*Czechoslovak*
*Academy*
*of*
*Sciences*

# MICROBIAL TRANSFORMATIONS OF STEROIDS

# BIOLOGIA

# ET

# INDUSTRIA

*EDITOR*

W. ROMAN, ADELAIDE

*CO-EDITOR*

L. GENEVOIS, BORDEAUX

# III

# MICROBIAL TRANSFORMATIONS OF STEROIDS

1966

Springer-Science+Business Media, B.V.

# MICROBIAL TRANSFORMATIONS OF STEROIDS

ALOIS ČAPEK
OLDŘICH HANČ
MILAN TADRA

ACADEMIA
PUBLISHING HOUSE
OF THE CZECHOSLOVAK
ACADEMY OF SCIENCES

PRAGUE 1966

*Scientific Editors*

ING. ŽELIMÍR **PROCHÁZKA,** C.Sc.

PROF. ING. DR. JOSEF **DYR,** D.Sc.

*Translated by*

DR. ODON **MACEK**

ING. DR. LUBOMÍR **URBÁNEK,** C.Sc.

ING. DR. OLDŘICH **HANČ,** C.Sc.

ISBN 978-94-011-7605-7 ISBN 978-94-011-7603-3 (eBook)
DOI 10.1007/978-94-011-7603-3

Softcover reprint of the hardcover 1st edition 1966

# PREFACE

Fundamental studies of *steroid compounds* have begun approximately thirty years ago with the solution of structures of cholesterol and bile acids. This field has been receiving more attention when the structures of the principal steroid hormones, aglycones of heart glycosides, sapogenins, steroid alkaloids, toad poisons have been established. An intensive interest in these derivatives of cyclopentanoperhydrophenanthrene was determined especially by practical viewpoints, notably by attempts to prepare by synthetical economical procedures a series of *steroid hormones* and their biologically active derivatives, further by theoretical viewpoints.

Biochemistry of steroids was developing in its beginnings by studies of metabolites of steroid hormones in animal organism and soon has brought a number of practical results that could be used also in medical diagnostics. Considerably later, the metabolism of steroid compounds was followed also *in microorganisms,* again in connection with the practical problem to aid to economical syntheses of some steroid hormones. Two names are of primary importance for the pioneer works in microbial transformations of steroids, that of the Italian worker L. MAMOLI who first realized the microbial biosynthesis of testosterone from dehydroepiandrosterone, and of the American chemist and microbiologist D. H. PETERSON, who has contributed by his fundamental experiments in microbial hydroxylation of steroids to the position eleven, especially to the synthesis of cortisone. Microbial transformations of steroids nowadays offer a valuable tool in the preparation of series otherwise difficultly accessible steroid compounds.

The relatively new field of *microbial transformation of compounds, with which the microbial cell normally does not come into contact,* depends above all on the high adaptability of these cells and their enzymic systems for transformation of such substrates. Analogously as are metabolized various steroid compounds by animal cells, where they represent physiological components, they undergo structural changes in the microbial cell catalyzed by enzymes, most often of the induced type.

The microbial transformations of steroids have not been so far summed up monographically, although there exist partial treatises from various viewpoints, summarizing the findings, especially of American authors, in periodicals and annual reviews. This field of activity has been also dealt with in *the Research Institute of Pharmacy and Biochemistry in Prague* for about ten years, both from the viewpoint of practical contribution to the synthesis of corticoids, and systematical investigation of similarities in the enzymatic conversions of steroids in some genera of fungi and also to explain the mechanism of the influence of steroids on the microbial metabolism. This review summarizes the experiences and results obtained in this field within the general state of knowledge of microbial steroid transformations. It features also a list of references (over 1200 ref.) available before the end of 1963, further a list of used microorganisms tested for steroid conversions and a list of steroid metabolites.

*Prague, December 1963.* *Dr. Oldřich Hanč*

# CONTENTS

Preface 5

**1. Introduction** 11

Chemistry of steroids 11
Biogenesis of steroid hormones 13
Metabolism of steroid hormones 14

**2. Types of microbial transformations of steroids** 18

*A*) Oxidative processes 18

*a*) Oxidation of the secondary alcohol group into the keto group 18
*b*) Formation of secondary alcohol group on steroid skeleton 19
*c*) Formation of tertiary alcohol group on steroid skeleton 25
*d*) Formation of primary alcohol group in side chains of steroid molecule 28
*e*) Dehydrogenation of ring A of the steroid molecule in position 1-2 and 4-5 29
*f*) Aromatization of ring A of the steroid molecule 30
*g*) Oxidation of the methylene group into the keto group 30
*h*) Cleavage of the side chain of pregnane steroids on C-17 under formation of a keto group 31
*i*) Cleavage of the side chain of pregnane steroids on C-17 under formation of a secondary alcohol group 32
*j*) Cleavage of the side chain of pregnane steroids on C-17 and opening of ring D under formation of testololactone 32
*k*) Cleavage of the side chain of steroids under formation of carboxyl group 33
*l*) Dehydrogenation of ring B of the steroid molecule 33
*m*) Formation of epoxides 34

*B*) Reduction processes 35

*a*) Reduction of the keto group into a secondary alcohol group 35
*b*) Reduction of the aldehyde group of the steroid molecule into a primary alcohol group 36
*c*) Hydrogenation of ring A of the steroid molecule in position 1—2 37
*d*) Hydrogenation of ring A of the steroid molecule in position 4—5 and of ring B in position 5—6 37
*e*) Elimination of secondary alcohol group 38
*f*) Rise of Homo steroids of the androstane series from pregnane derivative 38

*C*) Hydrolysis 39

*a*) Microbial saponification of steroid esters 39

*D*) Esterification 40
*a*) Microbial acetylation 40

**3. Mechanisms of microbial transformations of steroids** 41

Solubility of steroids and enzyme action 41
Specificity of enzyme action 42
Mechanism of isomerization 46
Mechanism of 1—2 dehydrogenation 47
Reduction of the 20-keto group 48
Cleavage of the side chain of pregnanes 49
Mechanisms of the hydroxylation 50
Microbial degradation of steroid molecules 51

**4. The influence of steroids on microorganims** 55

Microbial binding of sterois 56
The influence of sterols 57
The influence of steroid hormones 57
Steroids and microbial metabolism 61

**5. Practical methods of fermentation** 65

Procedures for the transformations of steroids 65
The maintenance of microbial cultures 67
Practical examples 67

**6. Analytical evaluation of the fermentation course** 69

Paper chromatography 70
Thin-layer chromatography 72
Colometric methods 73
Spectrophotometry 75
Polarography 76

**7. Isolation and identification of steroid metabolites** 78

Isolation of steroids from the fermentation medium 79
Choice of solvent for extraction 81
Distribution (extraction) coefficient 82
Types of extractors in use 82
Emulsions and their breaking 83
Selective extractions 83
Preliminary purification of extracts 84
Drying of extract 84
Evaporation of extracts 85
Preliminary purification of evaporation residues 85
Direct crystallization of metabolites 86
Isolation of steroids from mycelium 87

Separation of metabolite mixtures 88
Fractional crystallization 88
Preparative paper chromatography 88

Preparative thin-layer chromatography 89
Column chromatography 89
Counter-current distribution 90
Chemical separation methods 91

Identification of steroid metabolites 92

Preliminary identification processes 93
Colour and fluorescence reactions 93
Paper or thin-layer chromatography 95
Determination of melting point 95
Optical rotation 95
Molecular rotation 96
Elementary analysis 96
Study of crystal structure 96
Optical rotatory dispersion 97
Nuclear magnetic resonance 97

Chemical constituion proofs 97
Esterification and saponification 98
Saponification of esters 99
Influence of strong acids and alkalies 99
Oxidation with chromium trioxide 102
Selective oxidation methods 103
Reduction of keto groups 105
Hydrogenation of double bonds 106
Removal of hydroxy group under formation of double bond 106
Reductive elimination of the hydroxy group with zinc in acetic acid 107
Preparation of acetonides 107
Dienon-phenol rearrangement 108

Identification of steroid metabolites by microbiological methods 108

**8. Transformations of steroids by various strains of microorganisms** 110

Type of reaction 110

*A*) Oxidations 110
*a*) Alcohol to ketone 110
*b*) Hydroxylations 111
*c*) Epoxidation 114
*d*) Dehydrogenation (new double bound formation) 114
*e*) Aromatization of ring A 115
*f*) Side chain cleavage and ring D-splitting 115
*g*) Ring B-splitting 115

*B*) Reductions 115
*h*) Reductions of keto group 115
*i*) Hydrogenation (double bound elimination) 116

*C*) Hydrolyses 116
*j*) Saponification of esters 116
*k*) Cleavage of glycoside bond 116

*D*) Esterification 116
*l*) Formation of acetate 116

**9. Applications for the preparation of steroid hormones and their analogues** 117

Preparations of androgens 118
Preparations of estrogens 121
Preparations of gestagens 123
Preparations of corticoids 124

**10. List of microorganisms performing transformations of steroids** 129

**11. List of steroid metabolites** 147

**12. Literature** 181

*a*) Reviews 181
*b*) Individual papers dealing with microbial transformations of steroids 183
*c*) Patent literature 205
*d*) Papers dealing with isolation, identification and analysis of steroids 224
*e*) Papers dealing with general problems in microbiology and fermentation 233

**Subject index** 235

# 1

# INTRODUCTION

## CHEMISTRY OF STEROIDS

Steroids are differently substituted derivatives of the fundamental four nuclei cyclopentanoperhydrophenanthrene skeleton containing 17 carbons (*sterane*). The system contains 6 asymmetrical carbons in all and it is therefore theoretically possible to derive from it 64 isomer forms. The natural steroids and derivatives obtained by partial synthesis from natural steroids all have, however, a structure of only one basic form (*trans, anti, trans, anti, trans*) according to the space coupling of the four rings of the carbon chain into chair shape.

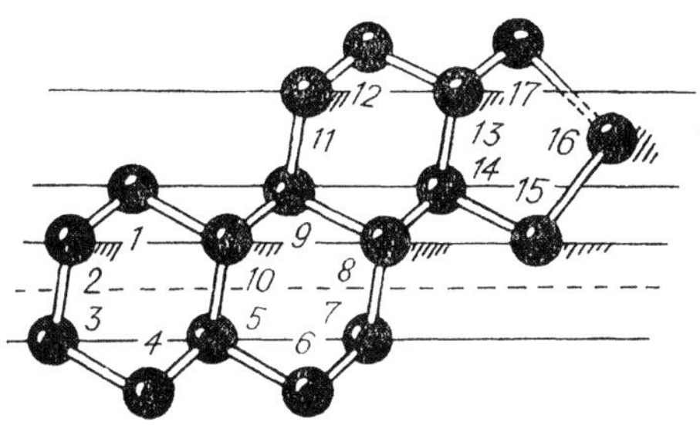


Fig. 1. *Sterane* (cyclopentanoperhydrophenanthrene).

The majority of steroids carries on carbon 10 and 13 a methyl group reciprocally oriented in *cis*-position. The space arrangement of substituents on the particular carbons of the skeleton is distinguished in case of *cis*-position towards angular methyls as $\beta$, i.e. by full bond, in case of *trans* as $\alpha$ by dotted bond. According to the total number of carbons in the skeleton, the basic steroid hydrocarbons are distinguished by means of the following nomenclature.

| Steroid hydrocarbon | Number of C | Substituents on carbon in position 10 | 13 | 17 |
|---|---|---|---|---|
| *Sterane* | 17 | H | H | 2 H |
| *Estrane* | 18 | H | $CH_3$ | 2 H |
| *Androstane* | 19 | $CH_3$ | $CH_3$ | 2 H |
| *Pregnane* | 21 | $CH_3$ | $CH_3$ | $\beta$-$CH_2CH_3$ + $\alpha$H |
| *Cholane* | 24 | $CH_3$ | $CH_3$ | $\beta$-$CH(CH_3).(CH_2)_2.CH_3$ + $\alpha$H |
| *Cholestane* | 27 | $CH_3$ | $CH_3$ | $\beta$-$CH(CH_3).(CH_2)_3.CH.(CH_3)_2$ + $\alpha$H |

According to the configuration of hydrogen on carbon 5 are distringuished hydrocarbons of series 5$\alpha$ and 5$\beta$.

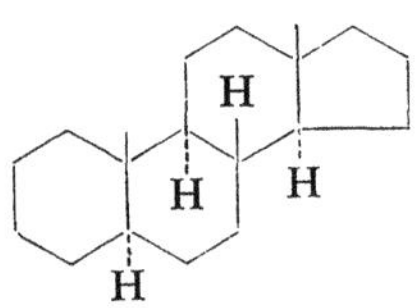


*5α-series*
*(earlier allo-series)*

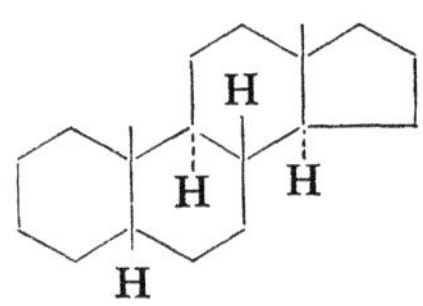


*5β-series*

By introduction of double bond between the carbons 4–5 or 5–6, the isomerism, however, is abolished in this centre.

The sterols, common costituents of the lipide fraction of the animal, plant and microbial cells are derived from the *cholestane* skeleton. Basic representatives are cholesterol in animal cells, sitosterols and stigmasterol in plant cells and ergosterol in microbial cells.

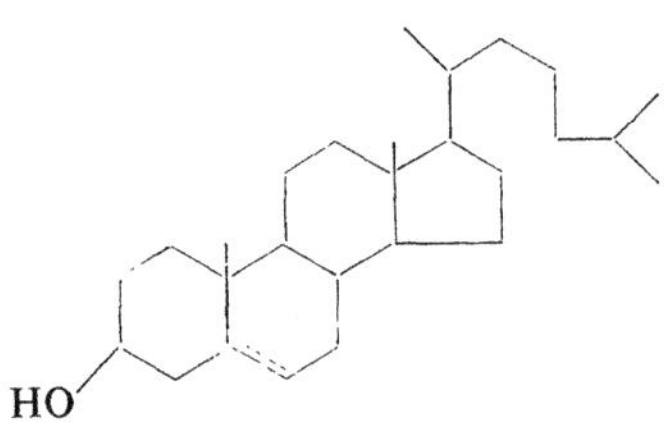


*cholesterol*

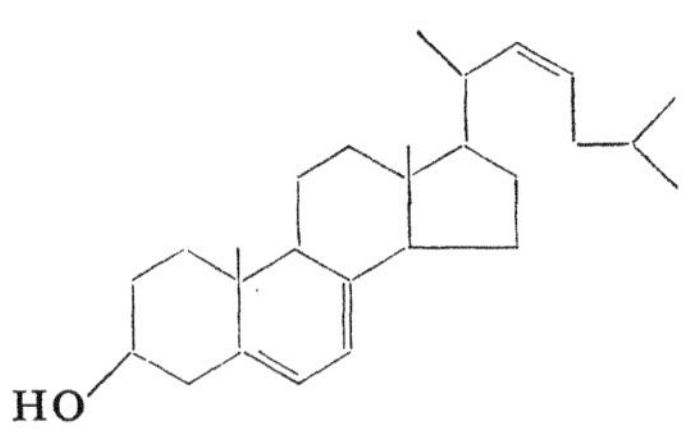


*ergosterol*

Cholesterol was the first raw-material for synthetic preparation of steroid hormones and in organism of higher animals it is also a possible precursor of these biologically imporgant substances. Another important group of steroid compounds, bile acids, are also formed from cholesterol in the liver. The sterols are synthetized in the organism from activated acetic acid via mevalonic acid and squalene; the genesis of the steroid molecule may nevertheless proceed also in other pathways than by cyclisation of squalene, as has been proved in experiments with isotopes labelled precursors.

## BIOGENESIS OF STEROID HORMONES

Schematically the *biogenesis of steroid hormones* in the organism of higher animals can be illustrated in the following way:

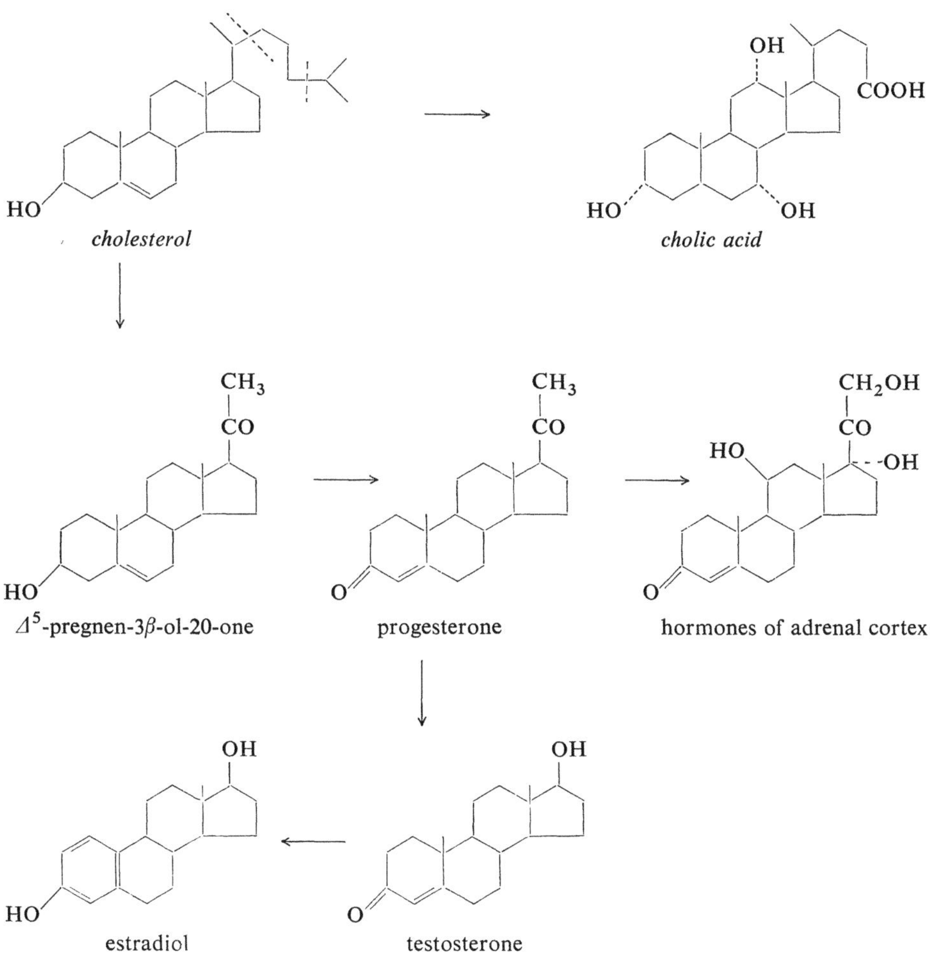


The steroid hormones are according to their origin and biological effect usually divided into *estrogens* or female sex hormones, chemically derived from estrane, *androgens* – male hormones, derived from androstane, *gestagens* – hormones of corpus luteum and *corticoids* – hormones of the adrenal cortex. Both latter groups chemically can be derived from the basic pregnane hydrocarbon. Considering the natural hormones and not their synthetic analogues with similar effect, we may schematically distinguish the above mentioned 4 group swith the following characteris ations:

| Groups of steroid hormones | Main representative | Basic hydrocarbon | Number of carbons | Structural characterisation |
|---|---|---|---|---|
| *Estrogens* | estradiol | estran | 18 | $\Delta^{1,3,5(10)}$-3,17 $\beta$-diol |
| *Gestagens* | progesterone | pregnane | 21 | $\Delta^4$-3,30-dion |
| *Androgens* | testosterone | androstane | 19 | $\Delta^4$-17$\beta$-ol-3-on |
| *Corticoids* | corticosterone | pregnane | 21 | $\Delta^4$-11$\beta$, 21-diol-3,20-dione |
| | cortisol | | 21 | $\Delta^4$-11$\beta$, 17$\alpha$, 21-triol-3,20-dione |
| | aldosteron | | 21 | $\Delta^4$-11$\beta$, 21-diol-18-al-3,20-dione |

In the biosynthesis of steroid hormones there occur besides the forming of the basic hydrocarbon skeleton enzymatically catalyzed oxido-reductive processes together with a splitting of the side chain on C–17.

*Processes in the biosynthesis of steroid hormones*

| Hormone | Precursor | Chemical transformations catalyzed by enzymes |
|---|---|---|
| *Progesterone* | pregnenolone | oxidation on C-3 shifting $\Delta^5$ to $\Delta^4$ |
| *Cortisol* | progesterone | oxidation (hydroxylations) in positions: 11$\beta$, 17$\alpha$, 21 |
| *Testosterone* | progesterone | oxidative splitting of *C-20* — *C-21* under formation of 17$\beta$-hydroxy group |
| *Estradiol* | testosterone | splitting $C_{19}$ and aromatization of ring A |

## METABOLISM OF STEROID HORMONES

By the enzymatic systems of animal tissues the steroid hormones are transformed by means of a series of oxido-reductive processes into derivatives mostly hormonally already inactive. The most intensive inactivating transformations of steroid hormones occur in the liver tissues. The steroid metabolites are eliminated in the urine either in the form of free alcohols or of glucuronides and esters with sulfuric acid. The estimation of some groups of these steroid metabolites in the urine is nowadays an effective diagnostic aid in endocrinological disturbances.

*Processes in inactivating transformations of steroid hormones in the organism*

| Hormone | Chemical transformations catalyzed by tissue enzymes | Final steroid metabolites |
|---|---|---|
| *Estradiol* | oxidation on C-17<br>oxidation on C-16 | estrone<br>estriol, 16-keto-estradiol, 16-keto-estron |
| *Testosterone* | oxidation on C-17, hydrogenation $C_4$-$C_5$ | androstandione |
| | oxidation on C-17<br>hydrogenation $C_4$-$C_5$<br>reduction of $C_3$ | androsterone |
| *Progesterone* | reduction of $C_3$ and $C_{20}$, hydrogenation $C_4$-$C_5$ | pregnandiols |
| *Cortisol* | oxidation C-11, hydrogenation $C_4$-$C_5$<br>reduction C-3 and C-20 | pregnantetrols |
| | cleavage of bond $C_{17}$-$C_{20}$, oxidation C-11 | adrenosterone |

*Sites in molecule of corticoid undergoing changes during the inactivating metabolism*

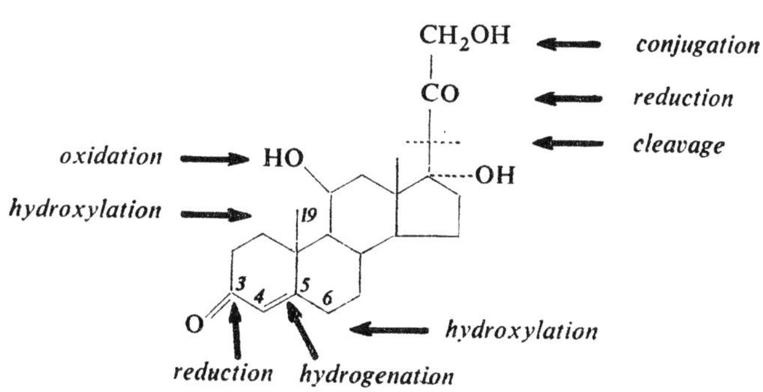

Some metabolites of steroid hormones still maintain a certain hormonal activity, though the most of them are already inactive. More than forty different individual steroids, most of them, however, precursors or metabolites of active hormones, have already been isolated from extracts of the adrenal glands. Of seven hormonally active steroids only three can be considered as true hormones.

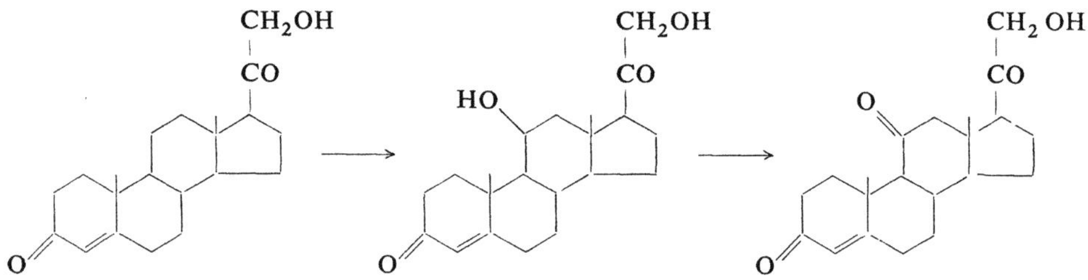

cortexone (*deoxycorticosterone*) | corticosterone | 11-dehydrocorticosterone

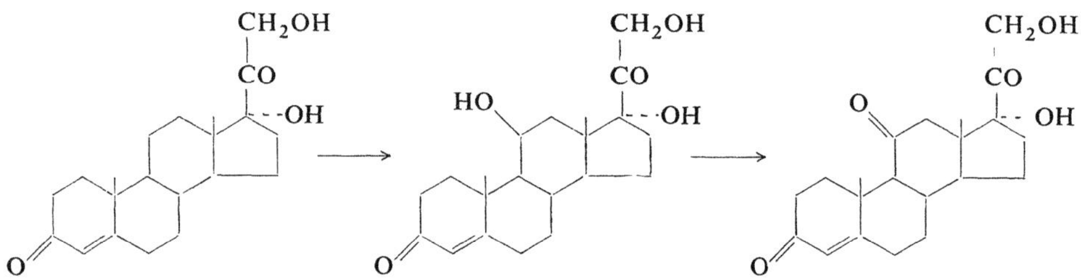

cortexolone (*compound S-Reichstein*) | cortisol (*compound F-Kendall*) | cortisone (*compound E-Kendall*)

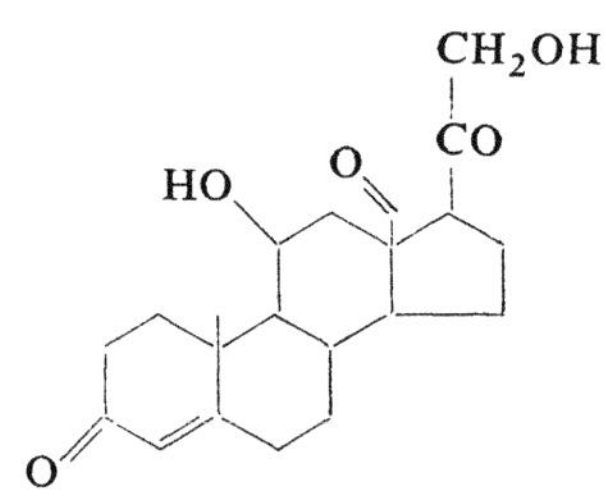

aldosterone

Perfusion methods of the individual organs, the techniques of tissue slices, homogenates and subcellular fractions in connection with the processes of chromatography and the use of steroids labelled by isotopes have facilitated detailed studies of the biosynthesis of steroids and of their further metabolism in the tissues of higher vertebrates. Recently attention has been drawn also to the enzymological aspects of the individual metabolic reactions. Analogous transformations of steroids have also been observed when *enzymatic systems of some microbes* were used (1–35). A number of oxido-reductions catalyzed by specific dehydrogenases is taking place not only in organ cells of higher vertebrates, but also in previously cultivated microbe cells after the adding of various steroids to the cultivation medium. In individual cases, oxido-reductions of steroid ketones and secondary alcohols have been carried out even by means of purified enzymatic preparations obtained from microbial cells (195–197, 361, 410, 416, 448–452).

Some microbial transformations of steroids may be used because of their specificity and high yields even in the synthesis of steroid hormones. Elsewhere the use of the enzymatic system of microbes for transformations of steroids may serve as a model experiment for the metabolism of the steroid in the cells of higher vertebrates. In some cases the ability of the microbes to perform certain transformations of steroids may even be used as a characterizing sign for species or strains. (73, 80, 235*a*).

In the following chapters first we shall deal with the *types* of the already established microbial transformations of steroids according to their chemical nature, further with the *mechanisms* of the transformations, as far as they are already known, we shall deal, on the other hand, with the *influence of steroids on the cells of microorganisms* and on their growth and metabolism, further with the methods used therein. Later we shall briefly discuss the results in a concise survey according to the microbes used and to the particular steroids.

# 2

# TYPES OF MICROBIAL TRANSFORMATIONS OF STEROIDS

The transformation reactions occurring in the steroid molecule under the catalytical effect of microbial enzymes range among *oxido-reductive reactions.* Besides esterification and hydrolysis a total of 12 oxidative and 5 reductive processes was proved, some of which are specific for the enzymatic system of the microorganism.

## *A*) OXIDATIVE PROCESSES

### *a*) Oxidation of the secondary alcohol group into the keto group

$$-\underset{\substack{|\\ OH}}{CH}- \longrightarrow -\underset{\substack{\|\\ O}}{C}-$$

In such way oxidation by means of various strains of aerobe bacteria (267) was achieved, further by means of *Flavobacterium dehydrogenans* (4) and *Proactinomyces erythropolis* (478) dehydroandrosterone into $\Delta^4$–androstene–3,17–dione. The same metabolite was formed when dehydroepiandrosterone was oxidized using the strain *Alcaligenes faecalis* (138).

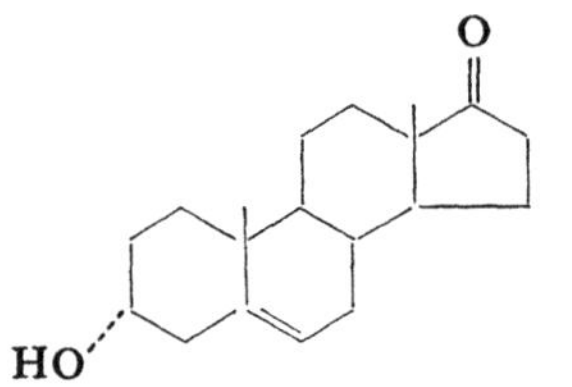

dehydroandrosterone
($\Delta^5$-androsten-3α-ol-17-one)

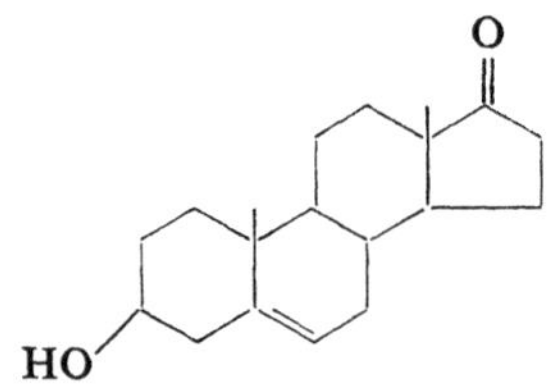

dehydroepiandrosterone
($\Delta^5$-androsten-3β-ol-17-one)

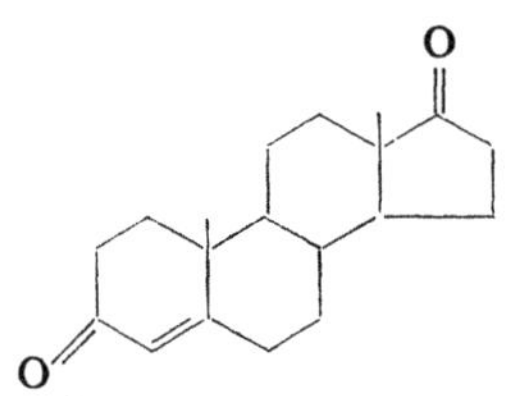

$\Delta^4$-androstene-3,17-dione

WELSCH (510) by means of the strain *Actinomyces albus* and TURFITT (479) by means of the strain *Proactinomyces* sp. have found in a high, nearly a hundred per cent yield, an analogous reaction on C-17 in estrogens.

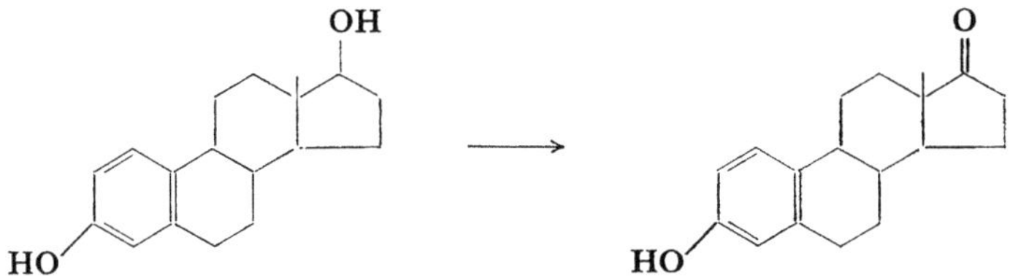

estradiol → estrone

By means of the strains *Azotobacter* sp. (194), *Proactinomyces roseus* (231), *Proactinomyces erythropolis* (478) and *Flavobacterium maris* (44) C-3 with the hydroxy group in $\beta$–position has been successfully oxidized into the respective keto group even on the molecule of cholesterol.

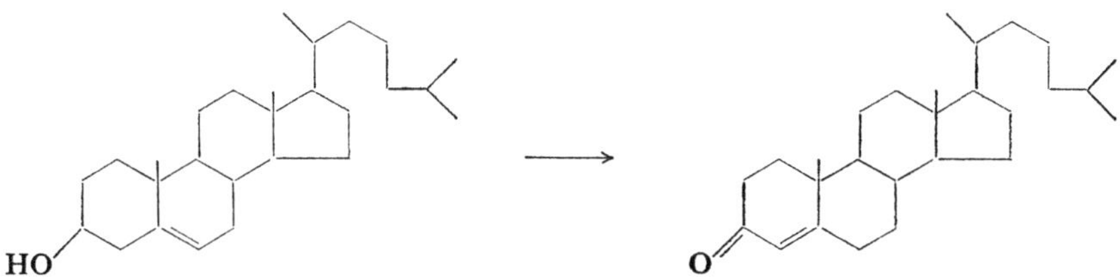

cholesterol → cholestenone

We have proved this type of reaction on the molecule of pregnenolone in all strains and species of *Aspergillus* (73) and *Actinomyces* tested by us.

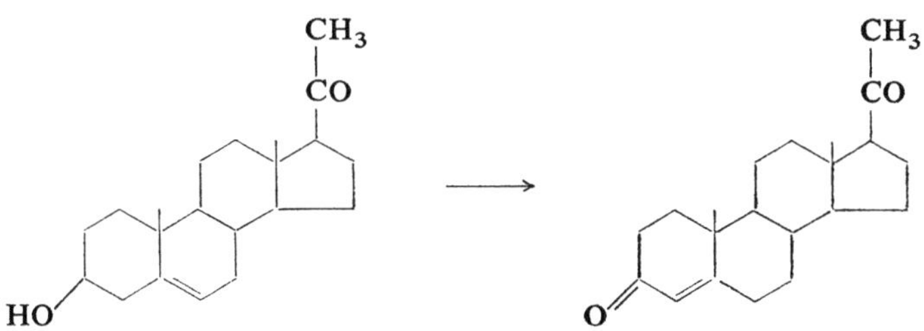

pregnenolone → progesterone

## *b*) Formation of secondary alcohol group on steroid skeleton

$$-CH_2- \longrightarrow \underset{\displaystyle OH}{\overset{|}{-CH-}}$$

The microbial hydroxylation of this type on the steroid skeleton was proved in positions 1α, 1β, 2β, 6β, 7α, 7β, 11α, 11β, 12β, 15α, 15β, 16α and 16β.

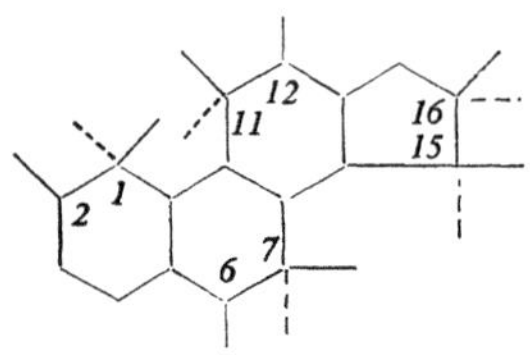


YOSHIO NOZAKI (320) has proved the hydroxylation of digitoxigenin by the strain *Absidia orchidis* in position 1β.

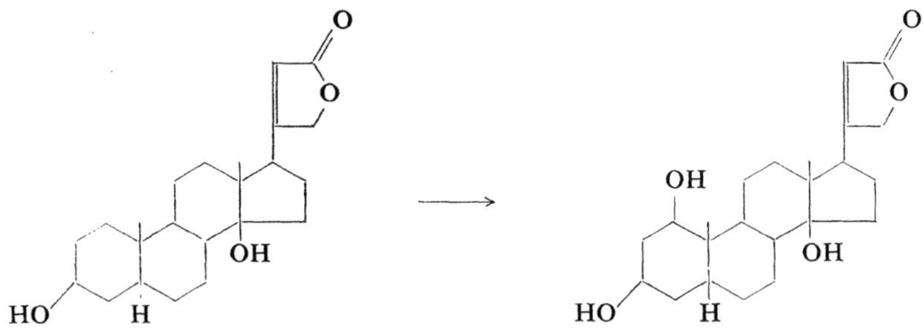


digitoxigenin

1β-hydroxydigitoxigenin

DODSON (105) has found by means of the strain *Penicillium* sp. the hydroxylation of $\Delta^4$-androstene-3,17-dione into position 1α and 2β.

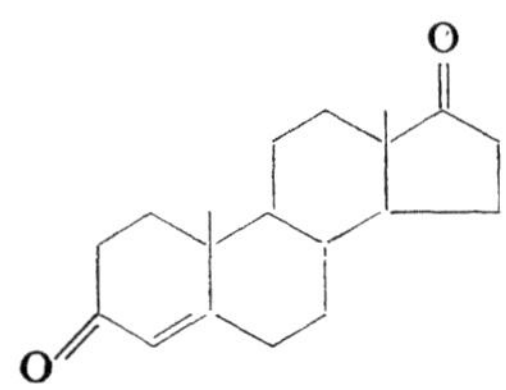


$\Delta^4$-androstene-3,17-dione

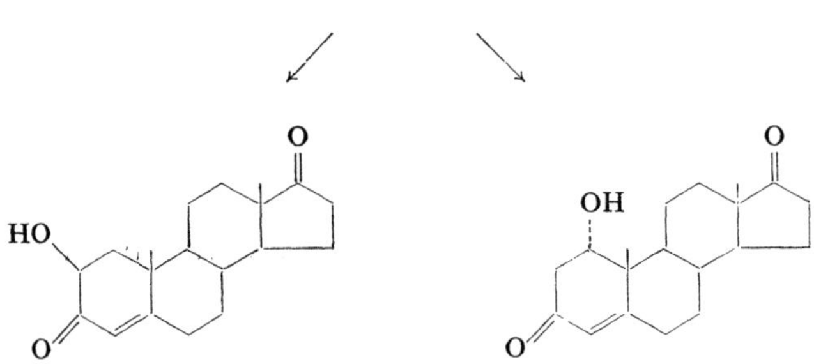


$\Delta^4$-androsten-2β-ol-3,17-dione

$\Delta^4$-androsten-1α-ol-3,17-dione

By means of the species *Rhizopus reflexus* and *Rhizopus arrhizus* EPPSTEIN (121) has proved the hydroxylation of $\Delta^4$-androstene-3,17-dione and of testosterone into position 6$\beta$.

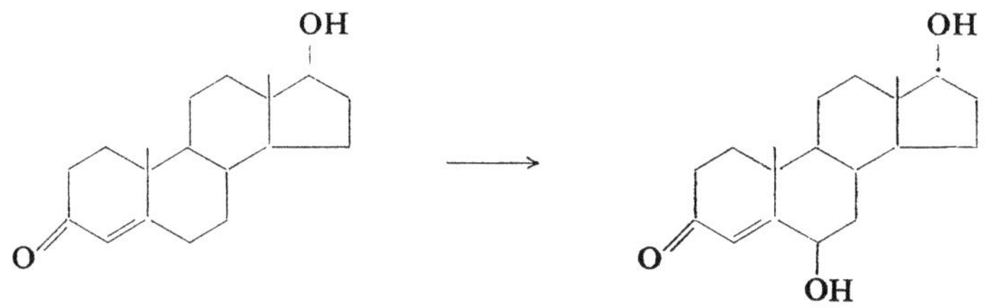


testosterone — 6$\beta$-hydroxytestosterone

FRIED (8) used the strain *Streptomyces auerofaciens* for introduction of hydroxyl into position 6$\beta$. Hydroxylation into position 6$\beta$ was proved in steroids with 18, 19, as also 21 carbons, but the formation of the 6$\beta$-hydroxy derivative is conditioned by the presence of the $\Delta^4$-3-keto configuration.

Hydroxylation into position 7$\beta$ is now less frequent reaction in the study of the microbial transformations of steroids. The formation of 7$\beta$-hydroxy derivative of cholesterol by means of the strain *Proactinomyces roseus* has been described by KRÁMLI (231).

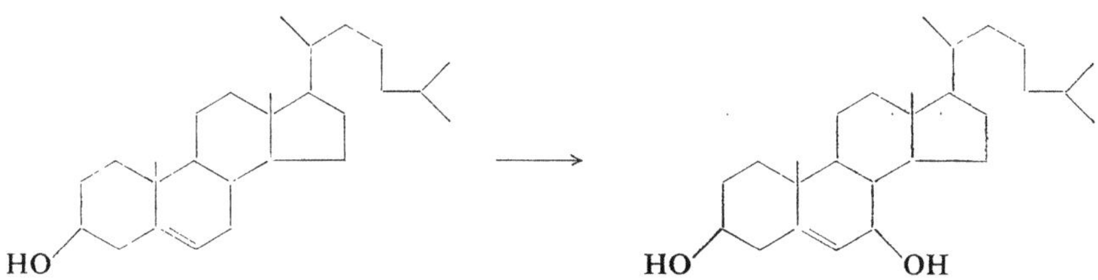


cholesterol — 7$\beta$-hydroxycholesterol

The Japanese authors ISHII H. (206) and NOZAKI Y. (319, 321) have proved the hydroxylation of digitoxigenin into position 7$\beta$ by means of the strain *Rhizopus nigricans, Rhizopus arrhizus* and *Absidia orchidis.*

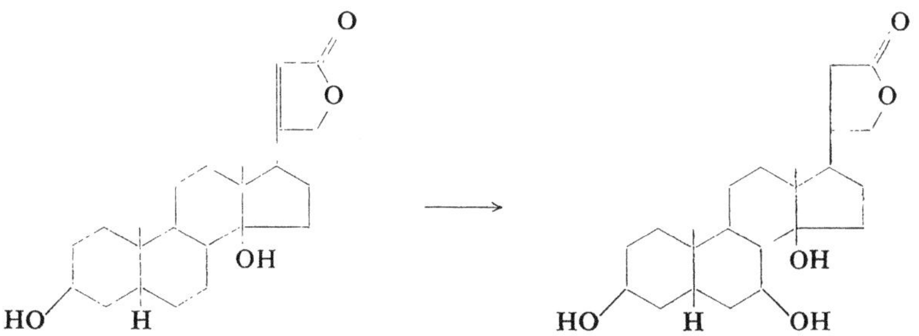


digitoxigenin — 7$\beta$-hydroxydigitoxigenin

McAlleer (259) has found, besides the 7β-hydroxylation of progesterone and of cortexone by means of species of genus *Cladosporium* also 7α-hydroxylation of the same steroids by means of species of genus *Helminthosporium*.

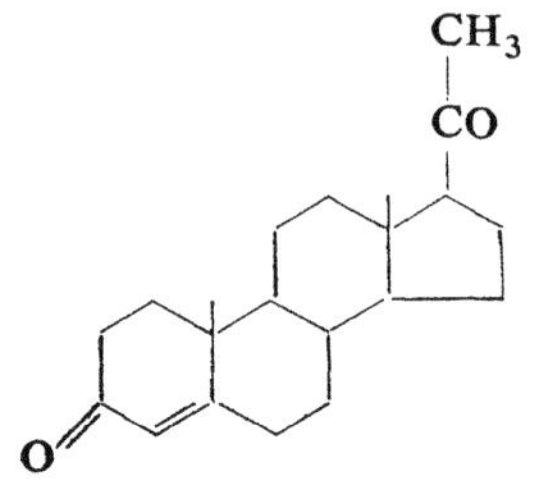


progesterone

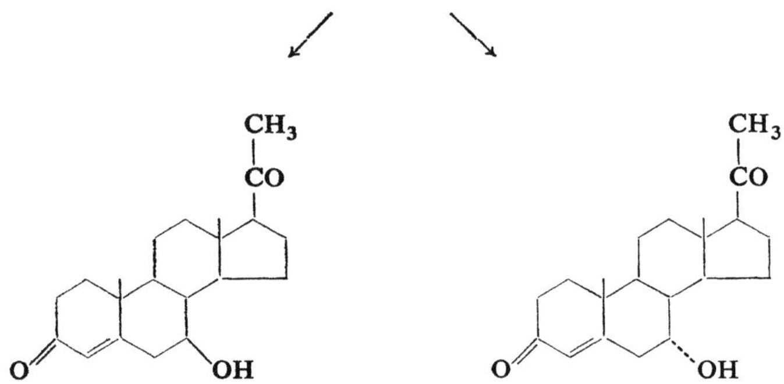


7β-hydroxyprogesterone    7α-hydroxyprogesterone

Likewise, species of the genus *Curvularia* (299) and *Peziza* (368) hydroxylate cortexone and 14α-hydroxyprogesterone into position 7α.

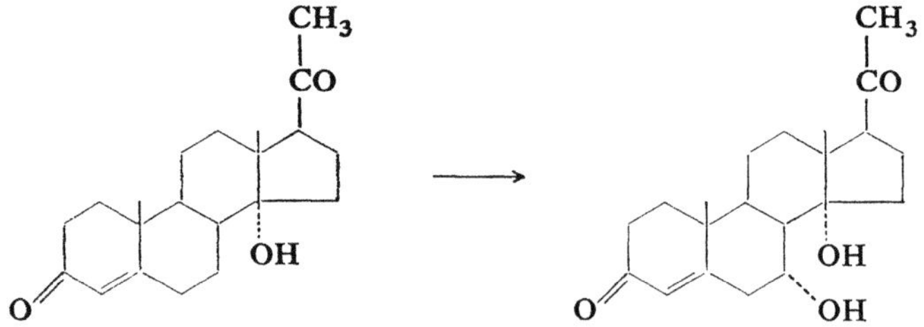


14α-hydroxyprogesterone    7α,14α-dihydroxyprogesterone

Another carbon that is hydroxylated according to the mentioned reaction type is C-11. The hydroxylation into position 11α was proved in a considerable number of strains and species of the genus *Rhizopus* (74, 119, 331, 474), *Absidia* (79), *Aspergillus* (141, 73, 531), *Penicillium* (80, 115), further in the species *Pestalotia foedans, Dactylium dendroides* (116), *Cephalothecium roseum* (297), and in some others.

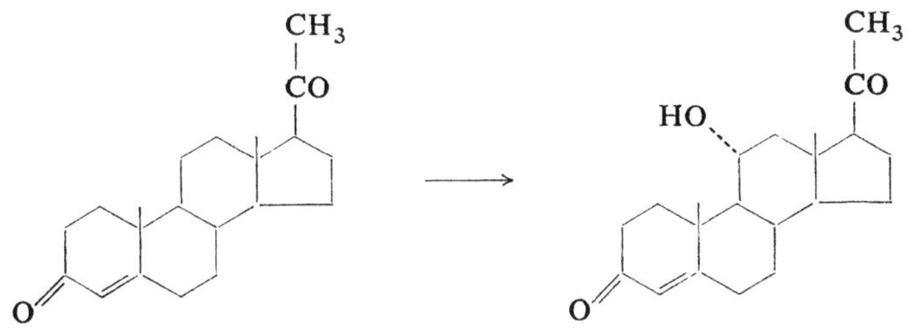

progesterone 11α-hydroxyprogesterone

Hydroxylation into position 11β was practically in no case proved on the steroid molecule as the unique transformation process. The formation of the 11β-hydroxy derivative is mostly simultaneously accompanied by the formation of the 11α-hydroxy derivative, eventually also by a further derivative, containing the hydroxy group in position 6β. So e.g. by means of species of the genus *Absidia*, there is formed from steroid "S" (Reichstein) besides cortisol also its 11α-epimer and by means of species of the genus *Cunninghamella* also the 6β-hydroxy derivative of steroid "S" (79).

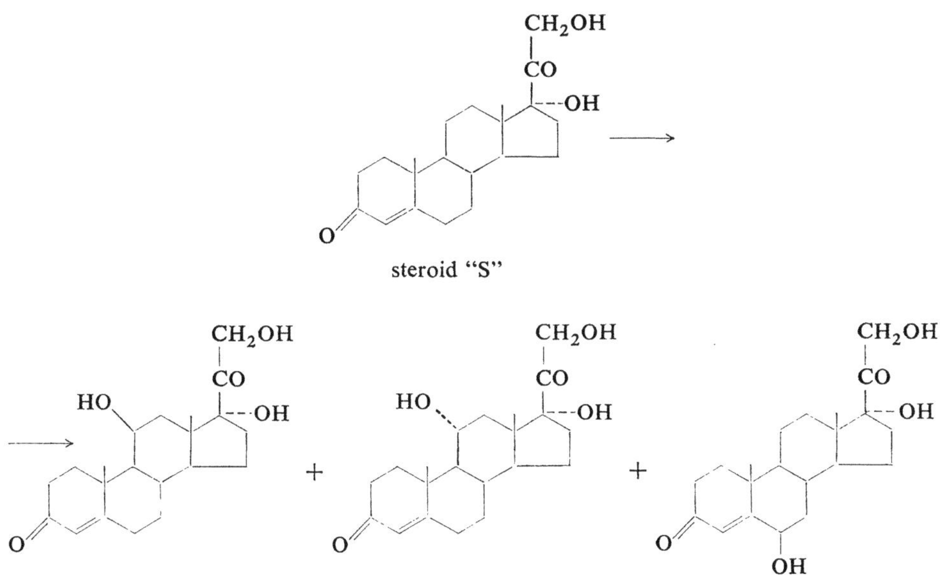

steroid "S"

cortisol epicortisol $\Delta^4$-pregnene-6β,17α,21-triol-3,20-dione

By means of the strain *Fusarium lini*, GUBLER (154*a*) has found the formation of the 12β-hydroxy derivative of digitoxigenin.

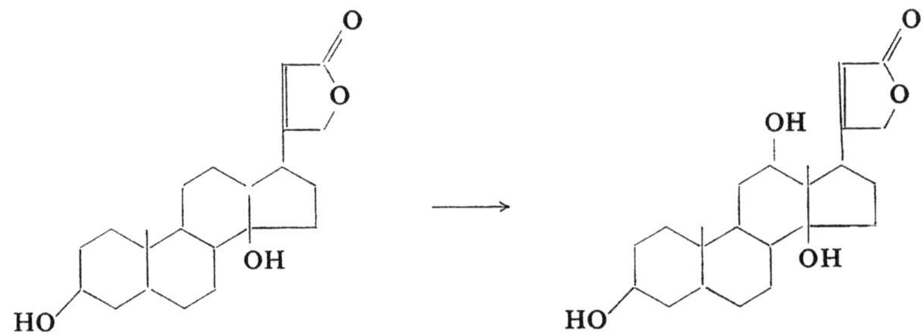


digitoxigenin → 12β-hydroxydigitoxigenin

Using the strain *Calonectria decora* (242) it has been possible to hydroxylate progresterone simultaneously into position 12β and 15β under formation of the respective dihydroxy derivative.

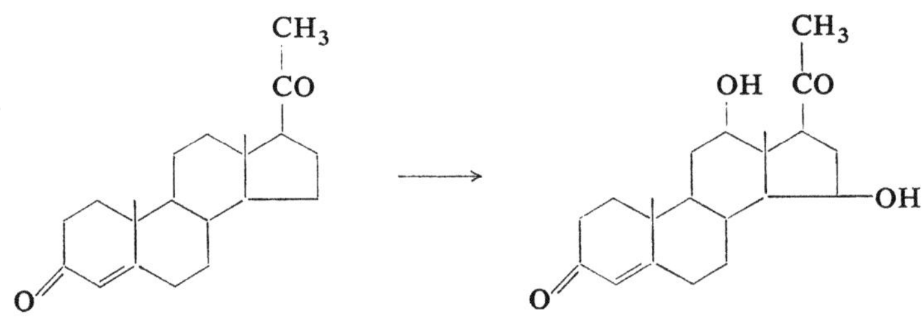


progesterone → 12β,15β-dihydroxyprogesterone

A relatively frequent hydroxylation process on the steroid molecule is the formation of 15α- and 15β-hydroxy derivatives. So e.g. the introduction of hydroxyl into position 15α on the molecule of $\Delta^4$-androstene-3,17-dione may be considered as a genus sign of the genus *Fusarium* (78).

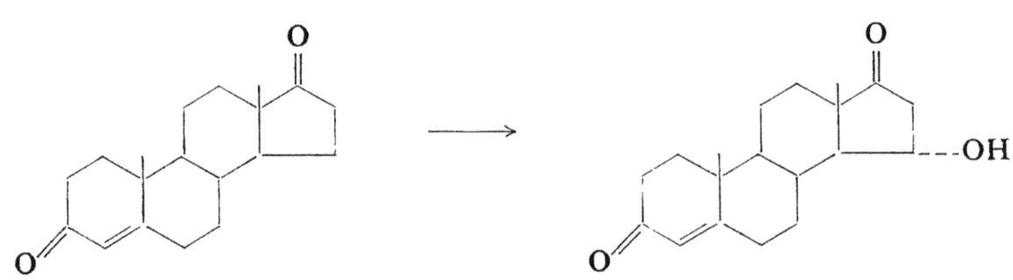


$\Delta^4$-androstene-3,17-dione → $\Delta^4$-androsten-15α-ol-3,17-dione

The hydroxylation into position 15β of progesterone was found by FRIED (8) by means of the strain *Phycomyces blakesleanus*. We have proved this course of hydroxylation in several species of genus *Aspergillus* and *Penicillium* (73, 80).

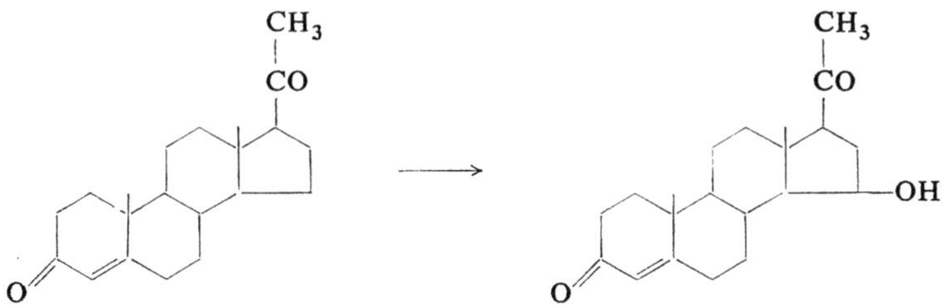


progesterone 15β-hydroxyprogesterone

The last carbon of the steroid skeleton which can be hydroxylated by microbial enzymes according to the mentioned reaction scheme is C-16. The hydroxylation into position 16α may also be used as one of the group signs of the genus *Actinomyces* (501).

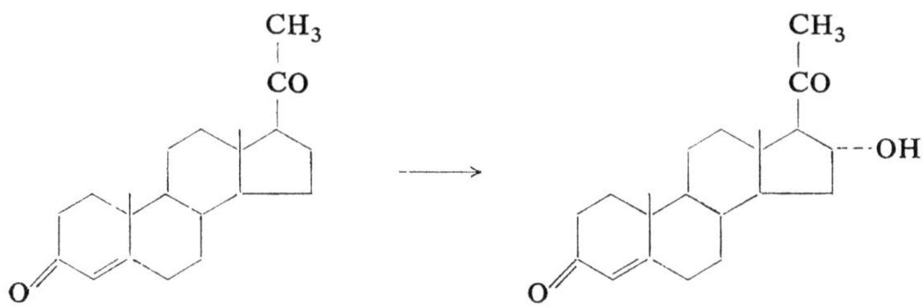


progesterone 16α-hydroxyprogesterone

The ability to hydroxylate the steroid molecule has been found on C-16 in the genus *Actinomyces* only. In bacteria, moulds and yeasts this ability has not been proved.

## *c*) Formation of tertiary alcohol group on steroid skeleton

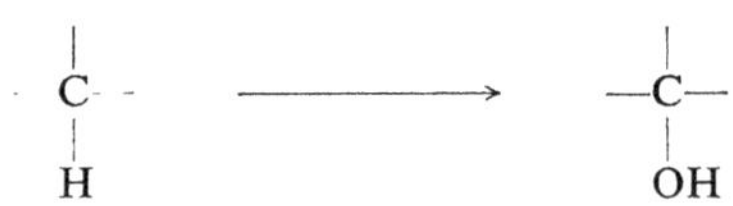


The microbial hydroxylation of this type was proved in position 5β, 9α, 14α and 17α of the steroid skeleton.

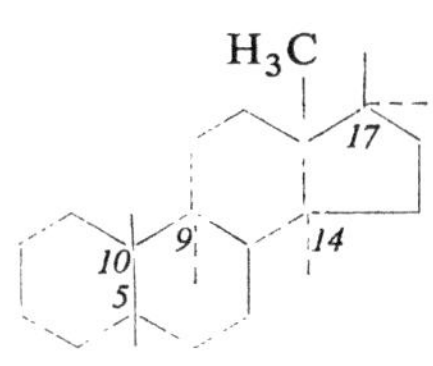

Ishii (205) has proved the hydroxylation into position 5$\beta$ by means of the strain *Mucor parasiticus* on the molecule of digitoxigenin.

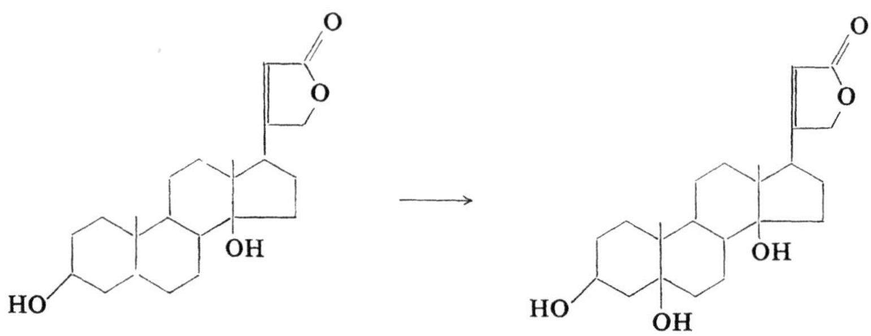


digitoxigenin 5$\beta$-hydroxydigitoxigenin

By means of the strain *Circinella* sp. (369) progesterone was hydroxylated into position 9$\alpha$.

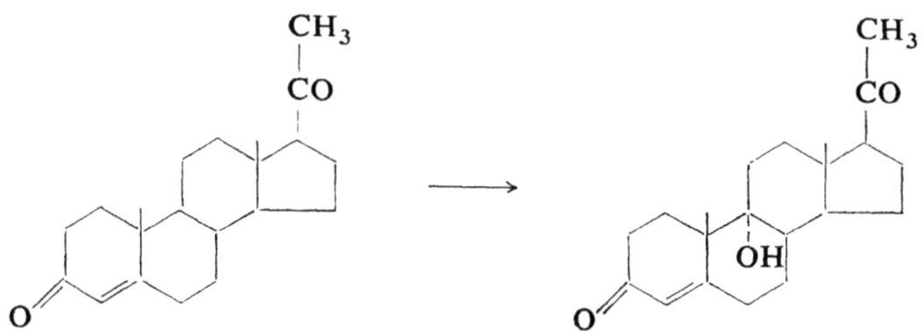


progesterone 9$\alpha$-hydroxyprogesterone

Hydroxylation on the same C-atom was proved by Sato and Hayakawa (355) by means of the strain *Helicostylum piriforme* on the molecule of solasodin.

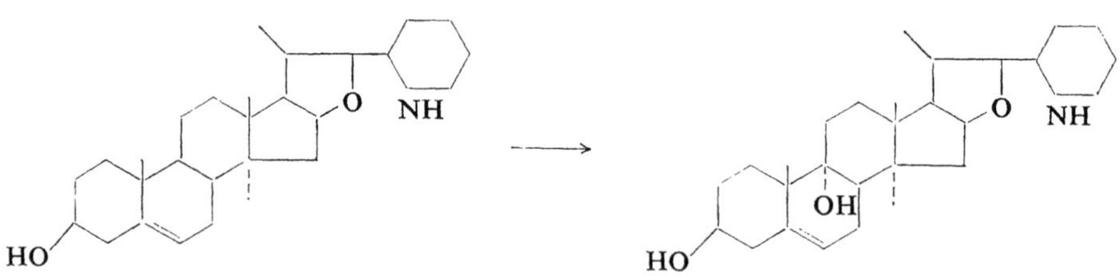


solasodin 9$\alpha$-hydroxysolasodin

Hydroxylation into position 10$\beta$ on the molecule of 19-nortestosterone (135) was carried out with the strains *Botrytis cinerea* and *Curvularia lunata*.

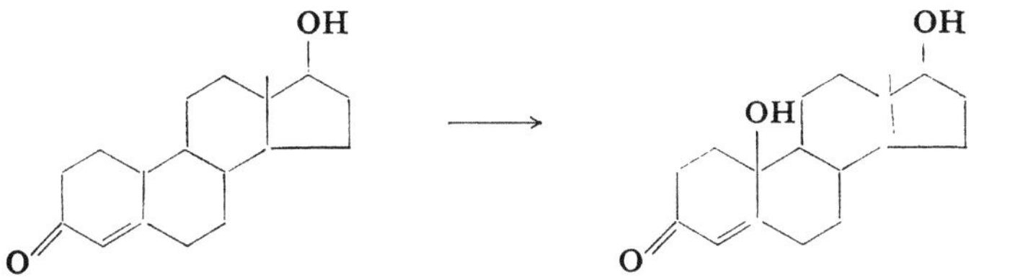

19-nortestosterone    10-hydroxy-19-nortestosterone

By means of various species of the genus *Mucor* (432, 538, 543), hydroxyl has been introduced into position 14$\alpha$ of the steroid molecule. As substrate testosterone, 10-nor-methyltestosterone, progesterone and cortexone have been used.

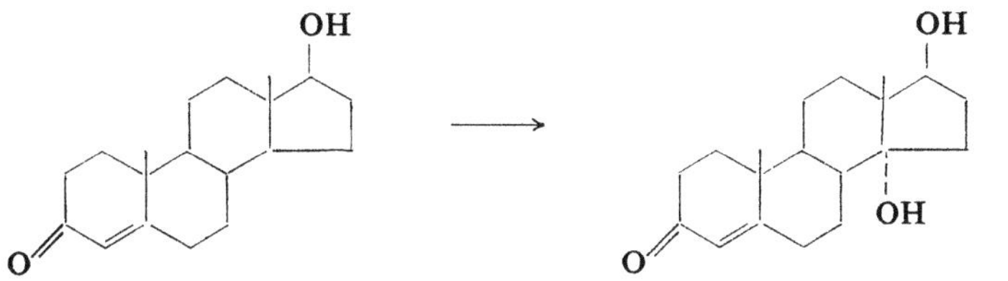

testosterone    14$\alpha$-hydroxytestosterone

The steroid "S" has been hydroxylated by the strain *Curvularia lunata* (38) simultaneously into position 11$\beta$, as well as into position 14$\alpha$ under formation of the respective dihydroxy derivative.

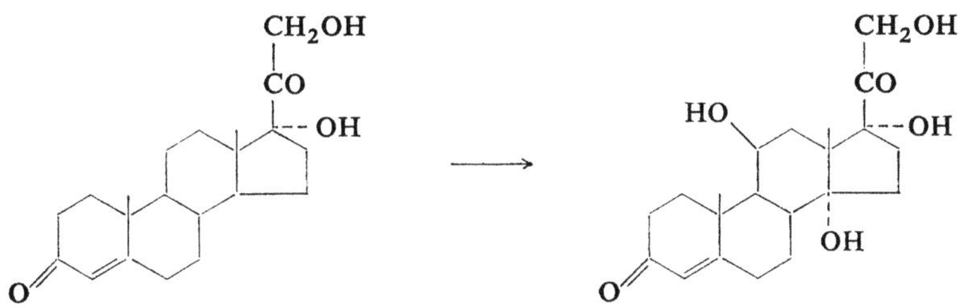

steroid "S"    14$\alpha$-hydroxycortisol

For introduction of hydroxyl into position 17$\alpha$ of the steroid skeleton the strains *Cephalothecium roseum* (181, 237, 299*a*) and *Trichoderma viridae* (256) have been used.

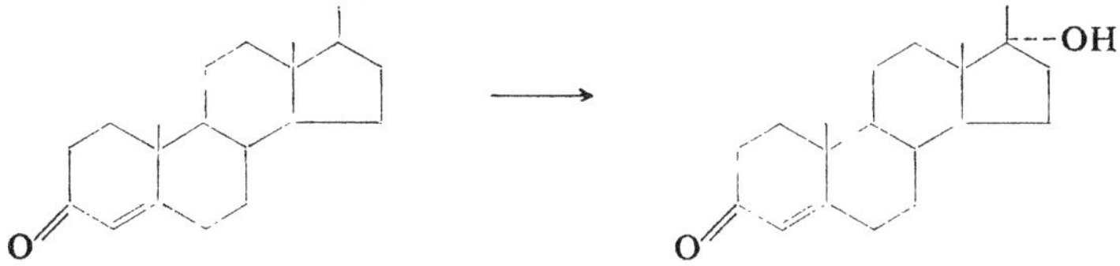

progesterone    17$\alpha$-hydroxyprogesterone

## *d*) Formation of primary alcohol group in side chains of steroid molecule

$$-CH_3 \rightarrow -CH_2OH$$

One of the very valuable transformation reactions is the substitution of hydrogen by hydroxyl on C-21 of the steroid molecule. The hydroxylation into position 21 was carried out by means of the strains *Ophiobolus herpotrichus* (299*a*, 580), *Aspergillus niger* (522) and *Wojnowicia graminis* (255).

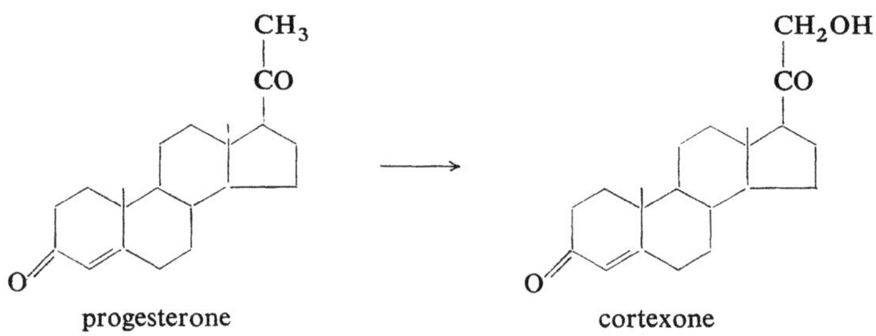


Some authors as WEISZ (509) have proved by means of the strain *Aspergillus niger* and RUBIN (351) with the strain *Curvularia lunata* the formation of dihydroxy derivatives of progesterone with the hydroxy group in positions 11α and 21 and further in positions 11β and 21.

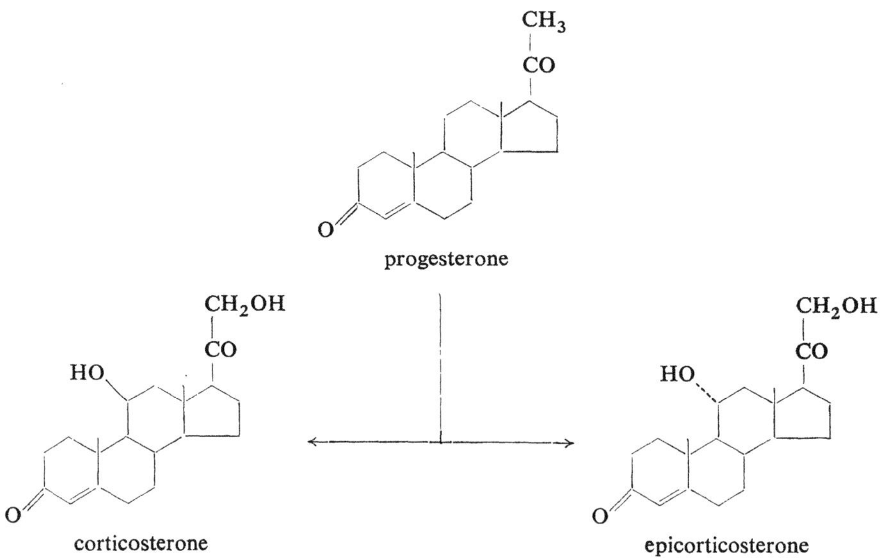

## *e*) Dehydrogenation of ring A of the steroid molecule in position 1—2 and 4—5

$$-CH_2-CH_2 \rightarrow -CH{=}CH-$$

This type of reaction belongs nowadays, besides hydroxylation into position 11$\beta$ and 16$\alpha$, to the most frequently studied reactions. The hydrogenation of ring A of the steroid molecule is, however, mostly accompanied either by the splitting of the side chain on C-17 under simultaneous opening of ring D or by reduction of the 20-keto group under formation of the respective 20$\beta$-hydroxy derivative (78, 81, 311).

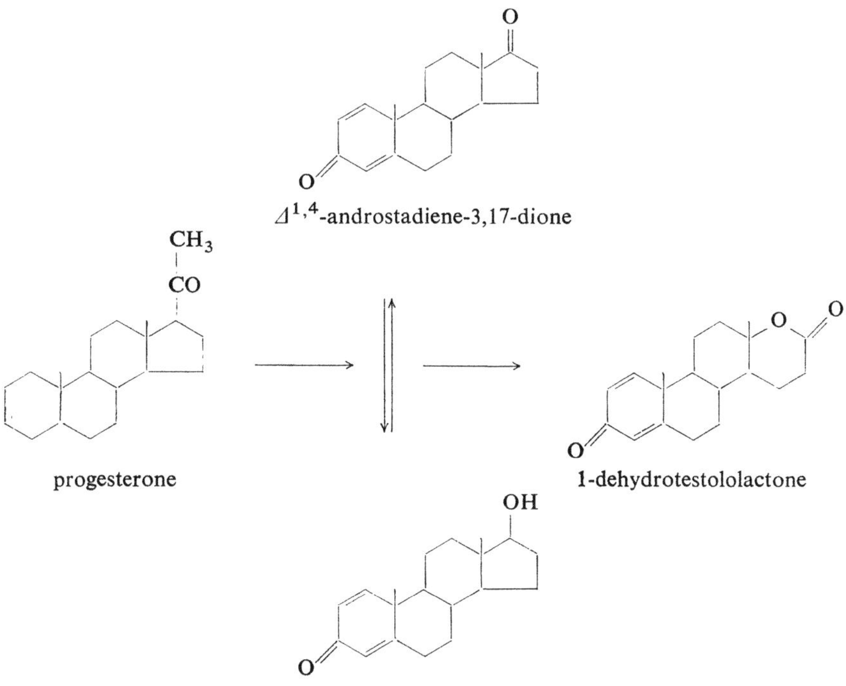

$\Delta^{1,4}$-androstadiene-3,17-dione

progesterone

1-dehydrotestololactone

1-dehydrotestosterone

The presence of hydroxy group in postion 17$\alpha$ prevents splitting of the side chain. In this case the 1-dehydrogenation is commonly accompanied by the reduction of the keto group on C-20.

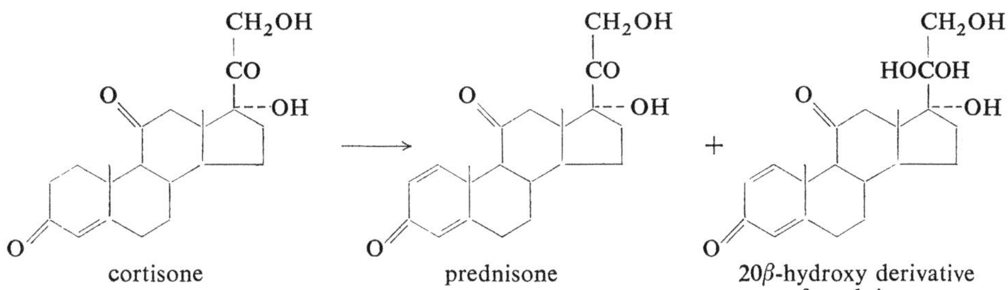

cortisone

prednisone

20$\beta$-hydroxy derivative of prednisone

Stoudt *et al.* (434) described, using the strain *Nocardia blackwelli*, dehydrogenation of the ring A of the steroid molecule simultaneously in the position 1–2 and 4–5.

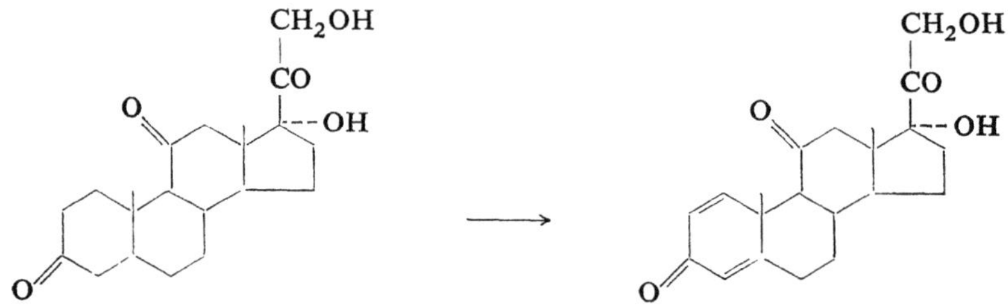


pregnane-17α,21-diol-3,11,20-trione — prednisone

## *f*) Aromatization of ring A of the steroid molecule

Dodson and Muir (106) succeeded in preparing estrone from $\Delta^4$-androsten-19-ol-3,17-dione by microbial aromatization of ring A.

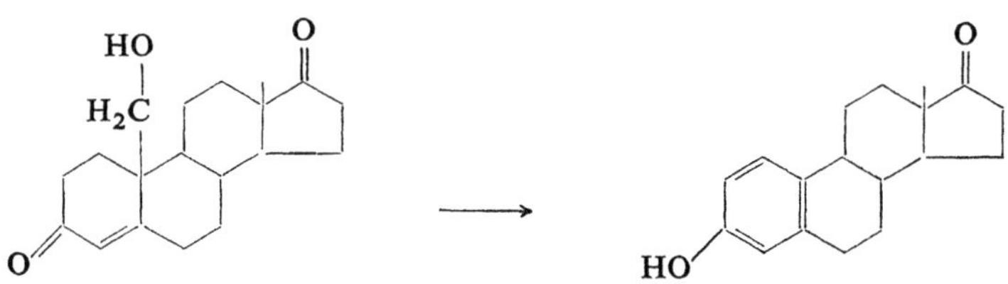


$\Delta^4$-androsten-19-ol-3,17-dione — estrone

## *g*) Oxidation of the methylene group into the keto group

$$-CH_2- \longrightarrow -\underset{\underset{O}{\|}}{C}-$$

This reaction type has been proved only in some species of genus *Rhizopus* and *Cunninghamella* (53, 655).

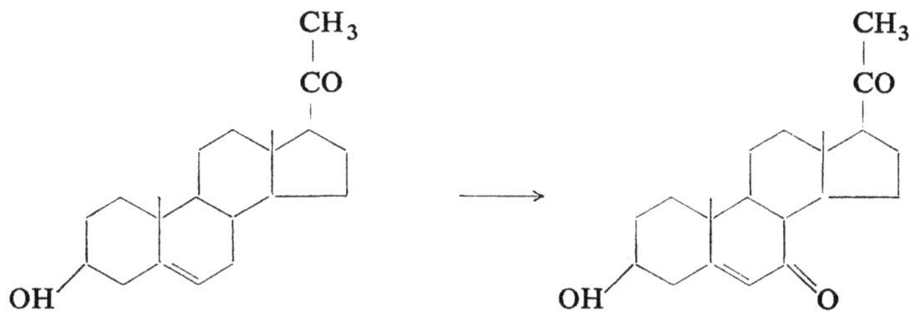

$\Delta^5$-pregnen-3$\beta$-ol-20-one $\quad$ $\Delta^5$-pregnene-3$\beta$,11$\alpha$-diol-7,20-dione

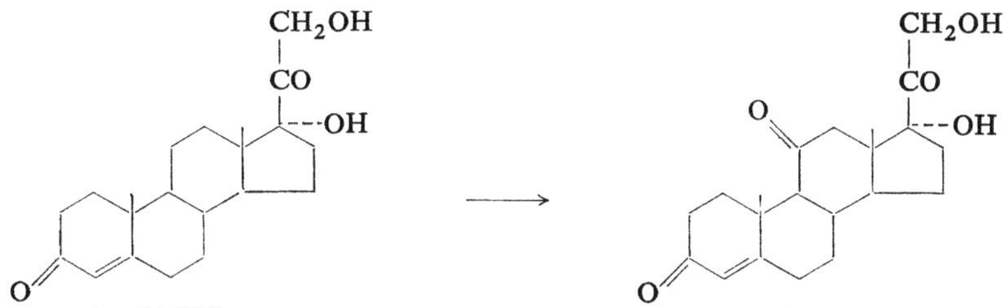

steroid "S" $\quad$ cortisone

In this second case cortisone is formed only as a by-product during the transformation by the strain *Conninghamella blakesleeana.*

## *h*) Cleavage of the side chain of pregnane steroids on C-17 under formation of a keto group

This oxidation in which the cleavage of the side chain on C-17 of the steroid molecule occurs under the formation of the keto derivative, is simultaneously accompanied by dehydrogenation, giving rise to the respective 1-dehydro derivative, or the formation of the keto group on C-17 is only an intermediate product for the subsequent oxidation degree, cf. the splitting of the side chain and the opening of ring D under formation of testololactone.

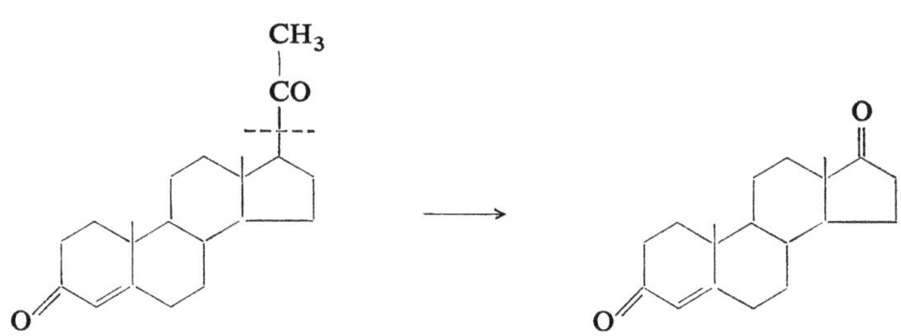

progesterone $\quad$ $\Delta^{1,4}$-androstadien-3,17-dione

This transformation has been carried out by means of the strains *Fusarium solani, Fusarium caucasicum* (493), *Fusarium lateritium* (78) and *Streptomyces lavandulae* (142).

## *i*) Cleavage of the side chain of pregnane steroids on C-17 under formation of a secondary alcohol group

This type of reaction was proved only in the genus *Penicillium* as a group sign (80, 160).

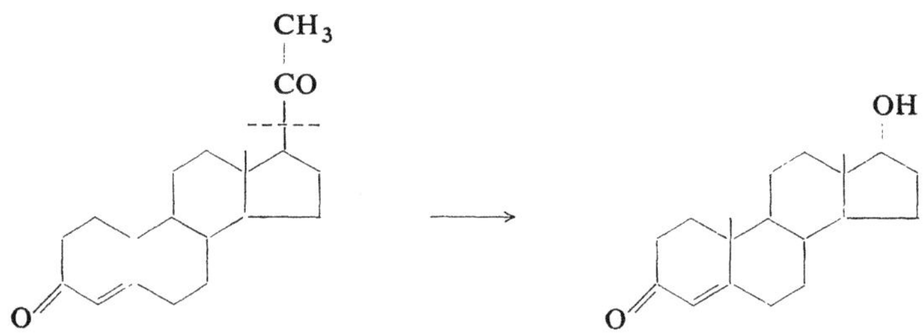

progesterone testosterone

## *j*) Cleavage of the side chain of pregnane steroids on C-17 and opening of ring D under formation of testololactone

In the pregnane steroids the cleavage of the side chain on C-17 occurs firstly under formation of $\Delta^4$androstene-3,17-dione and of testosterone as intermediate products which pass after the opening of ring D into testololactone as the final steroid product.

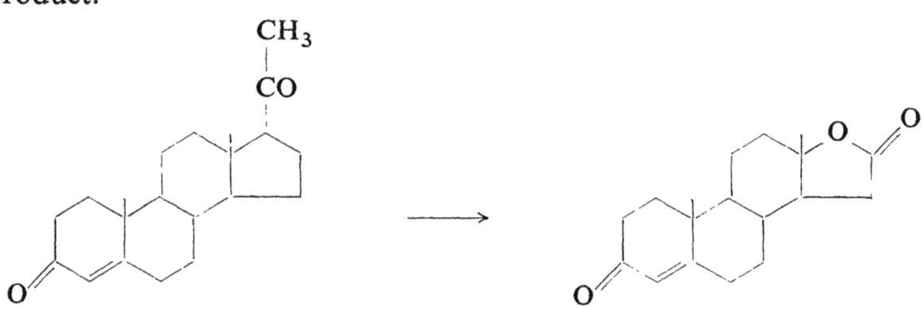

progesterone testololactone

This type of transformation is a relatively wide-spread group sign in the genera *Aspergillus* (74) and *Penicillium* (72, 73, 80, 142, 334, 336) and a species sign of genus *Fusarium* (78). In the latter case, however, the reaction is simultaneously accompanied by dehydrogenation of ring A in position 1–2 under formation of 1-dehydrotestololactone as the final product.

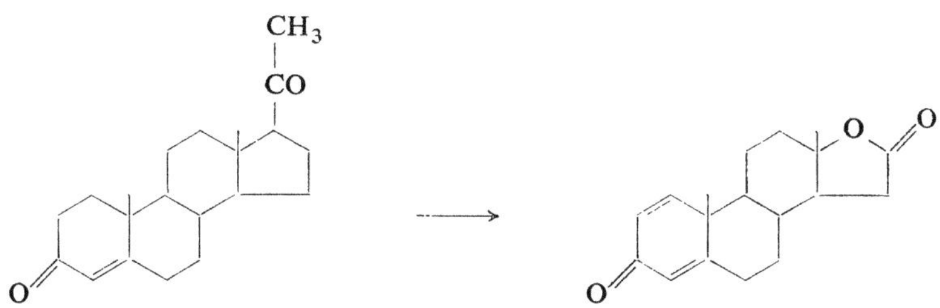


progesterone 1-dehydrotestololactone

## *k*) Cleavage of the side chain of steroids under formation of carboxyl group

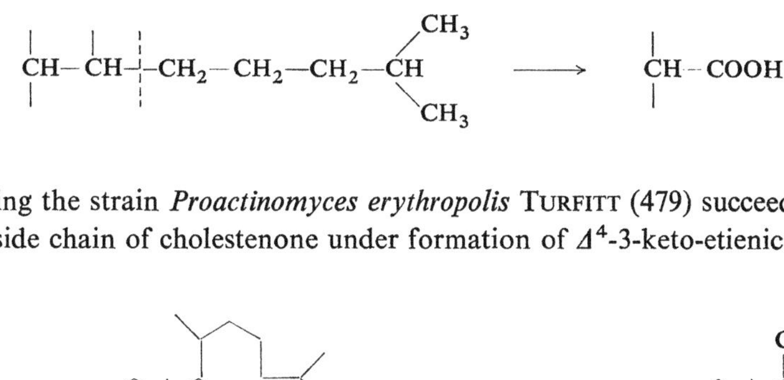


Using the strain *Proactinomyces erythropolis* TURFITT (479) succeeded in splitting the side chain of cholestenone under formation of $\Delta^4$-3-keto-etienic acid

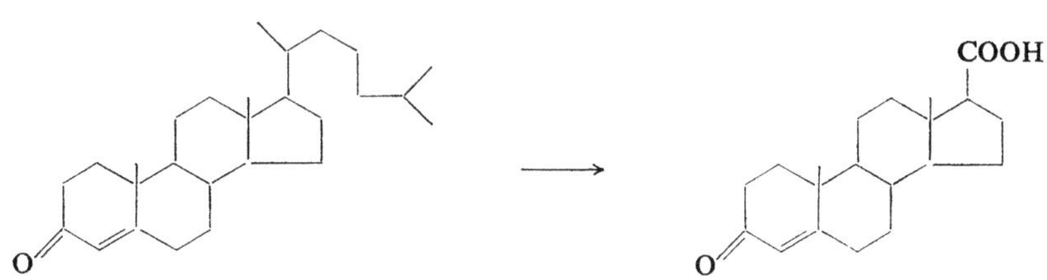


cholestenone $\Delta^4$-3-ketoetienic acid

## *l*) Dehydrogenation of ring B of the steroid molecule

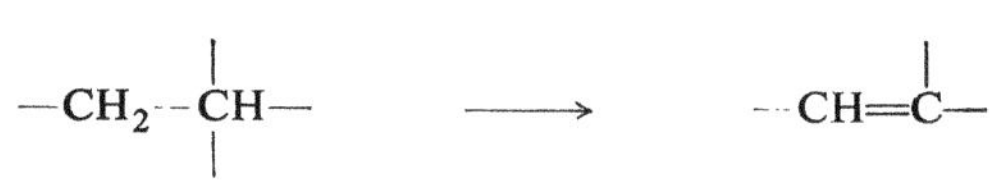

This relatively rare type of reaction has been proved by means of *Azotobacter* on the cholesterol molecule (194).

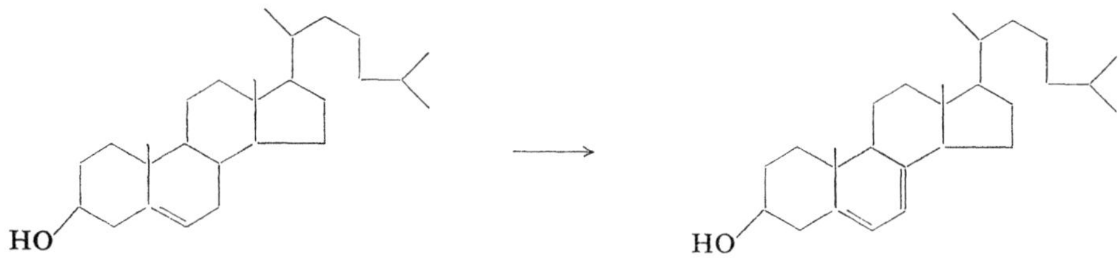

cholesterol → 7-dehydrocholesterol

## *m*) Formation of epoxides

In the strains *Curvularia lunata, Helicostylum pyriforme* and *Cunninghamella blakesleeana* (55) an interesting ability has been found to form epoxide in the site of a double bond in ring C and D of steroid molecule. The rise of epoxide in ring A and B of steroid molecule has not been proved, with the only exception of B-Nor-steroids.

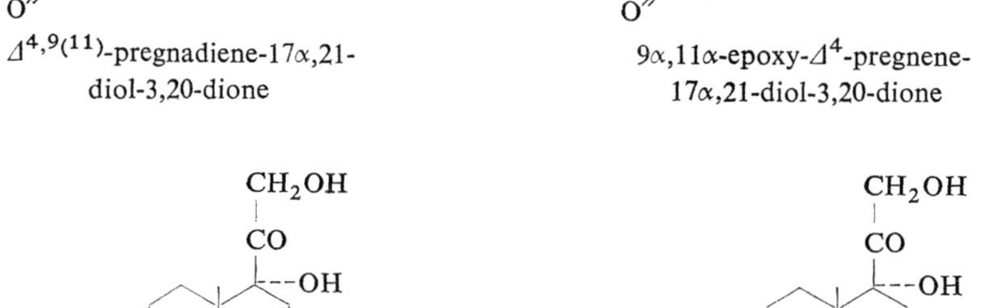

$\Delta^{4,9(11)}$-pregnadiene-17α,21-diol-3,20-dione → 9α,11α-epoxy-$\Delta^{4}$-pregnene-17α,21-diol-3,20-dione

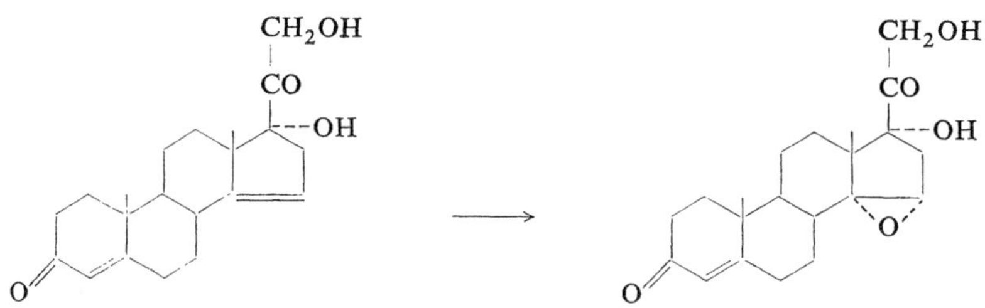

$\Delta^{4,14}$-pregnadiene-17α,21-diol-3,20-dione → 14α,15α-epoxy-$\Delta^{4}$-pregnene-17α,21-diol-3,20-dione

# *B*) REDUCTION PROCESSES

## *a*) Reduction of the keto group into a secondary alcohol group

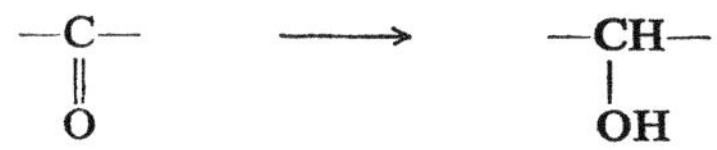


One of the works of the basic importance in the microbial transformation of steroids have been the papers of MAMOLI *et al.* (263, 264, 265) concerning reduction of the keto group on C-17 of androstane derivatives into the corresponding $\beta$-hydroxy derivative by means of *Saccharomycetes.*

$\Delta^4$-androstene-3,17-dione    testosterone

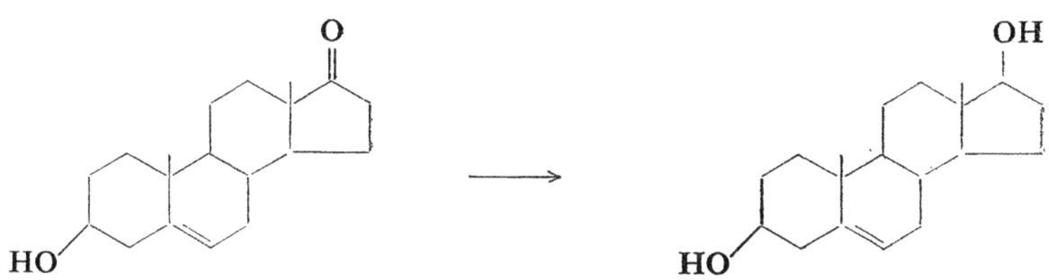


$\Delta^5$-androsten-3$\beta$-ol-17-one    $\Delta^5$-androstene-3$\beta$,17$\beta$-diol

Similarly proceeds the reduction of the keto group on C-3 and C-7 into the respective $\alpha$-or $\beta$-hydroxy derivatives. Thus, by means of the strain *Escherichia coli* (261) the reduction of the keto group on C-3 and C-7 into respective $\alpha$-hydroxy derivative has been successfully achieved.

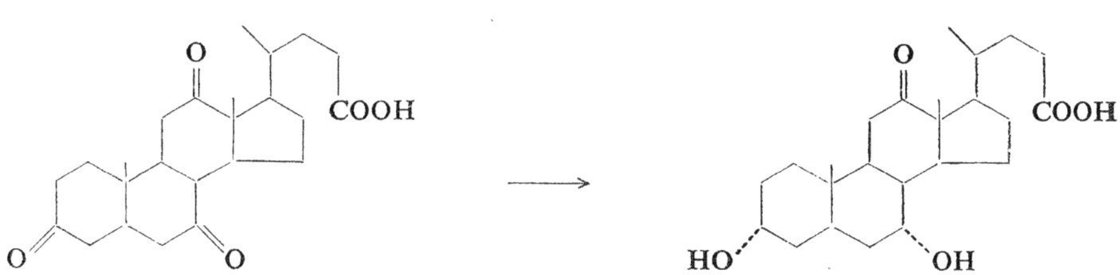


3,7,12-triketocholanic acid    3$\alpha$,7$\alpha$-dihydroxy-12-ketocholanic acid

By means of *Saccharomycetes* 3-keto group of the steroid molecule was reduced into the corresponding 3β-hydroxy derivative.

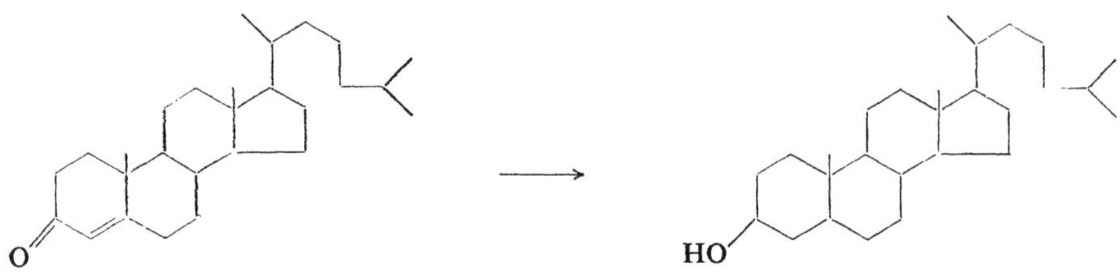


cholestenone → cholestanol

The last type of reduction of keto group proceeding on the steroid molecule is reduction of 20-keto group of the side chain of pregnene steroids into the corresponding 20β-hydroxy derivative (34, 78, 142, 150, 436).

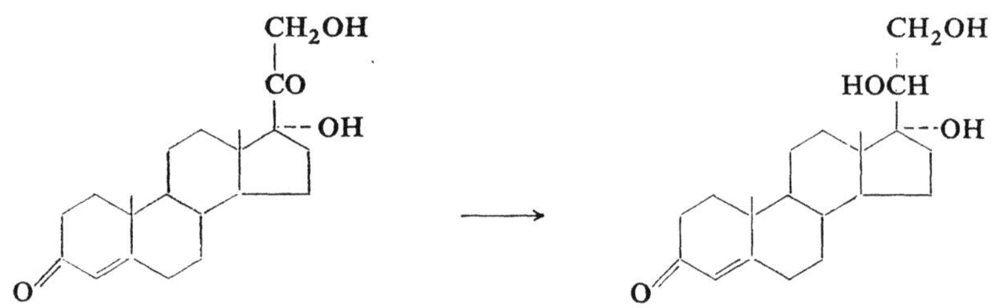


steroid "S" → $\Delta^4$-pregnene-17α-,20β,21-triol-3-one

In some cases the reduction of the 20-keto group does not proceed as the secondary reaction until after previous dehydrogenation of ring A in position 1–2.

The reduction of the keto group into a secondary alcohol group has only been proved on C-3, C-7, C-17 and C-20. On the other carbons of the steroid molecule, this type of reaction by means of microbial enzymes has not been proved.

## *b*) Reduction of the aldehyde group of the steroid molecule into a primary alcohol group

$$-CHO \longrightarrow -CH_2OH$$

This relatively exceptional transformation has been carried out by means of strains *Rhizopus arrhizus* (298) and *Penicillium lilacinum* (21).

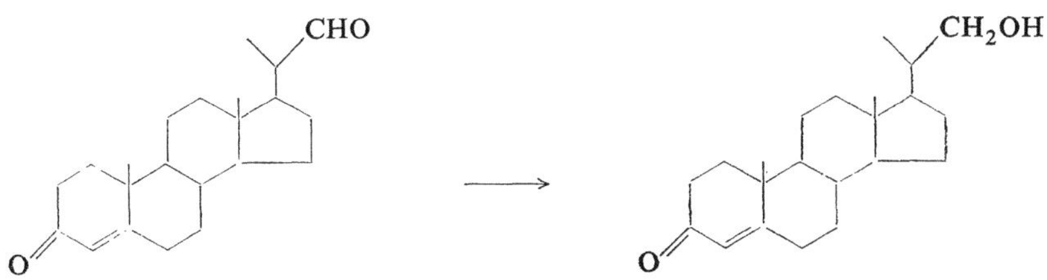

3-ketobisnor-$\Delta^4$-cholen-22-al → 3-ketobisnor-$\Delta^4$-cholen-22-ol

## *c*) Hydrogenation of ring A of the steroid molecule in position 1—2

$$-CH{=}CH- \longrightarrow -CH_2-CH_2-$$

The hydrogenation of ring A of the steroid molecule, i.e. the elimination of double bond in position 1—2 has been described by BUTENANDT using baker's yeast on the molecule of $\Delta^1$-androstene-3,17-dione. This process is simultaneously accompanied by the reduction of the 17-keto group into the respective $\beta$-hydroxy derivative.

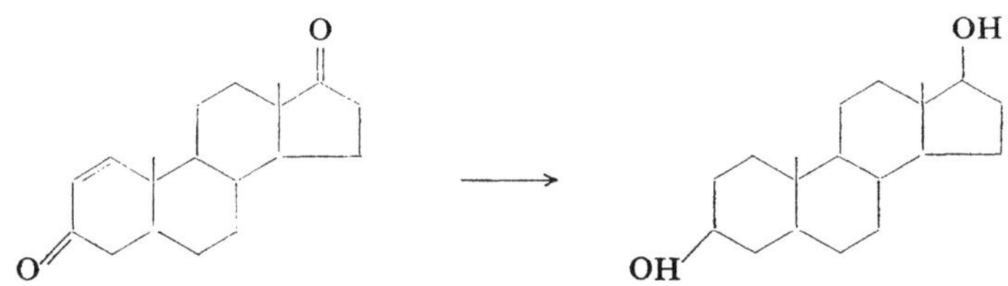

$\Delta^1$-androstene-3,17-dione → androstane-3$\beta$,17$\beta$-diol

## *d*) Hydrogenation of ring A of the steroid molecule in position 4—5 and of ring B in position 5—6

$$-CH{=}\overset{|}{C}- \longrightarrow -CH_2-\overset{|}{CH}-$$

By means of the strain *Clostridium lentoputrescens* MAMOLI (273) has performed hydrogenation of $\Delta^4$-androstene-3,17-dione and of testosterone into the corresponding androstane derivatives.

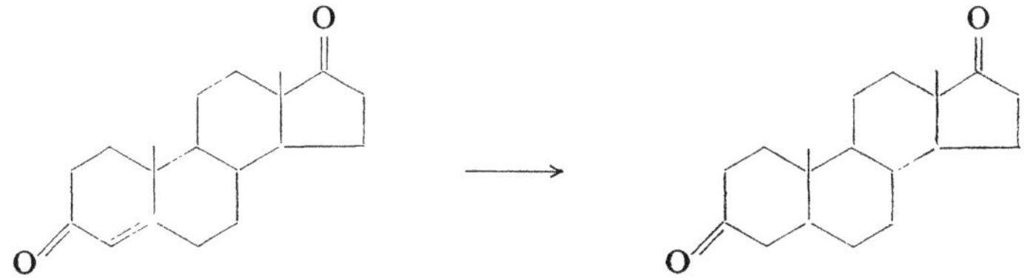

$\Delta^4$-androstene-3,17-dione → androstane-3,17-dione

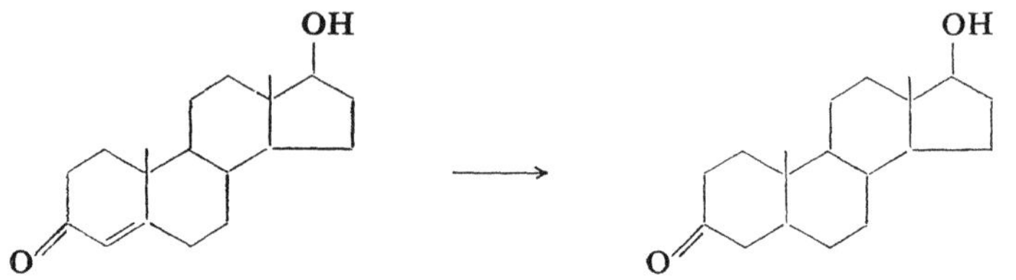

testosterone → androstan-17$\beta$-ol-3-one

The hydrogenation of $\Delta^5$-androstene-3$\beta$,17$\beta$-diol has been performed by MAMOLI using *Saccharomycetes.*

$\Delta^5$-androstene-3$\beta$,17$\beta$-diol → androstane-3$\beta$,17,-diol

## *e*) Elimination of secondary alcohol group

$$-\underset{\displaystyle OH}{\underset{|}{CH}}- \longrightarrow -CH_2-$$

By means of a non-identified bacterial strain it has been possible to substitute the secondary hydroxy group in the molecule of cholic acid by hydrogen.

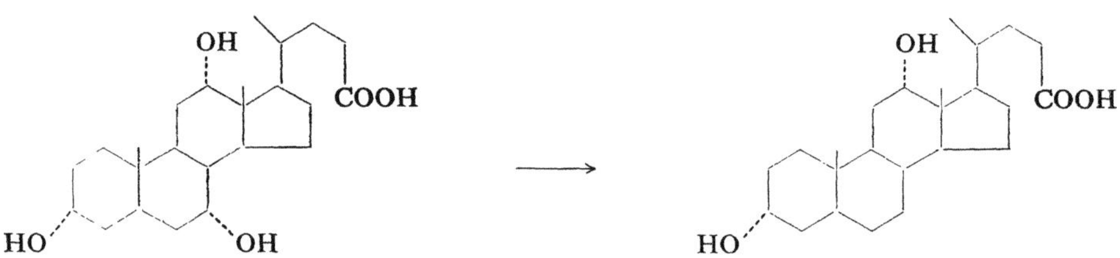

cholic acid → deoxycholic acid

## *f*) Rise of D-Homo steroids of the androstene series from pregnane derivative

The rise of 17 a $\alpha$-methyl-D-Homo-11$\alpha$-hydroxytestosterone from 17$\alpha$-hydroxyprogesterone by means of *Aspergillus niger* has been described (142).

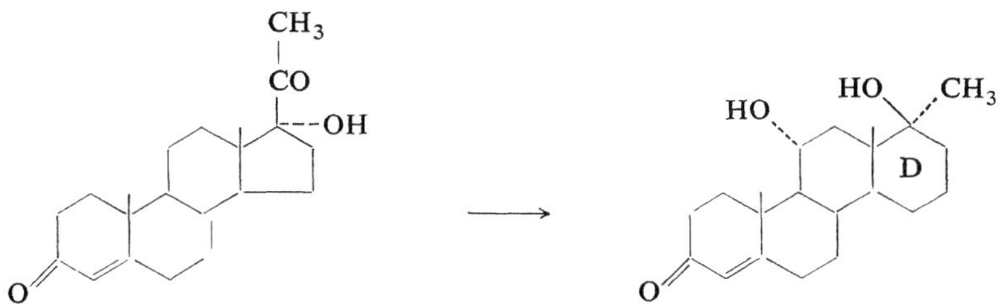

17α-hydroxyprogesterone → 17aα-methyl-D-Homo-11α-hydroxytestosterone

## *C*) HYDROLYSIS

### *a*) Microbial saponification of steroid esters

$$\text{R—COOR}' + \text{H}_2\text{O} \longrightarrow \text{R}'\text{OH} + \text{R—COOH}$$

By microbial hydrolysis of steroid esters free alcohol with the respective acid is obtained. CHARNEY *et al.* used the strain *Flavobacterium dehydrogenans var. hydrolyticum* (92) for the saponification of cortisol 11,21-diacetate.

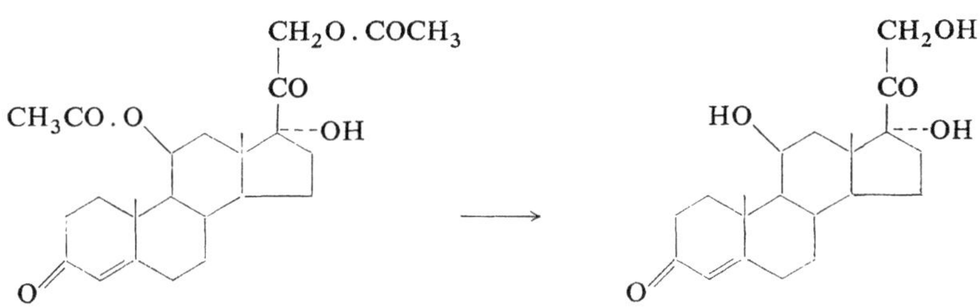

cortisol 11β,21-diacetate → cortisol

We have used the strain *Fusarium lateritium* (78) for the saponification of testosterone esters (formiate, acetate, propionate and benzoate).

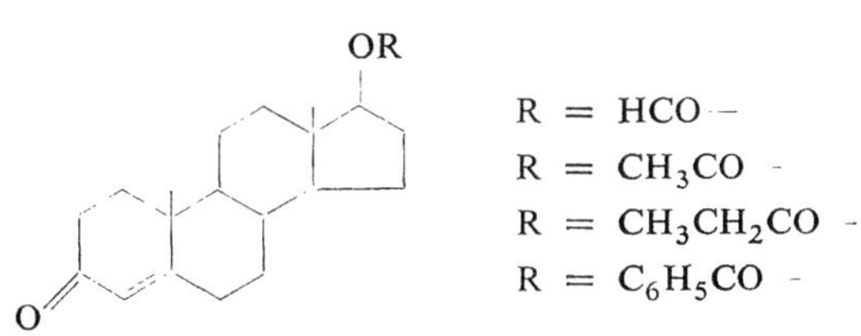

ester of testosterone

# *D*) ESTERIFICATION

## *a*) Microbial acetylation

$$CH_3-COOH + ROH \longrightarrow CH_3-COOR + H_2O$$

By means of the strain *Saccharomyces fragilis* McGuire (260) has proved the formation of the corresponding acetate from testosterone.

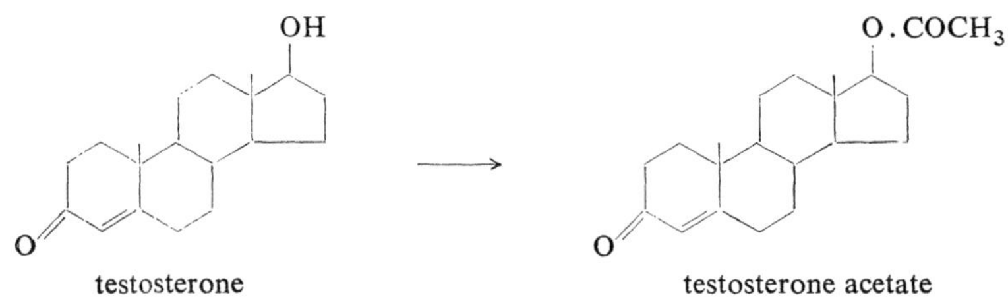


The ability to acetylate the steroid molecule has been proved a uniform sign of those strains and species of the order *Endomycetales* which are simultaneously able to ferment lactose.

# 3

# MECHANISMS OF MICROBIAL TRANSFORMATIONS OF STEROIDS

Various types of steroid transformations in microbes are very often combined so that the cells of microbes transfer the starting steroid into the end product possibly via several intermediate products, that are sometimes more or less easily distinguished by analysis or isolation. Each individual transformation step is catalyzed by a specific enzyme of microbial cells (12, 23, 33), sometimes of constitutive, sometimes of adaptative (inducible) character. In other cases we find among the end products of the transformation of a certain steroid several metabolites, or taken quantitatively, the main metabolite and several by-products. In addition to the current *oxido-reductions* of the steroid skeleton catalyzed by the individual enzymes, we witness, already in reactions of the *hydroxylation type,* the participation of the whole enzymatic system, which is partly insoluble and already inactive after desintegration of the cells. Otherwise the simple oxido-reductions, introduction of a new double bond especially in position 1 – 2 of the steroid molecule, have been already performed in cell-free systems with isolated enzymes (410).

An utter special position among the individual types of the transformations of steroids is held by the reactions *splitting the bond between the carbons* either in the side-chain in position 17 – 20 (160, 334, 477, 493, 388, 430), or in the ring skeleton mostly in position 2 – 3 of ring A (479), 9 – 10 of ring B (379, 382*a*, 417, 502*a*), or 13 – 17 of ring D (193*a*, 338*a*). The enzymic systems catalyzing this cleavage have been very little studied so far and are evidently more firmly rooted in subcellular structures of the microbe cells. These splittings of steroids leading to seco derivatives with altered solubility are the initial steps for the utilization of the steroid substrate as a source of carbon for certain microbes.

## Solubility of steroids and enzyme action

The *solubility of steroids* in water is usually very slight and is not in itself decisive for a possible conversion in a certain microorganism. In most cases steroid concentration used for transformations have been 0·05 – 0·1%, but the solubility of steroids in water is often essentially lower. Kondo and Masuo (226) described in

1 – 2-dehydrogenation of cortisol into prednisolone a special phenomenon of fermetation of steroids, which was called by them "*pseudocrystallofermentation*". The gram quantities of steroid (1 – 50 g/100 ml of medium) were directly without solution incubated in suspension with the medium and pre-cultivated microorganism and high yields of prednisolone were obtained. The added steroid seems to dissolve gradually, to react with the respective dehydrogenase of the microorganism and the arising prednisolone again cristallizes in the solution before an equillibrium, appropriate to low solubility, has been reached. The steroid pasees into solution gradually, after being in contact with the surface of the microbial cells and this process of subsequent cristallization of the metabolite occurs of course in the medium also in the presence of much lower concentrations (about 0·1% of steroid). A similar "*enzymatic transcrystallization*" has been described by KITAHARA *et al.* in the reaction catalyzed by fumarase of Lactobacilli in heterogenous phase (viz the rise of malic acid from fumaric acid).

## Specificity of enzyme action

The microbial enzymes catalyzing transformations of steroid often have a considerable *specificity* regarding the structure of the steroid *substrate* (414, 451, 456, 481, 485). Even though there are analogous groups in the same position in the steroid structure, not all such steroids are transformed uniformly by microbial enzymes. A role is also often played by the substituents in other positions of the steroid skeleton, often also by the length of the side chain in position 17. In contrast to that, the oxidation of the secondary alcohol group in position 3 of the steroid is often not specific for a certain group of these substances and they are quite commonly transformed into the respective ketones of the androstane, pregnane as well as cholestane series. Our strain *Penicillium notatum* transforming selectively and in high yields progesterone into testosterone (160), performed oxidations with cleavage of the side chain in all pregnene derivatives showing the conjugated structure 3-keto-$\Delta^4$ in ring A into respective androstene derivatives with hydroxy group in position 17$\beta$, as far the starting pregnene derivative had no hydroxyl in position 17$\alpha$. On the other hand, in the hydroxylation of cortexolone into position 11$\beta$ by means the enzyme system of *Absidia orchidis,* the hydroxyl group in position 17$\alpha$ was prerequisite. 11$\alpha$-Epimer was formed from cortexone instead of corticosterone. 11$\alpha$-Hydroxyprogesterone resulted analogously from the hydroxylation of progesterone.

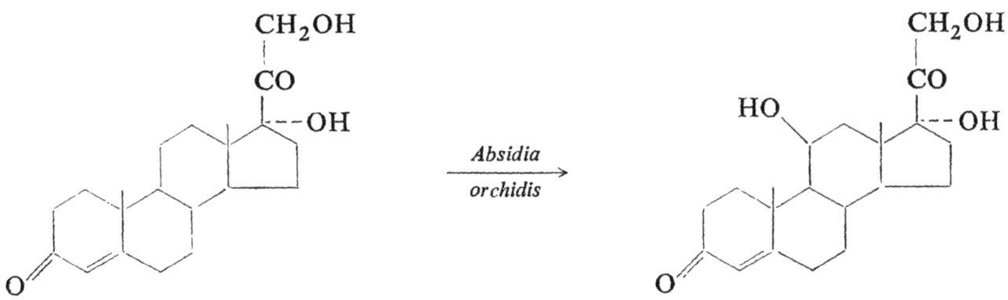


cortexolone cortisol

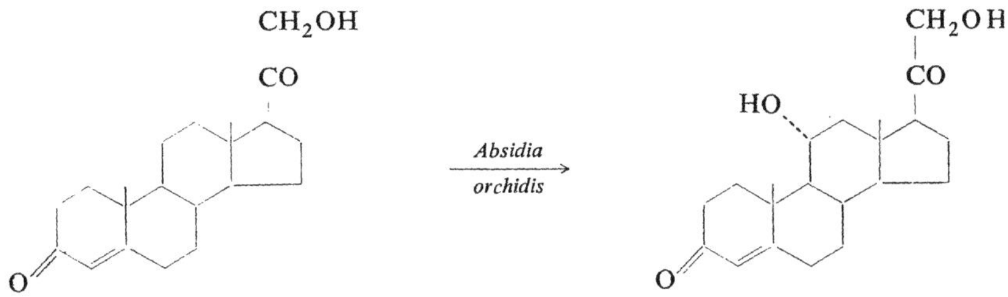


cortexone 11α-epicorticosterone

In conection with the position specificity are further some differences in the course of oxidation of alcohol groups in positions 3 and 17 of androstene derivatives (451), or in position 20 of pregnene derivatives (85). Also in the reduction transformations of the respective keto derivatives the position specificity of the substrate structure acts in a very expressive way. Several *Mycobacteria* (82) reduce for instance keto group in position 20 of the pregnene ring, the keto groups in positions 3 and 11 not being influenced. The introduction of a new double bond into the steroid by the action of dehydrogenation enzymes of microbes is often specific for position 1 – 2 and cannot be carried out by these enzymes into other positions of the steroid. Hydroxyl is introduced by some microbes into one or more positions of the steroid preferentially and several enzymes participate hereby. But even here, the substrate specificity plays a certain role, as has been demonstrated in the example of hydroxylation into position 11$\beta$ and 11$\alpha$ in the enzymes of *Absidia orchidis* (79).

The hydroxylation into positions 11$\alpha$, 11$\beta$ and 6$\beta$ by means of cultures *Cunninghamella blakesleeana* (79, 325) also depends to a considerable degree upon the structure of the steroid substrate. Steroids of different structure evidently take part in different ways in the induction of the respective enzymes. A. SCHUBERT *et al.* (25) described the hydroxylation into positions 12$\beta$ and 15$\alpha$ by means of the enzymes of *Calonectria decora* using differently substituted monohydroxyprogesterones. The enzymic system *Curvularia lunata* (371) hydroxylated cortexolone and progesterone into position 11$\beta$ (and 14$\alpha$), while 14$\alpha$-hydroxyprogesterone was hydroxylated into position 6$\beta$.

The hydrolytic enzymes of different microbes, too, show a different *position specificity* in the saponification of acetates of steroid alcohols. TAMM and GUBLER

(454) followed the action of *Fusarium lini* on various acetylated heart glycosides. The esterase of this microbe hydrolyzed the acetyl groups in positon 3 and 16, while the enzyme of *Giberella saubinetti* did so in position 3 only. The esterases of *Corynebacterium simplex* hydrolyzed acetate esters in positions 3 and 21 (313), the enzymes of *Flavobacterium dehydrogenans* in positions 3, 11α and 11β (92). The esterase of *Nocardia restrictus* hydrolyzed the acetoxy groups in positions 2β, 3, 6α, 15α, 16α, 17, 19, 20 and 21, while other acetates were resistant against hydrolysis (416). Neither were the esters of propionic, succinic and benzoic acid saponified into free steroid alcohols.

In a great many microbial enzymes catalyzing the transformations of steroids a considerable *stereospecificity* was observed as well with regard to the substrate as to the product of the transformation. VISCHER and WETTSTEIN (497) observed for the first time that if by transformation of a steroid a new centrum of asymetry is formed, usually only one of the possible epimers arises. If the microbial enzymes act on racemic steroid substrates, only the "natural enanthiomer" (*d*-form) is further transformed, while the antipode (*l*-form) remains unchanged. Thus was born a method for the resolution of steroid racemates obtained by total synthesis and the procedure was checked in a series of transformations, reduction of keto group in position 17, in hydroxylation in position 21 and dehydrogenation in position 1–2.

As an example of stereospecific reduction of 20-keto group of pregnane steroids may be considered the processes carried through with some *Streptomycetes*, leading to 20β-hydroxy derivatives. Using however, the enzymic systems of *Rhodotorula longissima* and *Rh. glutinis*, the respective 20α-hydroxy derivatives (444) were obtained. Another example of stereospecificity of microbial enzymes in transformations of steroid was presented by TALALAY *et al.* (448) by isolation and characterization of 3α, 3β, 17β-hydroxysteroid dehydrogenases in a species designated as *Pseudomonas testosteroni*.

We have examined the oxidation of 17-epimeric $\Delta^5$-androstene-3β,17-diols by means of cells of *Penicillium notatum* and *Actinomyces globisporus* (162). In the first case the secondary alcohol group is oxidized in position 3 into the respective 3-keto derivative while without oxidation in position 17 neither in the configuration of hydroxyl 17α nor in 17β. Out of this transformation there resulted either testosterone or its 17α-epimer, according to the starting steroid. Using the cells of *Actinomyces globispours*, both positions 3 and 17 were oxidized when the starting steroid had the configuration of hydroxyls β and there resulted androstenedione on the one hand, and 17α-epitestosterone on the other hand. When the starting steroids had configuration α in position 3, there occured no oxidation in this position even in this case.

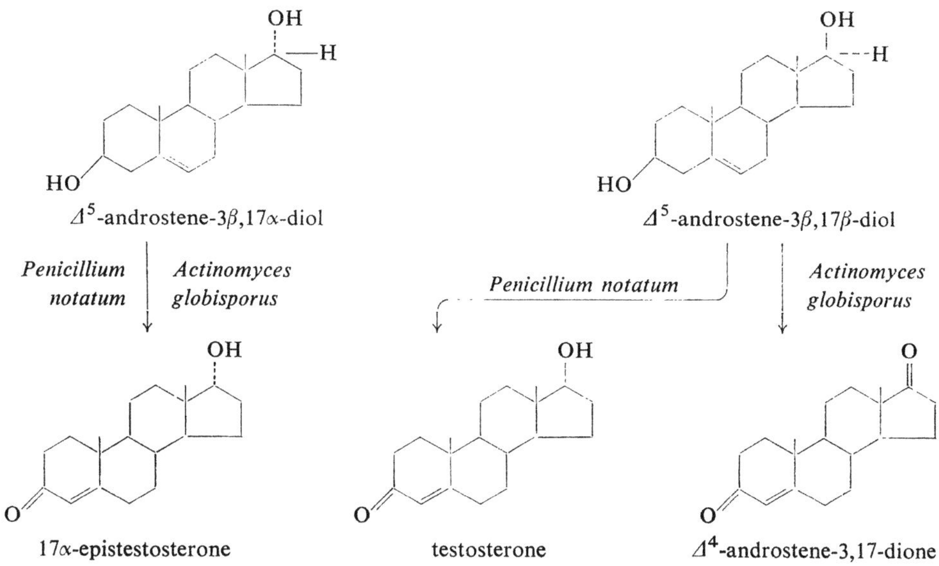


The qualities of *enzymes* catalyzing similar transformations have been studied especially by TALALAY *et al.* (478–452, 278) with special regard to the induction of biosynthesis of these enzymes, their isolation and purification.
Individually could be obtained:

3α-*Hydroxysteroid dehydrogenase* catalyzing the reversible transformation of 3α--hydroxy- and 3-ketosteroids saturated. $\Delta^4$-3-Ketosteroids, however, are not reduced.

*Dehydrogenases of* 5α – *and* 5β – *steroids,* catalyzing the introduction of double bond into ring A of the androstane skeleton.

*Isomerase of* $\Delta^5$-*3-ketosteroids,* catalyzing the *irreversible* transformation of $\Delta^5$-3--ketosteroids into $\Delta^4$-3-keto-konjugated isomer in the androstene as well as in the pregnene series.

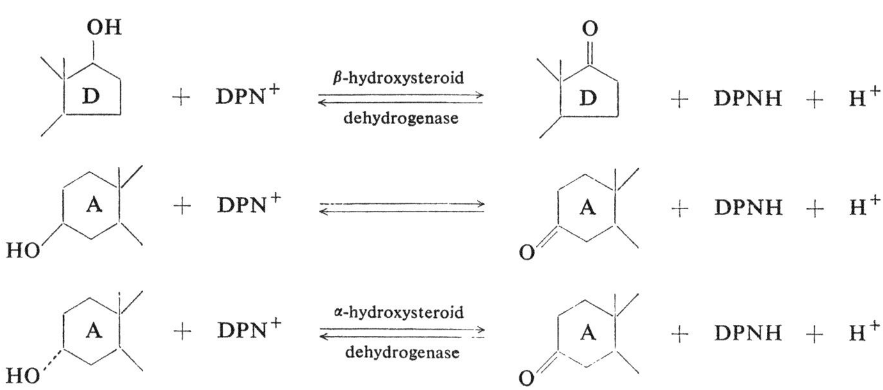

The reactions catalyzed by the 3α-hydroxysteroid dehydrogenase and 3-ketosteroid isomerase follow the Michaelis-Menten equation, whereas the β-hydroxysteroid dehydrogenases show considerable variations, when higher concentrations of the substrates had been used.

## Mechanism of isomerization

The ketosteroid isomerase of *Pseudomonas testosteroni* (211, 503) is one of the most active enzymes. Every molecule (mol. weight = 40 800) transforms 16 millions molecules of $\Delta^5$-androstenedione into the $\Delta^4$-isomer per minute (503). The cristalline isomerase catalyzes the isomerization of various 5–6 and 5–10 unsaturated 3-keto-steroids.

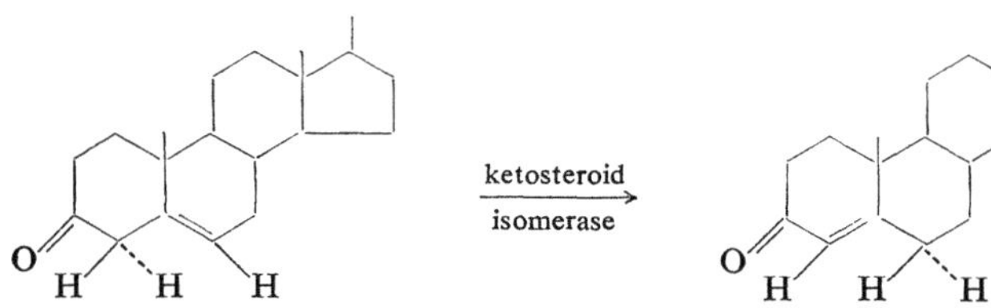


$\Delta^5$-steroid $\Delta^4$-steroid

The low concentrations of estradiol and of 19-nortestosterone competitively inhibit the action of the enzyme, forming molecular complexes with competively acting steroids. The isomerizations performed with the enzyme in a medium containing heavy water have shown that the process is carried through without incorporation of the isotope into the product. The reaction is therefore realized by a direct intra- or intermolecular transfer of proton (503). The chemical reactivity of the steroid itself and the character of reaction catalyzed by the enzyme suggests the probability of enolization together with formation of homo- or heteroannular diens or of a combination of these structures.

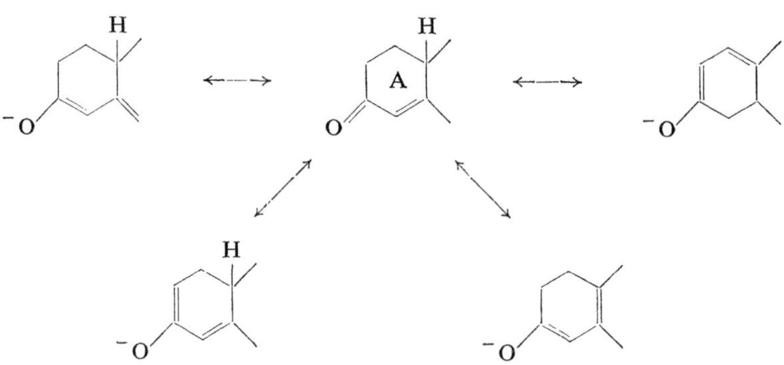


*Enol structures derived from 19-norsteroids*

TALALAY *et al.* (199, 449) have further proved that the steroids can act as *carriers of hydrogen* or coenzymes in oxido-reductions mediated between two types of pyridine nucleotides (TPNH and DPN). The basis of this finding is given by statement that the hydroxysteroid dehydrogenases reacting with estradiol or testosterone may perform transhydrogenations. The steroid hormones may, in this way, influence various metabolic reactions and thus various well-known biochemica effects of these hormones may be explained.

Recently TELLER and BONGIOVANNI (459) have described a mutant of *Pseudomonas testosteroni* with missing activity of 17$\beta$-hydroxysteroid dehydrogenase while the other enzymes were intact. The mutant may well be used in the analysis of steroids. The microbe enzyme catalyzes the oxidation of dehydroepiandrosterone into androstendione; it is, however ineffective upon testosterone. Similarly 20$\beta$-hydroxysteroid dehydrogenase from *Streptomyces hydrogenans,* isolated by HÜBENER (195–197) has been used for the identification and separation of steroids. The enzyme is a valuable tool in the microchemical estimations and a routine method for the identification of some steroids has been developped on this basis (185).

MONDER reported about the *role of DPN* in the transformation of steroids; he has found that pregnene steroids react reversibly with DPN forming products with characteristic optical qualities. Similarly react the analogues of DPN, too. The enolate anion of the steroid side chain on C-17 condenses with the quaternary pyridine base of DPN by an addition reaction which is reversible. The terminal oxidation, i.e. the transfer of $H^+$ up to O is mediated in moulds by an electron transfer system similarly as in animal cells. DOWLER *et al.* have carried through experiments in this field with cell-free extracts of various moulds.

## Mechanism of 1–2-dehydrogenation

A dehydrogenase introducting specifically a *new double bond* in the ring A of steroids in position 1–2 was isolated by SIH *et al.* (412) from *Nocardia restrictus.* It is a matter of an adaptive enzyme bound to intracellulary particles. The enzyme was partially purified and separated from 4(5$\alpha$)-dehydrogenase; it can utilize some oxido-reductive dyes as an electron acceptor and this is how this preparation differs from 1-dehydrogenase isolated later from the cells of *Pseudomonas testosteroni* as well (247–249). The enzyme contains the flavin prosthetic group catalyzing the direct dehydrogenation of the steroid. The participation of hydroxylated intermediary products in the dehydrogenation reactions is unlikely. On the basis of kinetic measurements the $\Delta^4$-3-ketosteroids can be also analytically distinguished from the $\Delta^{1,4}$-3-ketodienes and $\Delta^1$-3-keto-(5$\alpha$)-steroids (414). Many microorganisms dehydrogenating the steroid molecule in position 1–2 contain simultaneously the enzymatic system reducing the 20-keto group of pregnene steroids into the respective 20$\beta$-hydroxy derivatives. The different conditions in the phase of enzyme induction essentially change the relation of the dehydrogenation activity in position 1–2 and the reduction activity in position 20 of various steroid substrates.

The mechanism *1–2 dehydrogenation* of steroids, catalyzed by the dehydrogenase of *Bacillus sphaericus* has been studied by RINGOLD, HAYANO and STEFANOVIC (184, 346, 430*a*) in cells proper as well as in cell-free preparations in the presence of various electron acceptors as menadione (145), phenazinemethosulphate, dichlorophenolindophenol. In the absence of an electron acceptor the dehydrogenation proceeds only very slowly. In the experiments there were used steroids with an introduced isotope of hydrogen in positions 1 and 2 vith configuration $\alpha$ or $\beta$. Similar experiments were also carried out by HOLMLUND (192*a*) with an enzyme of species *Nocardia corallina.*

Similarly to what had already been proved by LEWY and TALALAY (249) on a soluble preparation of 1-dehydrogenase of *Pseudomonas testosteroni* as well as in the enzyme of *Bacillus sphaericus* the formation of hydroxylated intermediary products when introducing a double bond in position 1–2 is unlikely. The preferential transdiaxial loss of hydrogens has finally led to the suggestion of a two-step dehydrogenation process: The enolization of the ketone is followed by cleavage of activated hydride ion on C-1.

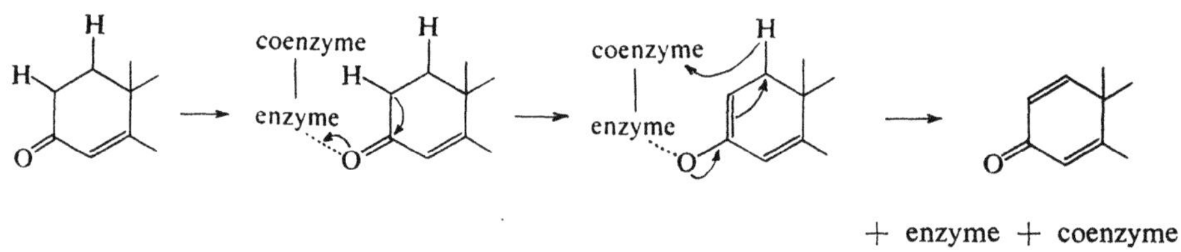


*Scheme of mechanism of 1–2 dehydrogenation of steroids*

## Reduction of the 20-keto group

A simultaneous *reduction of the 20-keto group* accompanies not only 1–2 dehydrogenation, but often also the cleavage of the side chain on C-17 in pregnene steroids under formation of androstene derivatives. Many *Penicillia* and *Aspergilli* e.g. splits progesterone via the 20$\beta$-hydroxy derivative into androstenedione and testosterone simultaneously opening ring D and giving rise to testololactone.

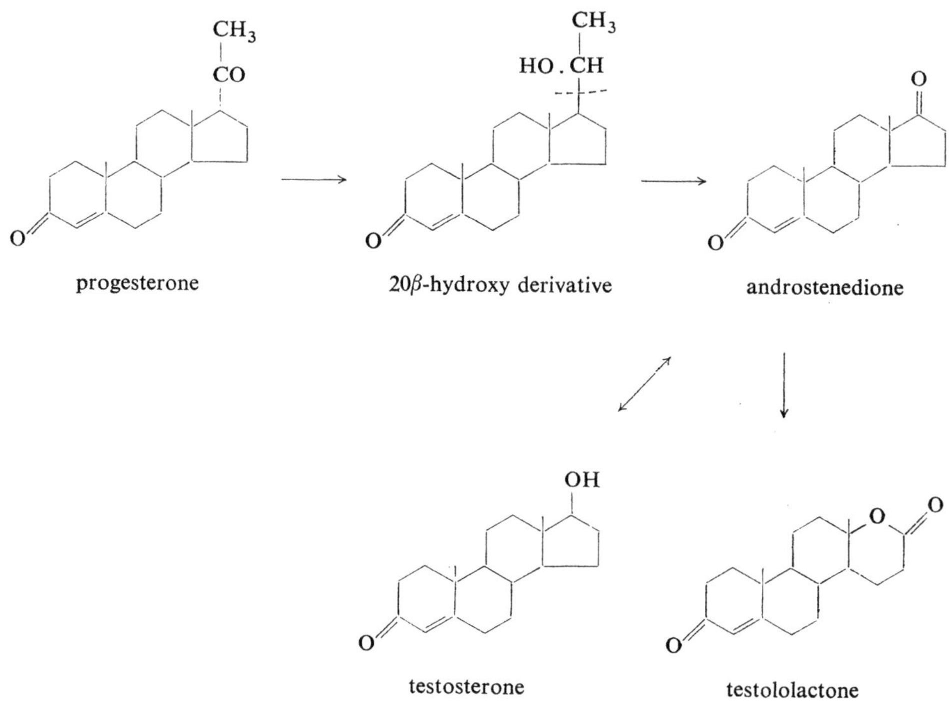


## Cleavage of the side chain of pregnenes

In some species of *Fusaria,* i.e. *F. lateritium, F. solani* and *F. caucasicum* a connexion of all these operations has been found: Dehydrogenation in position 1–2, splitting of the side chain of pregnene steroids in position C-17 a opening of ring D under formation of $\Delta^{1,4}$-androstadienedione and 1-dehydrostetololactone. Another mechanism of splitting the side chain was found in *Penicillium notatum* (160) under formation of a secondary alcohol group on C-17 with β-configuration of hydroxyl. Thus, over 60% of testosterone is formed from progesterone; analogously the respective derivatives of testosterone were obtained from 11-substitued derivatives of progesterone. Incubating progesterone with a culture of *Cladosporum resinane* f. *avellaneum,* FONKEN *et al.* (140) isolated besides androstenedione and testosterone also its acetate. The two-carbon residue split from the side chain of progesterone esterifies the molecule of resulting testosterone as was proved in experiments with progesterone labelled with isotope in the side chain. It is interesting that when using *Penicillia* for splitting the side chain from progesterone the formation of acetate of testosterone has never been observed; the result always being free alcohol only. The formation of acetate of testosterone has, however, been recorded also when reducing androstenedione in the yeast *Saccharomyces fragilis* (260).

## Mechanisms of the hydroxylation

As regards the mechanisms of *hydroxylation processes* two fundamental fact have been established. The oxygen of the introduced hydroxyl group originates from atmospheric molecular oxygen and not from water oxygen. This was found by the team of the *Worcester Foundation* (*Shrewsbury*) in USA and coll. (181, 239) in the hydroxylation into position 11, in cells of the adrenal cortex as well as in microbial hydroxylations of steroids into positions 6$\beta$, 11$\alpha$, 12$\beta$, 15$\alpha$, 17$\alpha$ and 21 using the isotope $^{18}O$. The second fundamental fact of the mechanism of steroid hydroxylation consists in establishing the character of the reaction which is a *direct substitution of hydrogen in the hydroxylated position* and not a Walden rearrangement with removal of hydrogen from the opposite position towards the hydroxylated one. These conclusions were made on the basis of experiments carried out with steroids labelled with tritium in certain positions (182, 183).

The labelled pregnanedione (T-11$\alpha$, 12$\alpha$) was e.g. incubated with a culture of *Rhizopus nigricans* and the obtained 11$\alpha$-hydroxyderivative contained only 31% of the original T of the starting substance. On the other side, the similarly labelled progesterone yielded after perfusion through beef adrenal glands the 11$\beta$-hydroxy derivative, which practically did not show any loss of isotope T. From the results of these studies performed by the *Worcester Foundation* and *Upjohn Company* (Kalamazoo) teams in USA (102, 182, 183), it is to be concluded that the enzymatic hydroxylations of steroids have mechanism of simple substitution of the hydrogen in a hydroxylated position. This conclusion was also additionally checked by an analogous microbial hydroxylation in position 12 in co-operation with the team of M. HAYANO and A. SCHUBERT (*Jena, East Germany*).

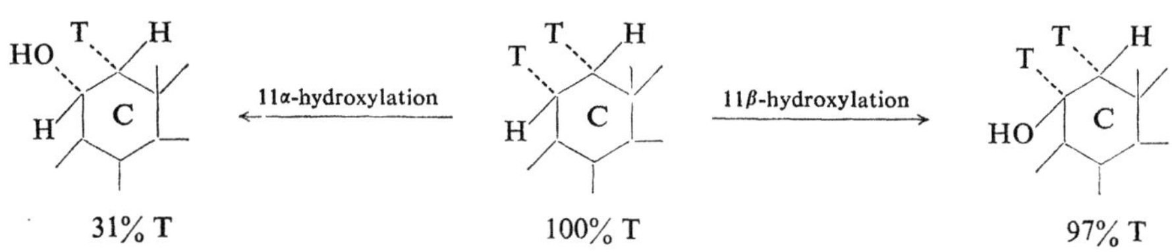


In the hydroxylations of cortexolone into the positions 11$\beta$ and 11$\alpha$ small amounts of cortisone were also isolated. It is interesting that this oxidation to a ketone has the character of a *direct* replacement of two hydrogens in position 11 by oxygen and not of a gradual oxidation via secondary alcohol.

Regarding the hydroxylation of steroids into *enolizable positions* e.g. into position 6 of ring B, the situation is somewhat more complicated considering the possibility of formation of enol intermediates before the hydroxylation proper. So there are two possible intermediary products in the hydroxylation of $\Delta^4$-androstene-3,17-dione by means of *Rhizopus arrhizus* according to whether 6$\beta$- or 6$\alpha$-hydrogen is liberated.

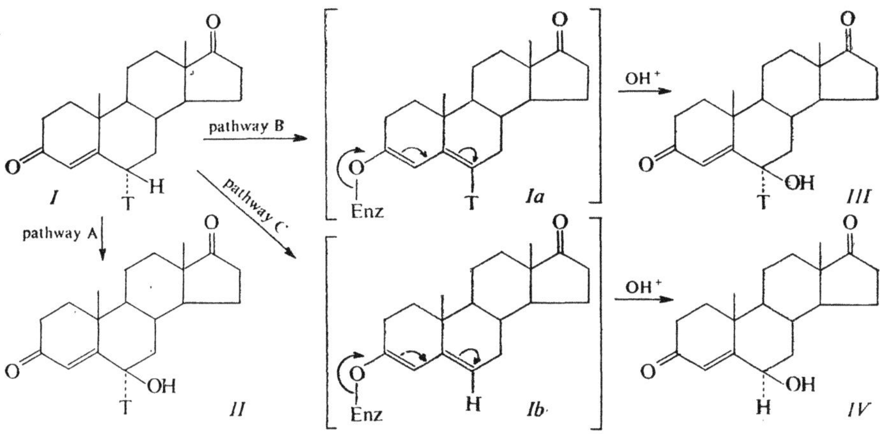


The *pathway A* represents the direct substitution of 6β-proton by hydroxyl. The *pathway B* and *C* assumes the enolization under participation of 6β- and 6α--hydrogen under formation of *enol Ia* and *Ib*. The results established by M. HAYANO and D. H. PETERSON *et al.* (47) have shown that the *pathway C* may be excluded as a possible hydroxylation mechanism, the *pathway A* and conditionally also *B* are the most probable ones, which fact indicates that *no keto-enol tautomeria occurs* in a considerable extent before the hydroxylation into enolizable positions in the steroid skeleton.

## Microbial degradation of steroid molecules

Most papers about microbial transformations of steroids deal with androstane and pregnane steroids with 19 and 21 carbon atoms in the molecule. Comparatively fewer papers discuss the transformations of *sterols* and *bile acids* under catalytic action of microbial enzymes. Cholesterol is very commonly oxidized by microbes into cholestenone. Some microorganisms perform a further oxidative cleavage of the cholesterol molecule, where in the side chain on C-17 of cholesterol is split, too. TURFITT (475–480) investigated the splitting of cholesterol with soil *Proactinomycetes* after preliminary blocking of the free hydroxy group in position 3 and he succeeded in isolation of 3-ketoetienic acid and A-seco-acid (Windaus' keto-acid) besides iso-caproic acid split from the side chain.

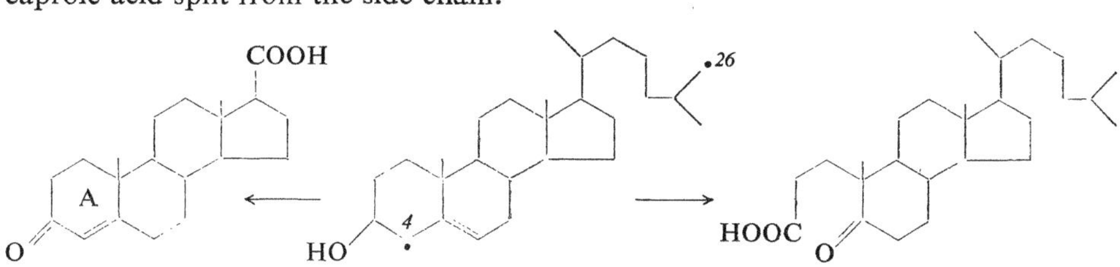


3-ketoetienic acid cholesterol Windaus' seco-keto-acid

Similarly HORVÁTH and KRÁMLI (194) reported, too, about the splitting of methylheptanone from the side chain of cholesterol by the action of certain microbes. When $^{14}C$ labelled cholesterol had been exposed to the action of soil *Mycobacteria* it was established by STADTMAN *et al.* (430) that carbon in position 4 of ring A is being oxidized into $CO_2$ four times faster than C-26 of the side chain.

In many microbial transformations of steroids also from the group of pregnane and androstane derivatives after longer incubation of the steroid with microbial cells, either analytically or after isolation of the metabolites, a decrease of the steroid fraction soluble in organic solvents has been observed. The oxidative changes in different positions of the steroid skeleton frequently serve as *initial steps for a further splitting* of the individual steroid rings. In the species *Nocardia restrictus*, C. J. SIH (413, 417, 502*a*) has found that the initial steps of steroid degradation are dehydrogenation in position 1—2 and hydroxylation in position 9α, followed by the splitting of ring B.

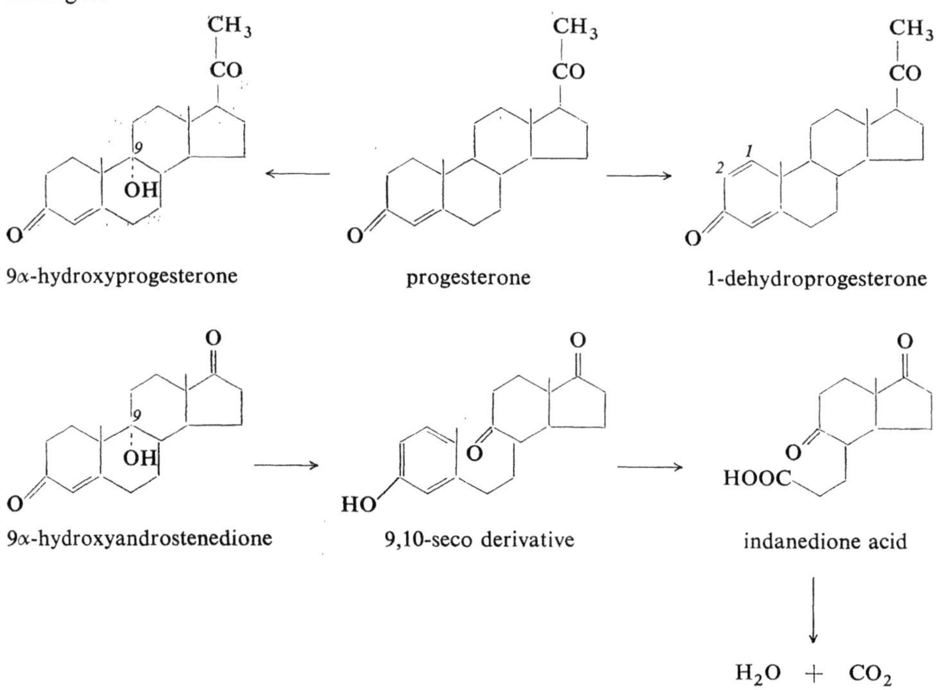


Similar splitting was found in progesterone as well as in androstenedione as the starting steroid. This mechanism is evidently common enough, since different microorganisms performing 1—2 dehydrogenation are also able to hydroxylate the steroid in position 9α. When position 9 and position 2 are blocked, the steroid cannot be split by oxidation using the mentioned microbe eventually into $CO_2$ and $H_2O$.

Most recently WANG and SIH (502*a*) isolated after incubation of testosterone with *Nocardia restrictus* additional 5 metabolites, which were identified as the corresponding 9,10-seco derivatives having hydroxyl in position 9 and further as lactones of 9-hydroxydicyclic acid. As concerns the *aromatization of ring A* after 1,2-dehydrogenation of steroids, SIH and RAHIM (418*a*) have found, in the study of the mechanism, that the action of steroid 1-dehydrogenase yields vinylogs of *β*-hydroxyketones which may undergo spontaneous nonenzymic rearrangement (reverse aldolization) to give their respective phenols.

Similarly the splitting of steroids with the use of the species *Mycobacterium smegmatis* has been studied by H. SOBEL (425) and K. SCHUBERT (375–382*a*). Various 17-ketosteroids were pre-dehydrogenated by the enzymatic system of the microbe in ring A and then split into 9–10-seco products with an aromatic nucleus A. After incubation with *M. smegmatis* the aromatic B-seco compound was isolated from progesterone, from which compound arises a ketoacid preserving ring C and D.

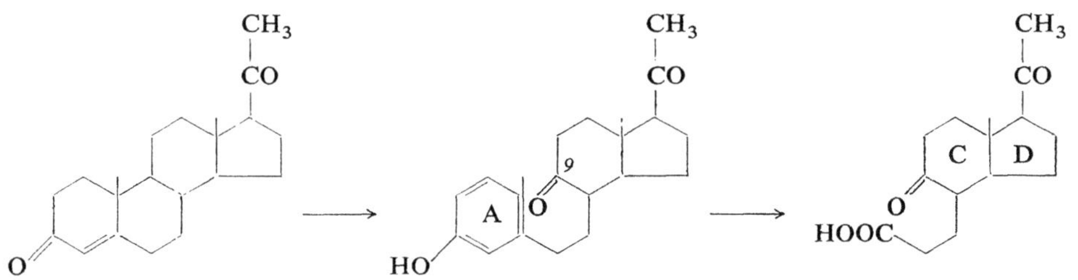


progesterone — aromatic B-seco compound — keto acid

From this *keto acid* 8 further metabolites were obtained by further incubation with *M. smegmatis* (K. SCHUBERT *et. al.*: Biochim. biophys. Acta 111: 529, 1965):

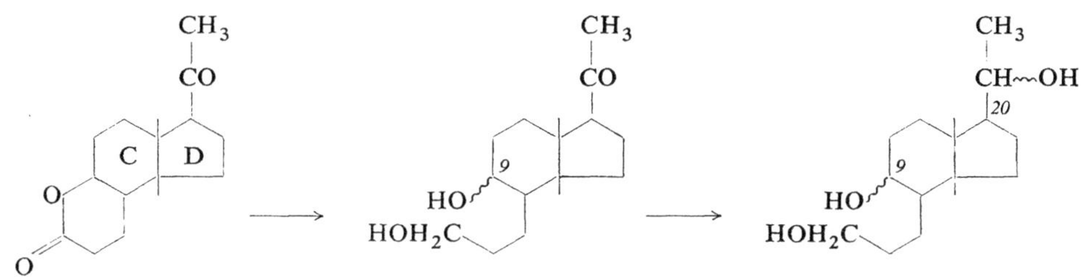


*δ-lactone* — *3 compounds* 9α-, 9β- and 9-keto- — *4 compounds* 9α, 20α-, 9α, 20β-, 9β, 20α-, 9β, 20β-,

Testololactone arising by the oxidative splitting of many pregnene and androstene derivatives in *Aspergilli* and *Penicillia* is a lactone of the D-seco acid being formed after opening of ring D (193*a*, 338*a*). Contrary to expectation, *testolic acid arises by enzymatic hydrolysis of the lacton and is not precursor in microbial formation of testololactone* (193*a*).

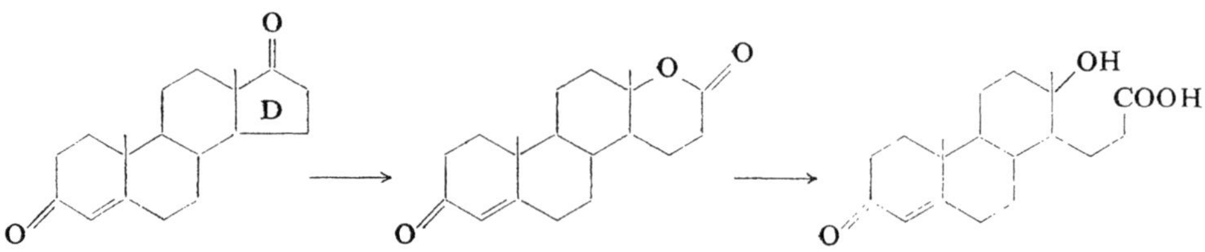


$\Delta^4$-androstene-3,17-dione testololactone testolic acid (D-seco acid)

The mentioned examples already show deeper oxidative interference with the steroid molecule with *degradation* of 4-member sterane skeleton. It cannot, however, be generalized that by adaptation all microorganisms would be able to use steroids as a source of carbon. The initial steps of oxidative conversions of steroids yield but relatively little energy to the microbe in comparison with other carbon-containing nutrients. By interfering with the metabolism of the microbes some steroids seem to compete with the current intermediates of the oxido-reductive transformations in bonds for the active site of the enzymes, inhibit utilization of sugars and the consumption of oxygen (74, 76, 77), mostly being transformed themselves, by the induced enzymes of microbial cells, into metabolites which do not posses these properties any more.

# 4

# THE INFLUENCE OF STEROIDS ON MICROORGANISMS

Microorganisms usually do not come into contact with steroids. These substances are not physiological substrates for microbes. An exception make some sterols, in particular ergosterol, which is synthesized, very commonly by microbial cells. The physiological importance of ergosterol with regard to the cells of microbes remains for the present quite unknown.

It has been already demonstrated in the preceding chapters, that microorganisms contain enzymatic systems capable of changing steroid compounds analogically as to the cells of higher vertebrates. Let us consider now the reverse aspect, namely, how do steroids affect the microbes themselves.

It appeared, that various steroid derivatives and also the proper steroid hormones influence microorganisms in different ways; they may be *growth factors* for some species of microbes (2a, 62, 348, 492). For other microorganisms some steroids may act as suitable *stimulators* (59, 61), but most frequently steroids in suitable concentrations function as *inhibitors* of the microbial growth (90, 100, 289). Nutritional requirements of steroids have been found in some strains of *Paramecium,* in *Trichomonades,* further in pleuropneumonia-like organisms (*PPLO*) (118) and in the yeast.

Microbes well known from various biochemical and genetical studies have often been chosen for the purpose of testing the influence of steroids on the growth. LESTER (244), for instance, has employed the microorganisms *Neurospora crassa,* which on the one hand synthesizes ergosterol and on the other hand transforms some steroid hormones. The aim of the studies following the effects of steroids on the growth of microbes had been most frequently the screening of substances from series of compounds with a more or less specific *antimicrobial* effect (90). Further on, an adequate testing object for the activity of steroid hormones has been looked for in the microorganisms, since this method of bioassay could replace an expensive biological titration of drugs on animals. Such experiments proved to be successful especially in the field of vitamines. Another aspect of investigating the influence of steroids on microorganisms was to find out the relations between the effects exerted by steroid compounds on the growth and metabolism of microbes on the one hand, and the ability of these microbes to transform steroid molecule on the other hand (76, 77).

## Microbial binding of steroids

Some microorganisms, in particular some fungi, have the ability of binding steroids, so can be set free by extraction by means of a solvent only after lyofilisation of the culture (170). For example in the species *Penicillium cannescens* even a relation has been found between the structure of the steroid and the avidity to bind a certain component of the microbe. Chiefly sterols with a long side chain, 3 beta-hydroxy group and a double bond in position 5–6 were found to be bound.

**Table 1.**

*The binding of β-sitosterol in various microbes*

(5 days' cultivation, 200 micrograms of steroid in 1 ml)

| Species | % of bound sterol | Species | % of bound sterol |
|---|---|---|---|
| *Penicillium canescens* | 65 | *Streptomyces levandulae* | 3 |
| *Gliocladium catenulatum* | 84 | *Nocardia erythropolis* | 0 |
| *Aspergillus flavus* | 45 | *Microspora* sp. | 23 |
| *Cylindrocephallum aureum* | 17 | *Mycobacterium laticola* | 16 |
| *Chaetomium funicola* | 60 | *Corynebacterium flaccumfaciens* | 12 |
| *Hormodendrum olivaceum* | 35 | *Pseudomonas aeruginosa* | 16 |
| *Rhizopus nigricans* | 35 | *Bacillus cereus* | 6 |
| | | *Staphylococcus aureus* | 8 |

In addition to sterols there has been examined the binding of all types of steroids in rinsed cells of the species *Penicillium canescens.*

**Table 2.**

*The binding of various steroids by Penicillium canescens*

| Steroid | % of bound steroid |
|---|---|
| β-sitosterol | 65 |
| Cholesterol | 69 |
| Stigmasterol | 49 |
| Dehydroepiandrosterone | 24 |
| $\Delta^5$-pregnen-3-ol-20-one | 19 |
| $\Delta^4$-androstene-3,17-dione | 21 |
| Cortisol | 3 |
| Testosterone | 7 |

The sterols were evidently bound in the highest extent. During incubations, the presence of oxygen was not necessary for the binding of sterol. When an alcoholic group in the position 3 of the steroid was oxidized in a keto-group, the binding of the steroid with the microorganisms decreased.

## The influence of sterols

RAAB (339) examined the effect produced by some sterols, particularly by cholesterol and ergosterol, on the growth of *Mycobacterium tuberculosis* and *Staphylococcus aureus.* A marked inhibition of the growth of microbes has been observed in certain concentrations of the employed sterols. SQUIRE (428) found similar action of cholesterol on 15 different strains of aerobes and anaerobes. ADELBERG (37) attempted to adapt some microorganisms to steroids, and succeeded in obtaining mutants in the species *Escherichia coli* partly adapted to ergosterol. MATKOVICS (283, 284) published very extensive studies on the action of cholesterol and ergosterol on microbes, especially also on the moulds. He followed on the one hand the influencing of the growth of microbes by various concentrations of sterol, and on the other hand the utilization of nutrients from the medium. Ergosterol produced an inhibitory effect on *Penicillium chrysogenum,* but stimulated the other moulds. No effect has been produced on the growth of cells of *Neurospora crassa* (244). The sterols change the surface tension of the medium and through it also the permeability of the cell walls for different compounds. The direct action of sterols on the dehydrogenase systems of microbial cells has been also taken into consideration. A mention has been already made of the stimulating effect of cholesterol and ergosterol in the organisms *PPLO* (423).

## The influence of steroid hormones

From the androgen steroids, the influence of dehydroepiandrosterone on the growth of *Escherichia coli* in the presence of inhibitory doses of sulfanilamides and in the presence of pteroylglutamic acid was examined. GAINES (144) states, that androgen reverses the inhibition caused, for instance, by 4-aminopteroyl-glutamic, or pteroyl-aspartic acids, analogically, as the inhibition is reversed by pteroyl-glutamic acid. The androgens stimulate growth of the alga *Euglena gracilis* (61) already in the concentration of the medium $10^{-4}$ mg/l. On the other hand, MAXWELL (289) found, that $\Delta^4$-androstene-3,17-dione inhibits almost completely the growth of the yeast *Saccharomyces fragilis* in the dose of 130 micrograms/ml. In the same way acted also some other hormones, as progesterone and cortexone, although their very close derivatives, on the contrary, exerted no influence on the growth.

**Table 3.**

*The influence of various steroids on the growth of Neurospora crassa*
(250 micrograms of steroid/ml.)

| Steroid | % of inhibition | Steroid | % of inhibiton |
|---|---|---|---|
| Ergosterol | 2 | cortexone-acetate | 5 |
| Cholesterol | 4 | cortexone-glucoside | 8 |
| Cholestenone | 4 | 5β-dihydrocortexone | 18 |
| Estrone | 7 | 5α-dihydrocortexone | 39 |
| Estradiol | 7 | 3α-hydroxy derivative (-5β) | 26 |
| Estriol | 2 | 3β-hydroxy derivative (-5β) | 8 |
| Testosterone | 26 | 3β-hydroxy derivative (-5α) | 22 |
| Androstenedione | 18 | 7-hydroxycortexone | 7 |
| Dihydrotestosterone | 6 | 11α-hydroxycortexone | 5 |
| Pregnenolone | 9 | 11β-hydroxycortexone | 4 |
| Progesterone | 22 | 14α-hydroxycortexone | 10 |
| 17-OH-progesterone | 9 | 15α-hydroxycortexone | 0 |
| Cortexone | 55 | 17α-hydroxycortexone | 3 |
| $\Delta^1$-dehydrocortexone | 2 | 11β,17α-dihydroxy-11-keto-cortexone | 5 |
| | | 9α-fluoro-11β, 17α-dihydroxycortexone | 3 |

Notably LESTER with coworkers (244) carried out thorough studies of the influence of various steroids on the growth of microbes. First of all the influence of various steroids on the growth of *Neurospora crassa* has been investigated.

Of 33 examined steroids a strong inhibitory effect was shown by cortexone, according to the weight of mycelium. The inhibitory effect is relatively specific since related steroids have only a small or no effect. The influence of cortexone on the growth becomes evident without the lag phase, so that cortexone acts directly itself and not after a metabolical change.

The inhibitory activity and solubility of the tested steroides have no mutual relation. For the time being it is not clear, whether the inhibition called forth by cortexone is an expression of interference with a function of endogenic steroids of *Neurospora crassa*. Some steroids, including ergosterol, can partly reverse the inhibition of growth caused by cortexone. The intensity of inhibition is dependent upon the sugar used as a source of carbon. It is unlikely, that it is a differential effect of cortexone on the cellular permeability. The results of measurements of respiration and of the consumption of sugar have shown that cortexone interferes with the penetration of sugar into the cell. Also the consumption of amino acids has been inhibited by cortexone. Analogous inhibitory effects on the respiration and incorporation of nutrients towards animal cells have been found in cortexone already earlier.

When the influence of cortexone was tested on the growth of grampositive and gramnegative bacteria, on the yeast and on the moulds, an inhibition has been found above all in grampositive microbes, whereas gramnegative microorganisms on the whole were not affected by analogous doses of steroid (see Tab. 4 on p. 60), TARBET with coworkers has found an inhibition of the growth of *Microsporum gypseum* by cortexone. MEIER (292, 293, 294) examined the influence of cortexone besides other hormones on the respiration of the yeast.

Among the microbes, whose growth was inhibited by steroids, also some *Dermatophyta* (343, 90) were described, which are interesting in this respect because there have been found changes in the occurrence of infections by this fungi in the time of the changed content of steroid hormones in the tissues. The infections called forth by *Microsporum* heal usually spontaneously in adolescence. Eruptions of the infection, brought about by *Trichophyton,* were recorded in connection with the occurence of Cushing's syndrome, and also during a therapy by means of cortisone. The steroids can directly, or indirectly influence the development of these infections. In a number of steroid hormones a fungistatic activity against fungus *Trichophyton purpureum* and *T. gypseum in vitro* has been tested (343). In the androgen series the activity has been found especially in 17α-methyltestosterone.

CHATTAWAY (94) examined 37 various steroids as regards the inhibition of the growth of the fungus *Trichophyton rubrum,* while the majority of compounds with a strong inhibitory effect were steroids of both, the androstane and the pregnane series, with a keto group in position 3. The reduction of keto groups leads to a decrease, or complete disappearance of the inhibitory effect. Double bonds in the ring A and B are not necessary for inhibitory activity of steroids. The presence of 17α-hydroxy group leads to a complete loss of inhibitory activity of steroids. (Tab. 5.)

A high inhibitory activity was shown by compounds with a keto group, β--hydroxy group, β-acetoxy group or β-acetyl group on the $C_{17}$ of the steroid skeleton. It is unlikely, however, that there exists a parallelism between the inhibitory activity against the fungus and the hormonal effect of steroids. Cortexone has a high inhibitory activity, 2α-methyl-9α-fluorocortisol, however, is without effect, although its corticoid activity is high. Likewise corticosterone has been found active as far as regards the inhibitory effect, whereas 16α-methyl-9α-fluoroprednisolone has been inactive. On the contrary, androstane-3,17-dione shows a strong inhibitory activity, though hormonally it is nearly without effect. CHATTAWAY (95) next found, that the inhibitory influence of cortexone on the growth, endogenic respiration and consumption of glucose in *Trichophyton rubrum* is similar to the influence of uncoupling agents on the metabolism of yeast. The similarity of action between dinitrophenol and cortexone, in this respect, has been pointed out already by CONNER (99, 100, 101) in *Tetrahymena piriformis.* Both these compounds inhibited growth, and the inhibition was reversed by stigmasterol. A similar effect of cortexone on *S. fragilis, E. coli,* and *B. subtilis,* was manifested also by an inhibition of cytochrome reductase and blocking of production of ATP, as found out by YIELDING and TONKINS. Also in

**Table 4.**

*The influence of cortexone on the growth of bacteria, yeasts and moulds*
(80—250 micrograms of the steroid/ml)

| Microbe | % of inhibition | Microbe | % of inhibiton |
|---|---|---|---|
| *(gramnegative)* | | *(yeasts)* | |
| *Escherichia coli* | <5 | *Torulopsis utilis* | 25—33 |
| *Aerobacter aerogenes* | <5 | *Saccharomyces cerevisiae* | 65—94 |
| *Proteus vulgaris* | <5 | | |
| *Serratio marcescens* | <5 | *(moulds)* | |
| *Pseudomonas fluorescens* | <5 | *Neurospora crassa* | 40—65 |
| *Pseudomonas aeruginosa* | <5 | *Aspergillus niger* | 56—76 |
| | | *Cunninghamella blakesleeana* | 38—63 |
| *(grampositive)* | | *Curvularia lunata* | 60—100 |
| *Micrococcus aureus* | 73—94 | *Penicillium lilacinum* | 71—86 |
| *Sarcina lutea* | 90—100 | *P. puberulum* | 76—100 |
| *B. megaterium* | 49—100 | | |
| *B. subtilis* | 50—100 | | |
| *Mycobacterium rannae* | 97—100 | | |

the fungus *Trichophyton mentagrophytes* cortexone similarly as 19-norprogesterone was a more effective inhibitor of growth among the tested steroids.

The inhibitory activity of testosterone in *Trichophyton rubrum* is similar to that exhibited against *Dermatophyta* (343), or against *Neurospora crassa* (244) and *Saccharomyces fragilis* (289). TARBET, however, has found a weak stimulating effect on the germination of macroconidia in *M. gypseum*. Besides, in a number of papers, in particular the influence of estrogens on the growth of microbes has been followed, as well as on the endogenic respiration of the yeast. In addition to the steroid estrogens also the synthetic analogues were tested, particularly diethylstilbestrol.

In our attempts to determine the influence of steroids on microbes we have pursued a somewhat different aim. During the study of oxido-reductive capacities of microbes in some steroids (78, 80) determination of steroid metabolites, mostly by paper chromatography, has served as criterion of the suitability of the employed strains. During our studies we soon found that the transformation of steroid is frequently in the quantitative and as well as in the qualitative respect different in the same strain, if the detection of steroid metabolites is carried out after a different period of fermentation. Furthermore, there have been cases, when even after an incubation of 30 days no changes have been found of the steroid added into the medium. For this reason we endeavoured to find a procedure, which would enable us in a short time to determine, whether the enzymatic system of the employed microbe in given conditions is capable of transforming the steroid or not.

**Table 5.**

*The influence of steroids on the growth of Trichophyton rubrum*
(250 micrograms of steroid/ml)

| Steroid | % of inhibition | Steroid | % of inhibition |
|---|---|---|---|
| Androstane-3,17-dione | 96 | Estrone | 11 |
| 1-androstene-3,17-dione | 96 | 5-androstene-3β,17β-diol | 10 |
| 19-nortestosterone | 94 | 17α-hydroxyprogesterone | 10 |
| Cortexone | 91 | Prednisolone | 7 |
| 4-androstene-3,17-dione | 89 | Androstane-3α,17β-Diol | 4 |
| 5β-androstan-17β-ol-3-one | 88 | Aldosterone | 0 |
| Calciferol | 79 | Pregnenolone | 0 |
| Testosterone | 77 | Cortisol | − 1 |
| 7-dehydrocholesterol | 77 | Cortisone | − 3 |
| 1,4-androstadien-17β-ol-3-one | 72 | Compound S (Reichstein) | − 4 |
| Progesterone | 70 | Lanosterol | − 5 |
| 5α-androstan-17β-ol-3-one | 62 | 2α-methyl-9α-fluoro-cortisol | − 9 |
| 11α-hydroxyprogesterone | 59 | 9α-fluorocortisol | −12 |
| Corticosterone | 44 | Stigmasterol | −14 |
| Lumisterol | 42 | Cholesterol | −18 |
| Estradiol | 39 | Sitosterol | −18 |
| Ergosterol | 28 | 17α-hydroxypregnenolone | −37 |
| 17α-epitestosterone | 20 | | |

## Steroids and microbial metabolism

We have already found earlier in the strain *Rhizopus nigricans* (74) that progesterone, added into the incubation medium, blocks the utilization of sugar, whereas the metabolite of progesterone, its 11α-hydroxy derivative has no influence in this respect. On ground of this discovery it has been interesting to examine to what extent is affected the consumption of oxygen by both mentioned steroids. In addition to the hydroxylation of progesterone in *Rh. nigricans,* we next followed from this point of view the oxidation of dehydroepiandrosterone to 4-androstene-3,17-dione in *Actinomyces globisporus* and reduction of 4-androstene-3,17-dione in *Saccharomyces cerevisiae* to testosterone. The chosen types of reactions were advantageous for that reason because they offer on the whole unambiguous metabolites without side reactions.

Fig. 2 shows the consumption of oxygen in *Rh. nigricans,* when a concentration of 62 micrograms of steroid in 1 ml of medium has been used. Steroid which is transformed by an enzymatic system of the mould does suppress in given concen-

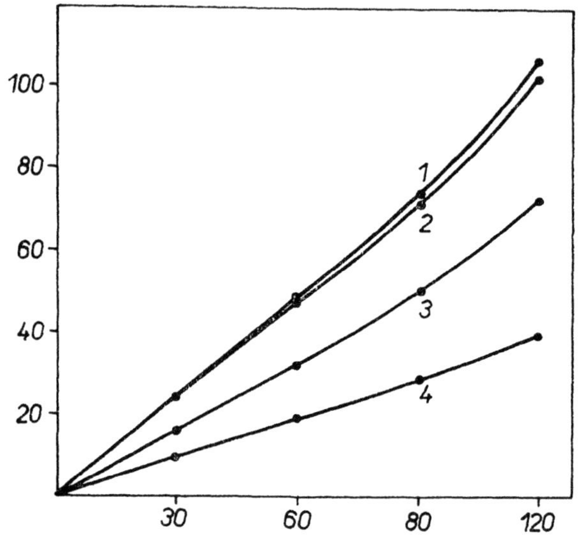


Fig. 2. The influence of progesterone (4), 11α-hydroxyprogesterone (2) and a mixture of both steroids in ratio 1 : 1 (3) on the consumption of oxygen in the strain *Rhizopus nigricans*. The culture without steroids (1). — *x*: time in minutes; *y*: microlitres $O_2$.

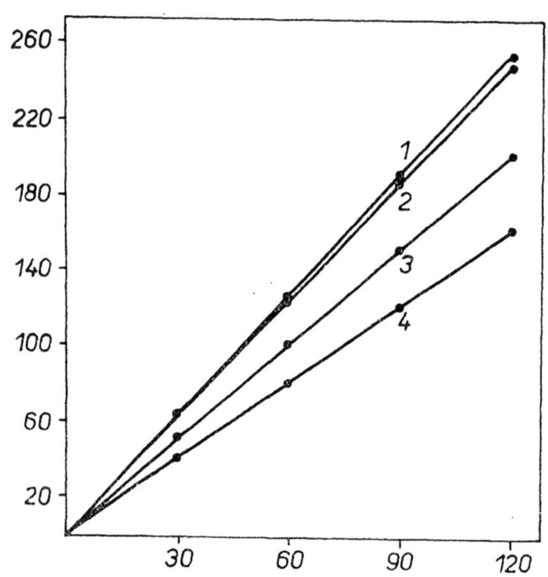


Fig. 3. The influence of dehydroepiandrosterone (4), androstenedione (2) and their mixture 1 : 1 (3) on the consumption of oxygen in the strain *Actinomyces globisporus*. Culture without steroids (1). — *x*: time in minutes; *y*: microlitres $O_2$.

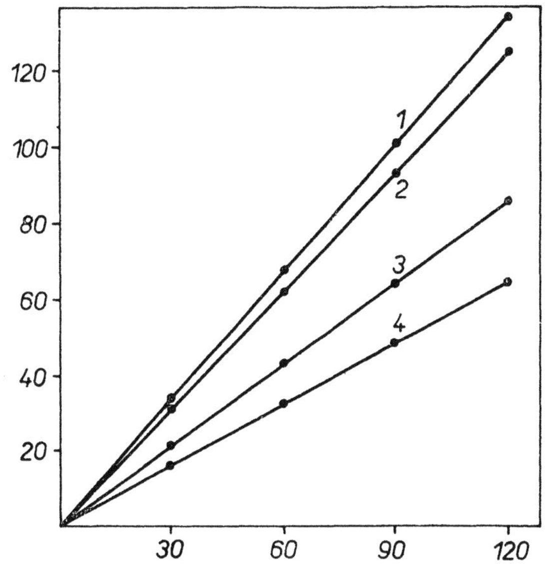


Fig. 4. The influence of androstene-3,17-dione (4) testosterone (2) and of their mixture 1 : 1 on the consumption of oxygen in *Saccharomyces cerevisiae*. Culture without steroids (1).
*x*: time in minutes; *y*: microlitres $O_2$.

tration the consumption of oxygen most intensively, whereas the originating metabolite exerts almost no influence at all.

In Fig. 3 is analogically illustrated the consumption of oxygen in *A. globisporus* in concentration of 62 micrograms of dehydroepiandrosterone, of androstene-3,17--dione, of their mixture, and in the control without steroids, here, too, the transformed steroid strongly suppresses the consumption of oxygen, whereas the originating steroid exerts practically no influence upon the consumption of the respective strain of microbe. Fig. 4 shows the rates of consumption of oxygen in *S. cerevisiae* under the influence of added steroids in concentration of 156 micrograms in 1 ml. Testosterone, originating through enzymatic transformation, has only slight influence upon the consumption of oxygen, whereas the added androstenedione strongly suppresses the consumption of oxygen by the yeast.

As shown in all quoted cases, the *transformed steroid always exerts an inhibitory effect on the consumption of oxygen by the microbe, while the metabolite, originating by transformation, practically does not influence respiration of the microbe any more.* For the present time it is not possible to determine whether the inhibition of respiration is of the competitive type with some intermediary product of metabolism of the microbe, or whether the inhibition of oxygen consumption is an expression of "*toxicity*" of the initial steroid, in respect with the microorganism, for instance by change of conditions of cellular permeability. It would then be possible to consider the transformation reaction of the steroid from that point of view of microbial metabolism as a *"detoxication" reaction,* during which a *disturbing compound is being changed to an "inert" compound.*

From this becomes evident a clear relation between an inhibitory action of the steroid and the ability of the microbe to transform the steroid in principal. A steroid that does not have an inhibitory effect on the respiration of microbes is also not metabolised by this kind of microorganism any more. For the present, with the exception of this dependence, no further relations of steroids to the growth of microbes have been found which could be generalised in the wide extent, be it for the genera of microbes, or on the contrary, for quite definite steroids structures. First of all, it is left to find out whether the steroids stimulating the growth of some microbes are changed in any way by their enzymatic systems in their structure, or whether they remain untouched. It is probable that for the present it is not possible to interpret uniformly the existing results of the influence of steroids on growth and upon the metabolism of some microbes. We can only conclude that steroids can act, and in some cases it has been already proved, in a purely physical-chemical way on the cellular membranes of microbes. But in other cases the steroids interfer with enzymatic processes in the cell either as their moderator in the sense of stimulation, or inhibition, and as a substrate which itself is subject to metabolic changes. In some cases the transformation of a steroid by the action of a microbe enzyme system proceeds in an increased extent, provided the sensitivity of the microbe to the added steroid substrate is increased in the sense of an induction of activity of the respective enzyme system. This sensitivity of the microbes, howerer, cannot be raised by means of current adaptation methods, but the above-mentioned phenomenon has rather the nature of a sort of "negative adaptation".

# 5

# PRACTICAL METHODS OF FERMENTATION

The entire fermentation procedure of the microbial preparation of steroid derivatives can be practically divided into two individual processes, viz:

a) Cultivation of the microorganism, namely the accumulation of the necessary enzymatic system;

b) The proper transformation of the steroid molecule (159).

The cultivation of the microorganism at all events takes place in the optimum cultivation medium. Its composition, namely the sources of N and C, and the presence of inorganic ions, is dependent on the used microorganism. Corn-steep liquor is used most frequently as an economically advantageous source of N and of additive factors. Of sugars, especially glucose, sucrose, dextran, molases are used as the source of C. Considering, that the desired chemical reaction is not realised until in the second part of the fermentation, when the steroid is added as an unphysiological substrate to the grown culture, the composition of the medium for cultivation of microorganisms is from the point of view of transformation of the steroid less important.

## Procedures for the transformations of steroids

The proper transformation of steroids may proceed either in the cultivation medium if the steroid is added directly into the cultivation medium of the grown culture (*single phase fermentation process*), or after terminated cultivation the microbial cells are separated from the cultivation medium and transferred into a simple medium such as distilled water, buffer solution of a certain pH, physiological saline to which at the same time the starting steroid is added (*double phase fermentation process*) (84, 1014).

The double phase fermentation process offers especially in the isolation part a number of advantages in comparison with the single phase process. These consists in the absence of contaminating substances and of microbial metabolites from the cultivation medium, in the possibility to evaporate the fermentation solution and thus crystallize directly the steroid metabolites without an extraction procedure

(*Czech. Pat.* 110.097). Furthermore it affords an easier regulation of pH during the steroid transformation, a simpler analytical control of the course of transformation.

The steroid is brought into the fermentation medium directly in the form of a crystalline substance, or more often in acetonic, methanolic or ethanolic solution, either in a single dose or semicontinuously in appropriate time intervals (75, 1127). The semicontinuous way of addition has, especially in mould or yeast fermentations, certain advantages, compared with the single dose method. Considering the relatively low solubility of steroids in aqueous medium, the possibility of adsorption of steroid on the microbial cell is lowered when using the semicontinuous addition. Also the isolation procedure is thus essentially simplified, since the microorganism cells devoid of the steroid substrate need not be subjected to extraction and the steroid metabolites are then isolated only from the fermentation solution.

While transformation of steroids is carried out practically always under submersion conditions, it is sometimes of advantage to use the stationary method for cultivation of microorganisms. It has been found that after stationary cultivation of the strain *Mycobacterium flavum* on mould agar it is possible to use washed cells for the transformation of cortisone to prednisone. In this case distilled water was used as the medium. When using cells cultivated in this way, the rate of transformation decreased but the rise of the underirable 20$\beta$-hydroxy derivative of prednisone was prevented almost completely (81, 82).

Orientation experiments in which the ability of steroid transformation is examined in a number of microbial strains are carried out on shaking apparatus throughout the entire fermentation (73, 78, 80). The qualitative course of transformation is followed by means of paper chromatography or thin-layer chromatography. On the basis of these tests a certain number of microorganisms is chosen in which then is studied the influence of external factors upon the course of steroid transformation. For this purpose (study of the influence of pH number of rotations, airation, temperature amount and kind of the used solvent, concentration of added steroid and the manner of addition) it is of advantage to employ the double phase fermentation procedure with cultivation of the microorganism on shaking apparatus, and subsequent steroid transformation by means of separated cells situated in buffer solution in a laboratory fermentation tank (165). Also here the course of steroid transformation is followed by paper chromatography or thin layer chromatography. After the elution of metabolites is carried out quantitative evaluation by spectrophotometry.

Wix and Albrecht (1005) used for the study of mould steroid hydroxylation in position 11 the double phase fermentation procedure in which globular mycelium is brought into the chromatographical column as a filling. Through the column is passed the steroid solution which is hydroxylated during the contact with the mycelium.

On analogous principles would be based the transformations of steroids taking place by means of the so called "bio-exchangers". The cells of microorganisms containing the desired enzymatic system are after cultivation separated from the medium and firmly fixed on a carrier, such as agar, silicagel, gelatine and others. This material is used in form of little cubes to fill chromatographical columns through which the solution of steroid is passed. Transformation of steroids by means of these bio-exchangers may be carried out even in a fermentor at submersion conditions. The advantage of this procedure would be possible continuous arrangement for the process and easier isolation of desired steroid metabolites, since the contamination by products of the composition of the biomass is said to be negligible.

On production scale for the microbial preparation of steroid hormones is used the equipment common in the industry of antibiotics, fermentors with stirring and airation devices, vacuum rotating separators or centrifuges for separation of the cell mass (1226, 1230).

## The maintenance of microbial cultures

A prerequisitive of the production is the maintaining of microbial strains used for the steroid transformation processes. The commonest employed method of preservation of microorganisms is lyofilisation (1231), sand or soil preserves or cultures on agar slant immersed in sterile oil (1226). FABIÁN and ŠINDELÁŘ (1225) followed the decrease of water in lyofilised ampoules and proved that a slip of filter paper added to dried microbe suspension favorably affects the course of the process. MAZUR and WESTON (1228) described a new method of preservation based upon spraying and drying of spores in an atomizer. In various microbes dried on cellulose fibres and stored at 37° as well as at room temperature, the vitality was observed during a period of 3–6 years (1223). Mixtures of peptone + glucose, peptone + + sorbitol proved to be efficient preservatives for dry cultures stored at room temperature.

For well sporulating cultures of moulds preservation on loose carriers such as rice, wheat bran, millet and similar material gave excellent results (1224). Equally good results were obtained if the spores of aktinomyces and moulds were under sterile conditions brought into empty test tubes and the cotton-wool stoppers then impregnated with paraffine oil. In the strain *Absidia orchidis*, employed for hydroxylation of compound S in the production of cortisol (1012) kept on millet at –5°C there was observed neither a loss of sporulating ability, nor damage of the enzymatic system necessary for hydroxylation as long as after 4 years of storage.

## Practical examples

In preparation of cortisol. In the hydroxylation of cortexolone by *Absidia orchidis*, the hydroxyl group is introduced simultaneously into the position

11$\beta$ and 11$\alpha$. For the preparation of cortisol, a 10 l medium containing 3% dextrin, 0·3% corn-steep-liquor, 0·3% $(NH_4)_2HPO_4$, 0·04% KCl, 0·04% $MgSO_4$ and 0·001% $FeSO_4$ of pH 6·5, was inoculated in a rustless fermentor with a single millet preserve of the above-mentioned microorganism. The microorganism was cultivated with stirring (250 r.p.m.) and aeration (5 l air per min.) at 28° for 16 hours. The growth of the mycelium in the fermentor was accompanied by a decrease in the pH value up to 4·5. After 16 hours' cultivation, giving rise to a fine highly branched mycelium having a homogenous protoplasm without vacuoles, pH was adjusted back to the value 6·5 and 4 g cortexolone added into the fermenfor as 1% ethanolic solution. The transformation proper proceeded under the same submersion conditions for approximately 18 hours. During the process, pH was maintained in the range 5·5 to 6·5. After 12 hours of transformation, its course was followed in 2-hour intervals, on the one hand by paper chromatography, on the other hand spectrophotometrically after extraction with chloroform of a collected sample of the fermentation liquid. The transformation was interrupted at the moment when the starting steroid could not be proved in the fermentation liquid any more. From the filtrate after the separation of the mycelium, cortisol and its 11-epimer besides a small amount of cortisone, was isolated by extraction with butyl acetate and the subsequent separation carried out by fractionated crystallization or after conversion to acetates.

In preparation of prednisone. In the microbial 1–2 dehydrogenation of cortisone to prednisone, the strain *Mycobacterium flavum* was used and the cultivation of the strain carried out under submersion conditions (300 r.p.m., 5 l air/min.) in a laboratory fermentor 6–8 hours at 30° in 10 l MB-11 medium (0·5% cornsteep-liquor, 0·1% $KH_2PO_4$, 0·1% $Na_2HPO_4$, pH 7·0). The course of the cultivation was followed turbidimetrically and the cultivation terminated at a growth of the strain, corresponding to an ixtinction of 0·9–0·95. The contents of the fermentor were then centrifuged and the separated cells introduced into half of the original volume of distilled water (pH 5·8–6·0) and 4 g cortisone added as 1% acetone solution. The transformation proceeded at 30° under the same submersion conditions for 4–5 hours. The course of the transformation was followed polarographically from the decrease in the starting cortisone and the rise of prednisone. As soon as the starting cortisone could not be proved in the transformation medium, the contents of the fermentor were made acid with sulfuric acid to pH 2·5–3·0, filtered and prednisone isolated from the obtained filtrate by extraction with chloroform.

# 6

# ANALYTICAL EVALUATION OF THE FERMENTATION COURSE

In the study of microbial conversion of steroids analytical evaluation of the process is one of the most important conditions. At the time starting steroid is being altered under the catalytical effect of the enzyme system of the microorganism used, an important role is played by the rapidity of analytical evaluation of the transformation. Here the qualitative point of view ranks before the quantitative one, especially in preliminary experiments. For these evaluation purposes chromatography is used in the first place, both paper chromatography as well as thin-layer chromatography. Both these methods comply with the requirements as to rapidity and are suitable to follow the decrease of the starting steroid and of the metabolite formation. Chromatography has also proved reliable in the second part of the process, i.e. in the isolation of the formed metabolites, as a control of the separation and purification operations. The chromatographic methods are also used when characterising and evaluating the purity of the already isolated steroid metabolites.

To follow the course of transformation of the steroid in already developed fermentation processes, apart from chromatographic methods, other physico-chemical procedures are successfully made use of. They are e.g. colorimetry, spectrophotometry and polarography, which allow quantitative evaluation of the transformation course; this evaluation is carried out by determining the content either of the starting steroid or the arising metabolite.

Though a number of *group colour reactions* (e.g. for $\Delta^4$-3-ketosteroids, $\Delta^5$-3-hydroxysteroids, 17-ketosteroids, 20-keto-21-hydroxysteroids etc.) has been worked out, or even a number of specific reactions for the individual compounds (e.g. blue reaction with Lugol solution for 11α-hydroxyprogesterone, cortisone, dehydroepiandrosterone), isolation of the newly formed product and its identification are absolutely necessary in each case for perfect characterization of a new metabolite. A great number of characteristics is used for the identification, i.e. physical, eventually physico-chemical methods (melting point, specific rotation, ultraviolet and infra-red spectra, estimation of the molecular rotations, elementary analysis etc.), chemical methods (acetylation, oxidation, reduction, hydrogenation, dehydration, cleavage of side-chain etc.), as well as microbiological methods, where we utilize the considerable transformation capacity of the microorganism as a supplementary feature for chemical and physico-chemical identification of the metabolite (hydroxylation into various positions, dehydrogenation in position 1 – 2, reduction of 17-ketogroup into appropriate 17$\beta$-hydroxy derivative, cleavage of the side-chain etc.).

An isolated steroid metabolite, whose $R_F$ value (at least in three different systems), colour reaction, melting point, specific rotation and infra-red spectrum entirely correspond to some steroid derivative already described, may be considered as identified.

## Paper chromatography

For quick evaluation of steroids conversion the course by means of paper chromatography, paper Whatman Nr. 4 has proved suitable. Using this paper and the Zaffaroni system, we find the delimitation of the metabolite spots achieved less sharp than on Whatman paper No 1, but the time necessary for the chromatogram is essentially shorter. For the identification of isolated derivatives we use the paper Whatman No 1, eventually separation systems according to Bush, where a better reproductibility of $R_F$ value is obtained.

Paper chromatography of steroids is nowadays sufficiently described in a number of monographies, where the individual papers concerning the chromatography of steroids, summary tables of the $R_F$ values in different solvent systems, and the detection agents are reported (1138, 1141, 1171, 1183, 1184). Therefore we confine ourselves to our own experiences obtained in analytical control and evaluation of the course of microbial conversions of steroids.

In preliminary transformation experiments, where it is necessary to follow a very great number of strains and microorganisms on one or several steroid substrates, we use paper bands of a length of approx. 30 cm for chromatographic control of the course of conversion. As stationary phase we use formamide in ethanolic solution (the concentration of formamide of 15 – 50% is dependent on the steroid substrate used) and as a mobile phase according to the polarity of the steroid: petroleum ether, benzene, chloroform and their mixtures (72, 73) are used. In detection of steroids 6 group-detection agents will be sufficient. They are: Lugol solution (1019, 1219, 1220), a mixture of p-fenylenediamine and of phthalic acid (1026), 2,4-dinitrofenylhydrazine (1035), blue tetrazolium salt (1044), saturated solution of $SbCl_3$ in chloroform (1142) and Turnbul's blue reaction with potassium ferricyanide and ferric chloride (1020).

Lugol's iodine solution gives a yellow-brown reaction with nearly all steroids. An exception is formed by progesterone derivatives with a hydroxy group in position 11$\alpha$, or 17$\alpha$, cortisone, dehydroepiandrosterone and some steroid acids, giving a specific blue colour.

An equally universal detection agent is 2,4-dinitrofenylhydrazine, which gives a yellow or orange colour practically with all steroid ketones. The sensibility of this colour reaction depends, however, on the position of the keto group in a steroid molecule. The steroids with a keto group in position 3, give the strongest reaction weaker in keto group in position 11 and weakest with a keto group in position 20.

A mixture of p-fenylenediamine and of phthalic acid is used for the detection of steroids, containing a group $\Delta^4$-3-keto, which yield with the above mentioned mixture yellow-brown to brown colour. The presence of a further double bond in position 1,2 (e.g. in $\Delta^{1,4}$-androstadiene-3,17-dione) weakens the colour reaction considerably. E.g. 1,4-androstadiene-3-17-dione, 1-dehydrotestosterone and 1-dehydrotestololactone do not present with the mentioned detection agent a yellow colour until after 24 hours. Prednisone practically does not react at all with the mentioned agent.

As a specific agent for α-ketol grouping in a steroid molecule we use successfully a blue tetrazolium salt, which with these substances presents an intensive blue colour (e.g. in corticoids with the 20-keto-21-hydroxy group or in some steroid metabolites with α-ketol grouping).

Very widely used is detection with a saturated chloroform solution of $SbCl_3$ which gives after spraying with most steroids containing a grouping of $\Delta^5$-3-hydroxy a orange to red colour reaction. In any case it is, howerer, possible to use the mentioned detection agent combined with observation in ultraviolet light detecting the spots of metabolites.

For the detection of steroids containing phenolic hydroxyl in the molecule it is of advantage to use a mixture of potassium ferricyanide and ferric chloride, which presents a blue colour (*Turnbul's blue*) with the mentioned compounds. For the detection of sterols (cholesterol, ergosterol and others) the Liebermann reaction (1142) or spraying with sulphur acid are used (1110), but we do not carry out the chromatography on paper, but on thin layers of aluminium oxide or silicagel.

For the control of the course of the conversion processes already studied, where the polarity of the resulting metabolite differs considerably from that of the starting steroid substrate (e.g. preparation of cortisol from cortexelone, 11α-hydroxylation of progesterone by the strain *Rhizopus nigricans,* saponification of cortisone acetate into free cortisone by the strain *Flavobacterium peregrinum* etc.), ascendent paper chromatography on thin paper strips approximately 10 cm long and 1 cm wide may be successfully used. The main purpose is to follow the decrease and disappearance of the starting steroid. The time needed for carrying out a complete chromatographic control, i.e. collecting of the specimen, extraction, evaporation, spotting, developing of the chromatogram and its detection range from 20–30 minutes. In the practical control of the fermentation course we collect from the fermentation medium a specimen corresponding to the quantity of 500–1000 nm calculated in the steroid used. Without separating bacterial cells or the mycelium of the microorganism used, the specimen is extracted with a 2–3 fold volume of chloroform. After separating and drying the extract with anhydrous sodium sulphate, the chloroform solution is evaporated and the residue is taken up in a small volume of ethanol and applied on chromatographic paper.

For quantitative evaluation of paper chromatograms we may use either photometric estimation of the intensity of the spot after detection, or spectrophotometric estimation of the content of the steroid after elution on the undetected chromatogram (165). In this case it is necessary to determine the position of the spots in ultraviolet light and to trace out the position of the searched metabolites on the chromatogram. For preparative purposes we have used mostly Whatman paper No 3, previously purified in advance by washing with ether and methanol.

Recently some further methods of chromatographic analysis have been mentioned in literature. Their aim is to speed up the course of the analysis, e.g. chromatography at a raised temperature, and in the centrifugal field (347, 1052, 1124).

## Thin-layer chromatography

Besides paper chromatography the thin-layer technique has recently been used, as reviewed by STAHL and RANDERATH (1163). In their monographs they deal in detail with the preparation of the adsorbent, of the layer application of the specimen, the choice of suitable solvents and of the detection agents. The advantage of this method of chromatographic analysis is the rapidity of its performance and the possibility of using drastic detection agents such as concentrated sulphuric acid, phosphoric acid etc., which can be used only with difficulty in paper chromatography. This procedure may also be used successfully for preparative purposes. A great advantage of thin-layer chromatography is the use of aluminium oxide without binder as an adsorbent for the chromatography of some steroids in the way, as worked out by MOTTIER and POTTERAT (1140*a*) in their method for the separation of organic dyes.

The influence of the functional groups and of their configuration on the adsorbitivity in 73 steroids of the cholestane series has been studied by ČERNÝ *et al.* (1043).

The separation of estrogens on thin layers of aluminium oxide without binder has been described by HEŘMÁNEK *et. al.* (1083). A number of papers in this field describe the use of other kinds of adsorbents such as silicagel, diatomaceous earth, powdered cellulose, powdered polyamide in layers without binder as well as in combination with a binder, which in most cases is plaster or starch (1023, 1125, 1162, 1192, 1210).

We have used this control method of the course of conversion especially in the study of microbial degradation of the cholesterol molecule and in that of microbial acetylation and reduction on C-17 by means of yeasts and yeast-like microorganisms. In both cases we have used concentrated sulphuric acid as the detection agent. After heating to 100°C the chromatogram shows the coloured spots of the metabolite.

## Colorimetric methods

For the quantitative evaluation of the course of microbial transformations of steroids, further for the determination of the purity of the isolated substances, eventually also for their identification colorimetric methods are used. Many of these analytical methods already have become popular for the facility and readiness of their performance on one hand, and for their selectivity on the other. In some cases, where the estimation is hindered by the presence of other steroid or non-steroid substances, it is possible to combine these methods with chromatographical purification or separation procedures, eventually to use selective extraction procedures in isolation of steroids from a fermentation medium (1136). Besides neutral steroids also phenolic steroids can be separated by means of ion exchangers (1186, 1187, 1188) in cases of determination of estrogens.

The basis of quantitative colorimetric estimation is the relation between the concentration and light absorption, formulated by the *Lambert-Beer law*. Extinction is measured at an appropriate wave-length, corresponding to the characteristical wave-maximum in the absorption spectrum. The concentration of the steroid is then determined by comparison with a standard solution or with a calibration curve. In the determination of steroids by colorimetrical methods the chromophore can either be a complex compound of the steroid with the agent used or a product of the conversion of the steroid itself e.g. by influence of an acid, alkalies, reducing or oxidating substances. In this case the chromophores are mostly unsaturated ketones, carbonyl groups, double bonds and conjugated systems.

In analytical control of microbial transformations of steroids either a decrease in the starting steroid, or the formation of the required metabolite is determined. Besides it is often necessary to follow the total sum of present steroids, i.e. of the initial substance and of the resulted metabolites in order to determine to what a degree the microorganism used cleaves the added steroid.

A review of colorimetric methods, used for the determination of steroids together with critical evaluation of the individual procedures is given by KAKÁČ and VEJDĚLEK (1106).

To the most frequently used methods belongs the reaction of steroids with sulphuric acid. The original *Kober reaction* (1110) has been many times modified, with the purpose to increase sensibility and selectivity (different concentrations of sulphuric acid, addition of further solvents, altering of reaction time, or of temperature etc.) (1022, 1150). This method has been used for the determination of estrogenes, of dehydroepiandrosterone and other $\Delta^5$-3-hydroxysteroids from the group of androgens and gestagens (1149), testosterone (1091, 1145) and corticoids.

MIZSEI and SZABO (1135) have used a reaction with sulphuric acid for the determination of $\Delta^{1,4}$-androstadiene-3,17-dione in microbial preparation from $\Delta^4$-androstene-3,17-dione, resp. from progesterone. For the determination of corticoids, mixtures of sulphuric acid with acetic acid are often used. BANES (1018) has worked out a method for determination of cortisol and corticosterone in addition to cortisone and cortexone. KAKÁČ (1104) uses an agent composed of a mixture of sulphuric and acetic acids in ratio 1 : 5 for the determination of prednisone in microbial transformation of cortisone and another modification for the determination of the 20β-hydroxy derivative of prednisone, resulting as an undesired by-product (1105). For the following of the microbial transformation of cortexolone into cortisol MIZSEI and SZABO have used colorimetric determination of the starting substance after selective extraction of the fermentation medium with a mixture of $CCl_4$–$CHCl_3$ 3 : 1 (1136). The same authors have used a reaction with sulphuric acid also for the determination of prednisolone besides cortisol in a fermentation medium (1137).

The *Zimmermann method* (1221), with *m*-dinitrobenzene in an alkaline medium has been successfully used for colorimetric determination of 17-ketosteroids from the androgens group as well from the estrogens series (1202). Some modifications may also be used for the determination of 3-keto- and 20-ketosteroids (1217). This method can also be applied for the determination of the so called *ketogeneous corticoids* after an oxidation cleavage of the side-chain on C-17 into the appropriate 17-keto-steroids by means of periodic acid (in types **I** and **II**) (1200) or by means of natrium bismuthate (in types **I**, **II** and **III**) (1147, 1178).

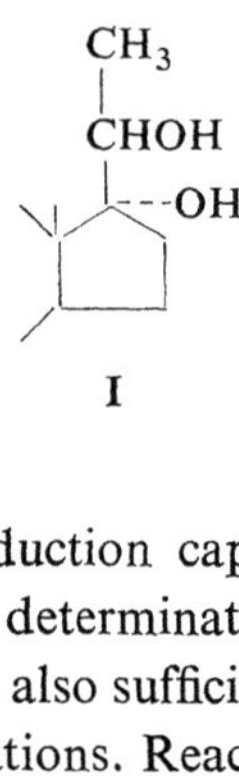


I

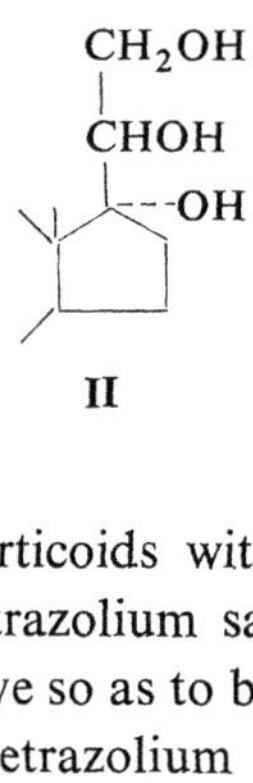


II

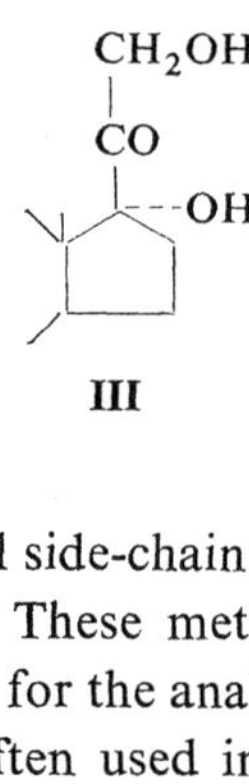


III

The reduction capacity of corticoids with a α-ketol side-chain is utilized in colorimetric determination with tetrazolium salts (1122). These methods are very sensitive and also sufficiently selective so as to be used also for the analysis of biological preparations. Reactions with tetrazolium salts are often used in combination with paper or column chromatography (1041, 1190). A modification of this method has been worked out by DOWLATABADI (1058), who used blue tetrazolium salt for the determination of $\Delta^4$-3-ketosteroids. This reaction, however, is not specific.

Another important reaction is the *Porter-Silber colorimetric method* (1159, 1185) with phenylhydrazine in diluted sulphuric acid for the determination of corticoids with a 17α-hydroxy group. WALSER and SCHLUNKE studied the mechanism of the Porter-Silber reaction (1212) and presume that in the reaction a 20,21-osazone is formed under simultaneous dehydratation into a 16-dehydro derivative. BARTON *et al.* (1021), however, have proved, that this reaction is an acid-catalyzed rearrangement of the side-chain of corticoids forming a 20,21-glyxoal grouping according to Mattox with subsequent formation of 21-monophenylhydrazone.

For the determination of corticoids with a 21-hydroxy group (so called *formaldehydogenic steroids*) an oxidation method with periodic acid or natrium bismuthate is used combined with colorimetric determination of the released formaldehyde by means of a reaction with chromotropic acid (1109).

Out of a number of further colorimetric methods we have to mention a method with 2,4-dinitrophenylhydrazine, suitable for the determination of ketosteroids (1076) with the hydrazide of isonicotinic acid (1046) and with 2,6-*tert*-butyl-*p*-cresol (1025) used especially for differentiation of $\Delta^4$-3-ketosteroids from $\Delta^{1,4}$-3-ketosteroids, further a colour reaction resulting at the conversion of an acetylated steroid hydroxy group into hydroxamic acid by means of hydroxylamine, which yields a colour reaction with ferric chloride (1067).

## Spectrophotometry

The use of spectrophotometry in the *ultraviolet* region has been limited to conjugated double bond systems, e.g. $\Delta^4$-3-keto, $\Delta^{1,4}$-3-keto, $\Delta^{4,6}$-dienes, $\Delta^{5,7}$-dienes, etc. The course of absorption spectra in theses cases is entirely characteristic and allows not only their identification on the basis of the position of the absorption maximum but also their quantitative determination on the basis of the extinction value (*absorbancy*). Marked differences in the course of the spectra allow to differentiate safely steroids containing a conjugated double bond system in two different rings (so called *heteroannular* dienes) from steroids having a conjugated double bond system in a single ring (so called *homoannular* dienes). The latter compounds present a considerably more active chromophor and absorb always at wave lengths which are higher by approximately 40 nm. The basic wave length value of the absorption maximum is subject to change in both cases owing to substitution, connection to a ring, etc. Similar relations are valid also for nonsaturated ketones. Steroids with aromatic ring A or B represent a special group as regards their course of absorption spectra. The former group is characterized by a relatively high absorption in the 210 nm region and a lower one in the 260 nm region. In the latter group, absorption is shifted into the 224 nm and 270 nm regions, as the result of higher substitution.

In the determination of related steroids, having an equal double bond system in the molecule and differing only in different, absorption spectrum unaffecting substituents, the combination of ultraviolet spectrophotometry with previous paper chromatographic separation (165) has given good results. In some special cases, however, even closely related compounds can be determined in a single solution. Thus, e.g. *Rössler* and *Brückner* (1174*a*) determined prednisone and cortisone on the basis of difference in absorption at 262·5 nm. Their method can be applied to the determination of other $\Delta^{1,4}$-3-ketosteroids.

Spectroscopy in the *infrared* region has been one of the most valuable physicochemical methods in the analysis of steroids. Its significance lies, on the one hand, in the identification of unknown substances through comparison with the spectra of known compounds, using especially the 700 – 1300 $cm^{-1}$ region in which the structural differences between the individual compounds are most markedly manifested, on the other hand, in the determination of structure from the position of absorption bands characteristic for the individual parts of the molecule, e.g. carbonyl, double bond, halogen, hydroxyl, etc. Application of infrared spectroscopy to the explanation of structure of steroid compounds and absorption values characteristic for the above-mentioned parts of the molecule are listed in numerous monographs and review articles*. In addition to these principal aspects, however, infrared spectroscopy has been finding application also as a control of isolation processes, especially in combination with chromatography, and in the purification of isolated steroid metabolites.

## Polarography

Polarographic methods are also sometimes used for quantitative determination of ketosteroids. The first paper in this field was published by AISENBRAND and PICHER (1064), who have determined testosterone, progesterone and cortexone in a 90% ethanolic solution of 0·1N-LiCl; they have established that the height of the polarographic wave is linearly dependent on the concentration of the steroids. Very extensive polarographic studies of $\alpha,\beta$-unsaturated 3-ketosteroids were carried out by WOLFE, HERSHBERG and FIESER (1218). Their measurements were performed in an alkaline solution of aqueous isopropanol at 1·1 volts. They have found, however, that saturated 3-ketosteroids, 17-keto and 20-ketosteroids cannot be determined under these conditions. Steroids with an isolated keto group can be determined as *Girard T-hydrazones* soluble in water. By preparation of these hydrazons the sensitivity of determination of $\Delta^4$-3-ketosteroids can be considerably increased. The *Girard hydrazones* of 17-ketosteroids and of 20-ketosteroids show a half-wave potential at −1·35 V. 20-Ketosteroids containing a tertiary 17α-hydroxy group cannot, however, be determined in this way. PRELOG and HÄFLIGER (1160) have found that also the *Girard T-hydrazones* can be determined in saturated 3-ketosteroids at pH 7·0 and − 1·44 V. HERSHBERG *et. al.* (1086) have worked out a method for the determination of $\Delta^5$-3$\beta$-hydroxysteroids after oxidation into $\Delta^4$-3-ketones according to Oppenauer

* Cf. e.g. DOBRINER K., KATZENELLENBOGEN E. R., JONES R. N.: Infrared absorption spectra of steroids, *Interscience* (New York), 1953.

JONES R. N., SANDORFY C. in the book WEISBERGER A. (Ed.): Technique of organic chemistry, Vol. IX., p. 247, *Interscience* (New York) 1956.

ROSENKRANTZ H. in the book GLICK D. (Ed.): Methods of biochemical analysis, Vol. II., p. 1, *Interscience* (New York) 1955.

ROSENKRANTZ H. in the book GLICK D. (Ed.): Methods of biochemical analysis, Vol. V., p. 407, *Interscience* (New York) 1957.

(Al-tert-butoxide). ZUMAN *et. al.* (1222) carried out direct colorimetric determination of the total content of $\Delta^4$-3-ketosteroids as well as the determination of these substances beside the 20-ketosteroids in 30–60% ethanolic solution of Britton and Robinson buffer. By means of the combined methods of polarographic determination together with a chromatographic purification the content of corticoids has been established as *Girard T-hydrazones* in the blood (1139, 1140). KABASAKALIAN *et. al.* (1100) carry out a direct polarographic determination of cortisone beside prednisone in a 50% methanolic solution adding an acetate buffer of pH 5·5. A half-wave potential of prednisone is measured at −1·30 V and of cortisone at −1·47 V against a normal calomel electrode. The same shift is to be found in the cortisone-prednisolone system. The influence of pH, of temperature, of concentration and of Hg pressure on polarographic waves were studied by KABASAKALIAN and MCGLOTTEN (1101). The determination of small quantities of cortisone beside prednisone (0·9–11·5%) were established by PICHER and HAJDU (1155). KABASAKALIAN and MCGLOTTEN (1102) have also worked out direct polarographic method for determination of isolated carbonyl groups in various positions of the steroid molecule in a 90% ethanolic solution of tetrabutylamonium chloride. GREENSPAN (151) has determined a 1-hydroxy derivative of cortexolone on the basis of the finding that in an alkaline solution there occurs dehydratation forming a double bond in position 1–2, and a shift from a voltage of −1·46 to −1·31 V is so brought about.

Polarographic methods can be successfully used in those cases of microbial transformations, where besides the starting steroid only a single steroid metabolite is present in the medium or if the resulting metabolites are so different structurally that they do not hinder the quantitative determination of the individual components.

# 7

# ISOLATION AND IDENTIFICATION OF STEROID METABOLITES

In microbial transformations of steroids the course of fermentation is most often followed by means of paper chromatography, or by thin layer chromatography. Even though the arising metabolites correspond, by their $R_F$ values, by the colour of spots after detection in ultraviolet light with the supposed substances, applied simultaneously as standards, it is not possible to consider this way of identification a sufficient and final proof. That is to say that in frequent cases a microorganism, which carrying out a wellknown transformation on one steroid substrate, carries out quite a different type of conversion with another steroid (stereospecificity of enzymes) (481, 1215).

For the purpose of qualitative and quantitative estimation of fermentation experiments when working with new strains of microorganisms or in changes of transformation conditions an isolation of arisen metabolites has to be carried out, it is necessary to prepare these substances in pure crystalline form and to perform their identification by means of physical, chemical eventually also microbiological methods. Since in microbial transformations there arise often two or even more metabolites, the mere isolation of individual substances (often of isomeres) is considerably time-consuming and tiresome. Besides the classical physical and chemical procedures a study of the spectra in ultraviolet and infrared range, nuclear magnetic resonance and other modern physical methods have been recently used more and more frequently for identification purposes.

Isolation of steroid metabolites from the fermentation medium is essentially easier than isolation from biological substrates e.g. from the urine, blood, tissues etc., especially so since in microbial transformations we can work with relatively high concentrations of steroids. Also the content of contaminating substances in fermentation processes is often essentially lower, since we work in very diluted solutions or even an enzymatic conversion proceeds in mere water. In this way many difficulties are eliminated, caused by the isolation of steroid substances from biological substrates e.g. formation of emulsions in the extraction by means of organic solvents. Even a small change in the composition of the fermentation medium can often essentially influence the isolation course. For extraction by means of organic solvents less suitable media, containing proteins soluble in water are e.g. yeast autolysate, bouillon, cornsteep and peptone, which are often the cause of unrequired emulsions.

Also oils, added into fermentation tanks, for the purpose of avoiding foaming sometimes render the isolation of less polar steroids considerably more difficult. Even that may be avoided by a little modification of the nutrient medium e.g. by mere dilution with water only. As a consequence of a very slow course of transformation (several days) an autolysis of a certain quantity of microbial cells may occur and the released proteins then cause some difficulties in the filtration as well as in the extraction of such substrates.

The identification of new steroid metabolites is carried out in cooperation with the specialists in the field of spectral analysis. In some cases an isolated steroid metabolite is returned to the microbiological laboratory for identification purposes, where it is submitted to further transformation e.g. microbial cleavage of the side-chain of C-21 steroids, introduction of a further hydroxy group, double bond etc. The most frequently used isolation and identification processes are illustrated as follows.

## Isolation of steroids from the fermentation medium

A condition of successful isolation of steroid metabolites is a fast processing of the material after finishing the fermentation. First it is necessary to stop any further action of enzymes of the used microorganism on already resulted steroid metabolites and herewith to prevent further unrequired transformations, eventually the cleavage of the steroid molecule. As it is mostly matter of endoenzymes a quick separation of the microorganism used will be sufficient. When the transformation proceeds very fast it is preferable to stop the action of the enzymes yet before the filtration e.g. by heating up to 60 – 80°C, by acidification of the fermentation solution or by adding a sufficient quantity of anorganic solvent. So e.g. the oxidation of cortisone into prednisone by means of some microorganisms is practically finished during 4 to 5 hours and the resulting prednisone is further transformed into the 20$\beta$-hydroxy derivative. This unrequired transformation can be prevented by an acidification with HCl to pH 3 – 3·5. In transformations, showing a slow course and where there is no danger of decomposition of the resulted metabolites a refrigeration of the fermentation solution to a temperature of 2 – 5°C is sufficient and the action of the enzymes is practically stopped. In any case it is absolutely necessary to prevent secondary infection of the finished fermentation process.

In microbial transformations of steroids the resulting metabolites are mostly of a more polar character that the used steroid substrate (saponification of esters, hydroxylation, rise of double bonds etc.). An exception form the recently described transformations when less polar steroid acetates (140, 192, 260) arise in such a way that the obtained metabolites are practically insoluble in water and are separated together with the microorganism used. It is therefore necessary, especially in isolation of new metabolites or when changing the concentration of steroid substrate used to carry out the isolation of steroids from the water fermentation medium on one hand and from the separated mycelium on the other hand.

From the point of view of isolation such types of transformation are the most advantageous, where practically one only metabolite arises and when it is possible to work in high concentrations. So e.g., when transforming progesterone into 11α-hydroxy progesterone WEAWER *et al.* (504) works with the strain *Aspergillus ochraceus* and uses concentrations of 20 – 50 g progesterone per 1 liter of fermentation solution. After performing the transformation over 90% of the present metabolite can be isolated by mere filtration. Another similar way of using high concentrations of steroids is described by KONDO and MASUO (226), who in such a way transform cortisol into prednisolone, while practically all steroids are filtered off after the transformation together with the microorganism used so that no isolation from the aqueous medium is carried out. Ways of transformations in so far incredibly high concentrations are, however, infrequent. The usual concentrations do not exceed 1 g of steroid per 1 liter of medium and then it depends on the polarity of the present steroids, if it is necessary to extract both the water part and the part of mycelium. In some cases it is possible to obtain crystalline steroid metabolites by evaporation of the fermentation solutions to a smaller volume, eventually by cooling to a low temperature. A condition for this manner of isolation is an suitable fermentation medium, preferably mere water in a two-phase fermentation technique and a good crystallizing ability of the steroid.

A relatively rarely used way of isolation of steroids is adsorption on charcoal or another suitable material. This process can be used successfully for the isolation of steroids from the androgenes and gestagenes series, while in labile steroids e.g. corticoids often considerable losses occur. JEANLOZ *et al.* (1098) and MAYER *et al.* (1130) have isolated metabolites of androstenedione and dehydroepiandrosterone on charcoal Darco G-60 with good results. We have used 10 g charcoal for the isolation of 1 g androstenedione after microbial oxidation of dehydroepiandrosterone by means of the strain *Actinomyces viridochromogenes* and the same technique also for the isolation of testosterone after microbial reduction of androstenedione by yeast. Part of these specimens was simultaneously extracted with chloroform and it was shown that the yields in both methods of isolation are completely equivalent. The adsorption proceeded during 30 minutes of stirring and after the separation the charcoal was dried and extracted with chloroform in a Soxhlett apparatus. Charcoal was also used for the isolation of corticoids where, however, the losses amounted to 25 and even more per cent. LEVY and KUSHINSKY (1118) have found that when isolating cortisone there occurs an oxidation cleavage of the side chain and they succeeded in isolating adrenosterone. A decrease of losses in the adsorption in labile steroids can be obtained by means of a special preparation of charcoal e.g. by boiling with HCl and by washing with KCN solution, whereby adsorbed oxygen and the metals catalyzing the oxidative decomposition processes will be removed. Conditions for the adsorption and elution of steroids from charcoal are reported by LOMBARDO *et al.* (1120) and by NEHER (1141). Recently some Soviet chemists (1112) studying the conditions for the isolation of corticoids on charcoal and various ways of elution have been drawing the attention to the adsorption methodology.

Other adsorbents e.g. aluminium oxide, silicagel, diatomaceous earth etc. have proved less suitable for the isolation of steroids from aqueous medium, sincle only a small part of the present steroids is recovered.

**Choice of solvent for extraction.** The kind of solvent used for the extraction of the steroid from an aqueous medium depends on the characteristics and polarity of the extracted steroid metabolite. Relatively nonpolar steroids, wherein e.g. dehydroepiandrosterone, androstenedione, testosterone, coitexone and others may be counted, can be extracted with ether (125, 268, 490). Hydroxy derivatives of androgenes and gestagenes are usually extracted with chlorinated hydrocarbons (chloroform, methylenechloride, dichloroethan) and for the extraction of very polar steroids e.g. corticoids and their hydroxy derivatives, ethyl acetate, butyl acetate or methylisobutylketone (233) are most often used. Nowadays the choice of solvents is mostly regulated according to the analytically determined exact values of the distribution coefficient. On the basis of these data is twas e.g. possible to prove the advantage in using ethyl acetate for the extraction of steroids with a higher number of O-containing functions (1032), which has been pointed out by REICHSTEIN (1166) previously. On the other hand, however, it is necessary, when using more polar solvents, to consider the fact that these solvents are less selective and a greater quantity of contaminating substances from the fermentation medium passes into the extracts.

A solvent designed to the extraction of steroids from an aqueous fermentation medium needs not be anhydrous, it has, however, to be pure and must not leave a perceptible evaporation residue. Ether must be free of peroxides, which in some cases can cause undesirable oxidations; e.g. oxidation of cortisone to adrenosterone was found (1129). Chloroform ought to be free of phosgene, alcohol of aldehydes, butyl acetate of butanol etc. The solvent used should have as low boiling point as possible, so that not too high temperature would be needed for evaporation of the extracts and it should be sparingly soluble in water. It is obvious that the solvent must not attact the isolated substance. Regarding extractions of greater volumes, further factors, especially the toxicity of the solvent, as well as its inflammability and economical advantages come into consideration.

Mostly one solvent only is being used for the extraction. To obtain a greater selectivity in extraction or to increase the extracting ability, however, a mixture of two or more solvents of different polarity may be chosen. For the extraction of some corticoids e.g. cortisol and epicortisol, we have used a mixture of chloroform with acetone in the ratio 9 : 1 or a mixture of butyl acetate with acetone in the ratio 9 : 1. These solvent systems are very effective, but more labile and sensitive to temperature. There is usually less danger of rise of emulsions when using these than in the extraction by means of one solvent only.

**Distribution (extraction) coefficient.** The distribution coefficient presents realiable information about the effectiveness of the extraction by means of a chosen organic solvent when the aqueous solution of the isolated substance is shaken with an equal volume of the solvent. In neutral steroids it is not dependent on pH, concentration and temperature under current working conditions (15–25°C). The transfer of the substance from one solvent into another occurs by diffusion, whose velocity has also to be established in advance; in steroids the equilibrium is established relatively soon, after approximately 5 minutes of shaking already. The distribution coefficient $K$ is calculated according to the following equation:

$$K = \frac{C_{\mathrm{org}}}{C_{\mathrm{aq}}}$$

$C_{\mathrm{org}}$ = concentration of steroid in organic solvent,
$C_{\mathrm{aq}}$ = concentration of steroid in water.

The estimation is usually carried out by shaking in separation funnels of a content up to 250 ml, preferably of cylindric shape with a scale, where the volumes of the solutions contained therein and eventual volume changes after shaking and the subsequent separation of both phases can be read directly. Very strong shaking is not of advantage, since sometimes very fine dispersions result, which then require very long time for separation. After perfect separation of both layers the contents of steroid is determined by any of the suitable analytical methods e.g. by colorimetry or spectrophotometry.

The literature quotes tables of distribution coefficients for binary up to quaternary systems of solvents determined on one hand for extraction purposes, on the other hand for the separation of steroid mixtures in countercurrent distribution (1032, 1038, 1068). A lucid summary of distribution coefficients of steroids was presented by MOORE and HEFTMANN (1138), ENGEL (1068*a*) and others.

**Types of extractors in use.** Current extraction methods are effective when the value of the distribution coefficients $K \geqq 1{\cdot}5$. In lower values of $K$ for lighter or heavier liquids perforators are sometimes used. This manner of extraction can, however, only be used when the isolated substances are thermostable, since the extraction lasts a longer time and the extract is constantly being heated to the temperature of the boiling point of the solvent used.

Extractors for the extraction of a liquid by a liquid can be divided, according to the course of the extraction into discontinuous and continuous (countercurrent). Small fermentation experiments are usually extracted by means of mere repeated shaking in separation funnels, or by means of stirring both phases with a mechanical stirrer or compressed air, eventually with inert gas.

For the extraction of larger volumes continual extraction columns are used. For laboratory purposes e.g. a descendent column, where both liquids are left to flow against each other without any outside mechanical interference (without stirring). The contact of the liquids is strengthened by filling the column with glass balls or with Raschig rings. The Rometsch column (1172) is equipped with a central stirrer device with perforated discs and solid perforated plates and the spaces between the stirring devices of the similarly constructed Scheibel column (1177) are filled up with porous material, where a temporary separation of both phases takes place. One of the most effective columns are the pulsation columns (1028) fitted with solid perforated plates, where the mixing of the liquids is supported by means of pulsation created by a pulsation device located outside the column. Vibration columns are also very effective.

Extractions of large volumes are carried out in centrifugal extractors of the Podbielniak (1157) and Luwesta type (1065), where both phases are moving against each other in the form of a narrow film. The function of these extractors, which are of a high efficiency when working correctly, is sufficiently known from the production of antibiotics.

All mentioned types of extractors have their advantages as well as their disadvantages. Column extractors require relatively slow work, centrifugal extractors have high performances, they cannot, however, be used for extraction with combustibles of the first class. The effectiveness of these devices depends especially upon maintaining of regular operation, upon the difference in specific weight of both solutions, upon the dispersion degree, speed of stirring etc. In pulsation and vibration columns the course of extraction depends on the amplitude, eventually on the pulsation frequency. A review of extraction methods, together with a description of various devices and theoretical calculations is given by TREYBAL (1206), ALDERS (1015) and others.

**Emulsions and their breaking.** Centrifuged or filtered fermentation solutions, if not quite clear, often form when they are extracted by organic solvents emulsions. Emulsions occur, however, even in completely clear media provided they contain a greater quantity of soluble proteins. Emulsions often result also in the extraction of alkaline solutions, or fermentation media, which were evaporated to a smaller volume before the extraction, which implied increased viscosity. Emulsions are sometimes very stable and do not separate even after 24 hours. Such emulsions can be broken by adding neutral anorganic salts (NaCl, $Na_2SO_4$), volatile solvents (ether, aceton, methanol) or by filtration over asbestos-cellulosic filtration plates, by a layer of diatomaceous earth, eventually by centrifugation. It is better to avoid these unfavourable phenomena by a careful filtration of the fermentation solutions, by adjusting the composition of the nutrient media or by using greater volumes of organic solvents for the extraction.

**Selective extractions.** A mixture of steroids, contrasting very distinctly by their distribution coefficients can in some cases be separated by means of selective extraction, using successively organic solvents with ascending polarity (671). TALBOT (1201) and later EIK-NES (1063) have used the selective extraction with benzene for the division of cortexone and dehydrocorticosterone from cortexelone, cortisone and other more polar steroids. The separation of steroid mixtures in the mentioned way is mostly not so sharp. MIZSEI and SZABÓ (1136) have obtained good separation of cortexelone from cortisol and epicortisol in medium extraction by means of a mixture of $CCl_4 - CHCl_3$ (3 : 1) and have used this method to follow the course of microbial transformation analytically. The possibility of using selective extraction for the division of a steroid mixture mostly results from the data of the distribution coefficients and is of outstanding importance in countercurrent distribution.

**Preliminary purification of extracts.** Extracting a steroid by means of organic solvents, nonmixible with water, a relatively selective separation of steroids from other organic as well inorganic components of nutrient medium can be obtained. In spite of that it is usually necessary to set free these extracts of contaminant subances, which can unfavourably influence the isolation of steroid metabolites. Organic and inorganic impurities of acid character are removed from the extracts by washing with a 2–5% aqueous solution of $NaHCO_3$, further with destilled water until neutral reaction, eventually also with a diluted solution of mineral acid and water. Phenolic substances can be removed by shaking up with a 1N- to 2N-solution of alkaline hydroxide, where a considerable decolourisation of the extract can be seen. Purification by means of solutions of alkaline lyes can, of course, be carried out only in such steroids and solvents, as are stable even in a considerably alkaline medium. No solution of lyes is used in steroids, where there is a danger of an isomerisation of the side chain on $C_{17}$ and especially in corticoids, which are considerably sensitive towards alkali in the presence of air oxygen. In extracts of very polar steroids a saturated solution of natrium chloride sometimes is used for purification instead of water, whereby the danger of losses decreases. Another problem usually presents itself in the decolourisation of very dark extracts. A good result often can be obtained by filtration of the extract through layer of an active adsorbent e.g. aluminium oxide or magnesol. They are used in quantities of approximately 0·25 g per 1 g of the present steroids (331). The decolourizing effect of charcoal in solvents used in extracts is usually very weak. Specimens determined for chromatographic separation are mostly not previously decolourized.

**Drying of extract.** During the extraction of aqueous fermentation solutions water also penetrates into the organic solvents, in quantitities corresponding to the solubility of water in the solvent used, as illustrated in the following table:

| *Solvent* | *Solubility* of water in solvent at 20°C (weight %) | *Solubility* of solvent in water at 20°C (weight %) |
|---|---|---|
| Ethyl ether | 1·47 (25°C) | 6·59 |
| Benzene | 0·06 | 0·18 |
| Methylenechloride | 0·14 | 2·00 |
| Chloroform | 0·10 | 0·82 |
| Tetrachloromethane | 0·03 | 0·08 |
| 1,2-dichloroethane | 0·15 | 0·81 |
| 1,1,2-trichloroethylene | 0·02 | 0·10 |
| Ethyl acetate | 3·25 | 8·70 |
| Isopropyl acetate | 1·8 | 2·9 |
| *n*-butyl acetate | 1·2 | 0·68 |
| *n*-butyl alcohol | 20·0 | 7·8 |

The table also shows the percentage loss of solvents, caused by the solubility of the solvent in water. This value is considerably high especially in ether and ethyl acetate and plays an important role when large volumes are extracted.

As a drying means anhydrous sodium sulphate or magnesium sulphate is mostly used. For the drying of 1 liter of chloroform extract e.g. 3 $^1/_2$ g of anhydrous $Na_2SO_4$ (331) is sufficient. During the drying process the solution often becomes clear. The solutions are set free of the drying agent by means of filtration and by washing with an appropriate anhydrous solvent.

**Evaporation of extracts.** Dried extracts are liberated from the solvent by vacuum distillation. The solutions of labile steroids are evaporated, if possible, at low temperatures in the atmosphere of inert gas, mostly of nitrogen, which can be directly introduced into the solution. The evaporation is carried out in circulation or rotation evaporators, larger volumes in film evaporators. Evaporation of the main quantity of the solvent takes place in the mentioned devices, whereas the final evaporation is carried out in previously weighed glass-flaks, where the extracts are evaporated to dryness. Only in the case when the steroid metabolites crystallize already during the evaporation, the evaporation is interrupted and the crystalline portion isolated.

Evaporation of the solvent yields water free evaporation residues. Either in the form of oil or of sirup or in the form of a semi-crystalline mass. When only one main steroid metabolite results from microbial transformation, the evaporation residue is mostly crystalline, or will crystallize during several hours. In these cases the crystalline products are submitted directly to an isolation.

**Preliminary purification of evaporation residues.** The oily and sirupy residues, before further processing are submitted to preliminary purification, consisting in the removal of lipide portions, eventually in further decolourisation. The separation of lipide portions is mostly carried out by shaking with petroleum ether (eventually with pur pentane or hexane) from 70–85% alcoholic solution. CARTLAND and KUIZENGA (1040) were the first to show the possibility of removing the lipide fractions with petroleum ether extraction. Using this method of purification it is often of advantage to dissolve the water-free evaporation residue in absolute alcohol, to add a lipophilic solvent and then to dilute the resulting homogeneous solution with 20% of water and to shake. So precipitation of the lipide portion in the form of flakes, following the dilution of the alcoholic solution with water, will often be avoided.

The mentioned method of removing lipide portions has to be carried out very carefully in steroids of low polarity, which can also partly pass into the petroleum ether extracts so that this operation sometimes becomes impossible. As to these substances, we have succeeded in removing the lipide fractions from the solutions of 70–80% acetic acid on shaking with petroleum ether. Under these conditions androstenedione did not pass into the lipophilic solvent.

In some cases the lipide portion, which sometimes amounts up to 50% of the total crude extract, can be removed by mere digestion of the water-free evaporation residue by means of pentane and decantation of the extract (113, 543). Sometimes the removal of the lipide portions in the above mentioned way can be combined with decolourisation. It has to be noticed that the solution of steroids in 80% alcohol can be sufficient well decolourised by means of charcoal (10–30% per the present steroids). Good stirring, eventually heating up to 40–50°C and filtration often yield good results.

The solutions of steroids in a dilute alcoholic solution purified in this way are again evaporated in vacuo to dryness or are only condensed to a smaller volume (evaporation of organic solvent), are diluted with water and extracted into chloroform or ethyl acetate. At this stage the frequently occurring emulsions are avoided by using for the dilution a saturated solution NaCl or $Na_2SO_4$ instead of water. After drying in the usual manner the extracts are evaporated to dryness.

**Direct crystallization of metabolites.** The purified and evaporated extracts often crystallize during a short time. The crystalization is facilitated by dilution by means of a suitable solvent, eventually by seeding. Such "universal" solvents for the crystallization of less polar steroids are most frequently ether or petroleum ether, in more polar steroids: acetone, ethyl acetate or alcohol, eventually their mixtures with ether, petroleum ether etc.

Steroid metabolites unwilling to crystallize can be frequently forced into crystallization by seeding and by cooling to low temperature. If no seeding material is available, we mostly proceed in that way that we add a less polar solvent (ether, pentane) dropwise into the saturated solution of the steroid until the first durable turbidity appears, which is then removed by gentle heating or by a drop of a more polar solvent. By stirring, friction (best with the sharp edge of a freshly cut glass rod) and by cooling the solution is often forced into crystallization.

The following technique is used for the crystallisation of some more polar steroids, e.g. corticoids: extracts of steroids in a water-inmiscible solvent (chloroform, butyl acetate, methylisobutyl ketone) are evaporated to a concentration 100–300 mg steroid/1 ml and an approximately equal volume of phosphate buffer of pH 8–8·5 is added. The resulting two-phase mixture is cooled to 0–10° and stirred several hours. The described crystallization procedure is carried out e.g. in cortisol, epicortisol, prednisone and prednisolone (1077*a*, 1077*b*).

**Isolation of steroids from mycelium.** Neutral steroids are mostly sparingly soluble in water. Their solubility in water increases with the number of oxygen functions, whereas double bonds raise the polarity of the steroid very little, only. In microbial transformations arise, as already mentioned, mostly more polar metabolites i.e. more soluble in water than the starting steroid substrate (hydroxylation, dehydrogenation, saponification of acetates, rise of D-Homosteroids etc.) Sometimes, however, formation of very little polar metabolites occurs e.g. when acetyl derivatives are formed.

It always depends upon the concentration, in which the transformation of the steroids is carried out and upon their polarity, whether it is necessary to isolate the products of microbial conversion also from the mycelium portion. Nearly always nonpolar and slightly polar steroids are adsorbed by separated mycelium, whereas more plar metabolites are in the solution. So e.g. in microbial oxidation of dehydroepiandrosterone into androstenedione in a concentration 0·4 g per 1 liter medium there are only approximately 30% of steroid in the solution, whereas the rest will be separated together with the mycelium. When transformating cortexelone to cortisol at the same concentration only the unchanged, slightly polar original steroid is adsorbed by the mycelium, whereas the metabolite is to be found in the solution.

Steroids adsorbed by the separated mycelium, are isolated by extraction by means of a suitable organic solvent e.g. methylene chloride, chloroform, ethyl acetate, acetone or methanol (260, 331, 543, 1087). When extracting by means of water-inmiscible solvents, the mycelium must be dried up previously in vacuum at 30–50°C or by azeotropical distillation with benzene or toluene; in labile steroids dehydrating with a solvent water-miscible e.g. methanol or acetone is used. When moist mycelium is dried under heat it sometimes agglomerates to hard horny material the crushing and extracting of which meets with great difficulties (especially in yeast microorganisms).

To prevent the difficulties connected with the extraction of the mycelium it is possible prior to filtration to solubilize the steroids undissolved in the fermentation medium by adding a sufficient quantity of acetone or methanol (456), eventually also to speed up the dissolving by stirring. After the filtration the solvent is usually evaporated. If the mycelium contains only a little quantity of adsorbed steroids, it is sufficient to wash the separated mycelium directly on the filter with a suitable solvent.

## Separation of metabolite mixtures

**Fractional crystallization.** Mixture of two or of more steroid metabolites, which very distinctly differ as to their polarity or crystallization ability, can be separated by fractional crystallization. So e.g. from the mixture of 11$\alpha$-hydroxyprogesterone and 6$\beta$,11$\alpha$-dihydroxyprogesterone, the dihydroxy derivative can be separated by crystallization from methanol (331), from a mixture of 6$\beta$-hydroxy-17$\alpha$-methyltestosterone and 11$\alpha$-hydroxy-17$\alpha$-methyl-testosterone the first mentioned substance can be separated by means of benzene (121). A fractional crystallization from chloroform was also used for the separation of prednisone from the present 20$\beta$-hydroxy derivative.

This method of separation of steroids cannot be, however, considered as a quantitative one and serves only to preliminary obtaining of a certain quantity of a relatively pure material. It happens frequently, especially in isomeres that crystallization is accompanied by the rise of mixed crystals, e.g. in a mixture of cortisol and epicortisol.

Sometimes the formation of crystalline hydrates or adducts with the solvent used is helpful in the separation by fractional crystallization. These crystalline forms frequently crystallize more easily and in coarser crystals than water-free substances. Epicortisol yields e.g. with chloroform an adduct that crystallizes very well and is suitable for the separation from cortisol; the latter again gives an adduct crystallizing very well with methanol.

**Preparative paper chromatography.** For the isolation of milligram quantities of steroids from a mixture preparative chromatography on thick chromatographic paper is sometimes used, e.g. Whatman No. 3. From this paper impurities must be removed before use by washing with organic solvents (ether, benzen, methanol) in a similar way, as the chromatography proper is carried out. Examples of the separation of steroid mixtures according to this method are quoted by KAHNT *et. al.* (1103).

The position of the individual substances on the chromatogram is established by detection of the marginal bands of paper by means of the usual detection agents or in ultraviolet light. The elution of steroids from the cut out bands of paper is carried out by extraction with suitable organic solvents on one hand, either by simple submersion of the paper in a vessel with the solvent for about 24 hours (per 1 cm$^2$ we take approx. 5 ml of solvent) or the extraction is speeded up by stirring, eventually by shaking. As more frequently used, and quantitative method of extracting steroids from paper is the elution in a chamber, whereas the eluate flows down into a flask placed below receiver. The solvents for the elution are chosen according to the polarity of the steroids, most frequently methylenechloride, ethyl acetate or methanol, eventually their mixtures. Very polar steroids are eluted by methanol diluted with 5–15% water. Details about the choice of solvents for the elution are given e.g. by BUSCH (1035) and SMITH (1089).

**Preparative thin-layer chromatography.** This method of semimicro-preparation is very often used besides column chromatography. Its advantage is fast performance, sharp separation and easy isolation from the chromatogram. The adsorbents that are most frequently used are aluminium oxide and silicagel, sometimes in a mixture with a binder (plaster). Separation on diatomaceous earth or Florisile is less sharp. The layers of adsorbents may be up to 5 mm thick (1090). HEŘMÁNEK *et al.* (1085) carried out the separation of various types of steroid substances (including esters) with three organic solvents only: petroleum ether, benzene and ethanol. Using a suitable mixture of two of the mentioned solvents they always obtained a very good separation. The designation of the spots of the separated steroids is carried out in a narrow band on the margins of the chromatogram by colour detection or in ultraviolet light. For preparative thin-layer chromatography in aluminium oxide detection with morin is also appropriate, as morin is not eluted from the adsorbent (1043). The separation of the band with the metabolite is carried out after careful drying of the chromatogram either by pouring or by suction. The steroids are then extracted from the adsorbent with an organic solvent either by stirring in a flask, or by chromatography after adjusting to a small column. The method of preparative thin-layer chromatography has been described in the monographies of STAHL (1196), RANDERATH (1163), BOBBITT (1025*a*), COPIUS PEEREBOOM (1048*a*) and others.

**Column chromatography.** One of the most reliable method of separation of steroid mixtures still remains column chromatography. In adsorption chromatography the separation of the mixture of substances proceeds on the column of active adsorbent by means of an eluotropic series of solvents with increasing polarity. This kind of chromatography is the oldest, but still often used especially for the separation of less polar steroids or even of esthers of considerably polar hydroxysteroids e.g. of corticoids (440). This method was developed by REICHSTEIN *et al.* (1168, 1170), the adsorbent most frequently being aluminium oxide, which shows the greatest adsorption activity and allows a reliable separation in considerably complicated steroid mixtures. As the alkaline and excessively active aluminium oxide is mostly unsuitable for the separation (hydrolysis of esthers, dehydratation and other undesirable secondary reactions) (1115, 1141, 1170), it has to be previously neutralized and desactivated by adding a certain quantity of water, whereafter it is standardized by means of fast azo dyes (1029, 1084, 1089). In labile steroids, instead of aluminium oxide, silicagel, Florisil, powdered cellulose and other neutral adsorbents are more frequently used. Even when silicagel was used the formation of artefacts has been observed (1119, 1128, 1205).

Instead of the eluotropic series of solvents, most frequently in the sequence of petroleum ether, benzene, ether, chloroform and methanol, the method of gradient elution is also often used, where the polarity is gradually increased by mixing two solvents of different polarity e.g.: benzene–ethyl acetate (481), benzene–ethanol (1108, 1113), chloroform–acetone (141, 299*a*), dichloroethane–acetone (119), chloroform–methanol (151).

Another technique, a little different, is the column distribution chromatography, used especially for the separation of more polar steroids. Instead of an adsorbent "a carrier of stationary phase" is used here, especially diatomaceous earth, Hyflo Supercel, Celite 535 and 545, further silicagel or even powdered cellulose. In the separation of polar steroids an aqueous stationary phase is usually used and the elution is carried out by some slight polar solvent e.g. pentane, hexane etc. In the separation slightly polar steroids ethylene glycol, propylene glycol or formamide serve as stationary phase.

In the separation of a steroid mixture by means of adsorption chromatography the ratio of steroid and adsorbent is usually 1 : 25 up to 1 : 50, while in distribution chromatography the ratio of steroid to carrier is much higher (1199). The use of adsorption chromatography is simpler and faster than in distribution chromatography, where much more careful work of preparation of the column is required. In adsorption chromatography the number of collected fractions is usually lower than 50, while in the second method the number of fractions reaches a multiple. The individual fractions of the eluate are evaporated and evaluated by paper chromatography photometrically and fractions containing pure individuals are combined, crystallized and further analyzed.

In column chromatography a mixture of two differently active adsorbents are also sometimes used for the purpose of attaining a certain separation ability e.g. a mixture of charcoal and diatomaceous earth (1078), charcoal–Celite 545 (531), Florisil–Celite 545 (328). Very important and decisive for a successfull separation of a mixture of steroids is the problem of purity of the solvents used. The usual techniques of purifying are quoted by LEDERER (1115), NEHER (1141) and by BUSCH (1035). In chromatography on aluminium oxide the use of acetone is not reccomended, owing to the rise of high-boiling condensation products, which cannot be easily destilled (diacetonealcohol, 160°C). For the separation of estrogenes, 17-ketosteroids and some $C_{21}$-steroids SEKI (1179, 1180, 1181) used chromatography on a column of partially esterified cation exchanger Amberlite IRC-50 and eluted with a mixture of alcohol and water of constant composition. Here the most polar steroids are eluted first of all.

Chromatographic separation of steroid isomeres, isomeric hydroxy derivatives, are illustrated in a review by JACQUES and KAGAN (1097).

**Counter-current distribution.** The method of separation of steroid mixture by counter-current distribution belongs to the most gentle methods of separation even in very labile steroid substances. It permits to separate even closely related compounds by their distribution coefficients and this high differentiating ability together with high purity of the isolated substances belong to main positive values of this method. The shaking is carried out either manually in separation funnels, or mechanically in apparatuses of different types with semiautomatic or automatic operation. Disadvantages of this method are the expensiveness and the dimensions of the equipment and considerably lengthy and exacting work. To reach a good separation it is necessary to maintain carefully the constant values of the distribution coefficient. The separation by means of counter-current distribution in neutral steroid substances is relatively very rarely used (1038, 1039, 1154), whereas this method was more frequently applied to the separation and identification of ionized steroids, e.g. estrogenes (1066, 1197).

**Chemical separation methods.** The mixtures of steroid metabolites in some cases cannot be reliably separated not even by means of column chromatography. It is a matter especially of the mixture of two or more metabolites, which, already when evaluated by paper chromatography, show very similar $R_F$ values, as it is e.g. in some epimeric hydroxy derivatives, or dehydro derivatives (pairs: cortisol and epicortisol, cortisone and prednisone, 4-androstenedione and 1,4-androstadienedione). In these cases satisfactory separation can be obtained only by chemical modifications.

The chemical methods of separation are used for such steroids derivatives, which may be easily changed back to the original substance, e.g. by saponification. Another way of separation of steroid hydroxy derivatives, frequently used, is esterification. This facilitates the separation of steroids with easily esterifiable hydroxyls (primary and secondary) from not hydroxylated substances, or from substances containing only tertiary hydroxyl groups. In this way the separation of mono from dihydroxysteroid is also facilitated, in general the separation of substances with a different number of hydroxyl groups. By chosing appropriate reaction conditions separation by selective acetylation can be reached. By esterification the stability of 21-hydroxysteroids is also essentially raised so that also the separation of corticoids on a aluminium oxide column is then made possible, which is not to be carried out in free 21-alcohols (440, 482) without danger of destruction. Acetates mostly crystallize well and their mixtures can be often separated by mere fractional crystallization (e.g. separation of cortisol 21-monoacetate from epicortisol 11α,21-diacetate (167, 251*a*)).

By help of anhydrides of dibasic acids it is also possible to prepare hemiesters, e.g. hemisuccinates or hemiphthalates, distinguished by their yielding salts with alkalies that are easily soluble in water. By shaking with an organic solvent it is then possible to separate a steroid fraction, which does not yield hemiesters (1056, 1156).

For separating ketosteroids from the non-ketonic steroid fraction the already classical method of Girard's P and T hydrazones is in use (1074, 1167). With these hydrazines react most keto groups except the 11-keto group, 20-keto group adjacent to 21-acetoxy group (1075) and 3-keto group between two double bonds, where steric hindrances occur (1152). Girard hydrazones are soluble in water, so that extraction by means of organic water-inmiscible solvent, makes possible separation of the non-ketonic portion of steroids. Even the hydrazones themselves can be sometimes separated by successive acidifying of the aqueous solution i.e. by fractional decomposition of the hydrazones into free ketosteroids (1167). EVANS *et al.* (1069) has reached separation of cortisone acetate from dihydrocortisone acetate by means of successive decomposition of Girard P hydrazones, at various pH values in the presence of formaldehyde. This way of separation of steroid ketones is sometimes not feasible inlabile steroids because of the danger of destruction, eventually because of acetylation. TAYLOR (1203) recommends in these cases the preparation of hydrazone in neutral alcoholic solution in the presence of Amberlite IRC-50, whereupon he cleaves the resulting Girard complex with formaldehyde.

A way already relatively rarely used nowadays, of separating of 3-hydroxysteroids is a preparation of insoluble compounds, which these substances yield with digitonin (1037). Of practical importance are some chemical processes described lately, in which addition or complex compounds form, as e.g. the molecular complexes of 1,4-androstadiene-3,17-dione with some phenols ($\alpha$- and $\beta$-naphtol, resorcinol etc.), insoluble in dichlorethane solution (1045) and enabling the separation of this substance from a steroid mixture. LONG and WYMAN (1121) describe the preparation of bisulfite compounds of $\Delta^1$-3-ketosteroids, which are soluble in water and therefore can be separated from the isomeric $\Delta^4$-3-ketones and $\Delta^{1,4}$-3-ketosteroids. SMITH *et al.* (421) shows the possibility of a separation of cis-$\alpha$-diols from other steroids by converting them into a cyclo-borate complex, soluble in water (e.g. in 9$\alpha$-fluoro-16$\alpha$-hydroxy derivative of cortisol).

## Identification of steroids metabolites

For a final analysis and identification the isolated steroid metabolites have to be first perfectly purified by repeated crystallization until a constant melting point is reached and the substance appears chromatographically uniform. Apart from few exceptions, steroid compounds are well crystallizing substances, whose melting point is mostly above 100°C (with exception of some esters). For recrystallization we usually choose a solvent with a steep solubility curve i.e. such a solvent, where the difference of heat and cold solubility is marked. Sometimes, especially in crude substances, the resulting solutions have to be decolourized, eventually mechanical impurities removed. In this case a considerable excess of solvent is used to prevent untimely crystallization during the filtration. Such a solvent is chosen, especially for the final recrystallization to yield a water-free crystalline steroid; most frequently ether, ethyl acetate, or acetone. Besides volatility these solvents have the advantage of not yielding with steroids hydrates nor adducts, which are sometimes considerably thermostable and support even the temperatures over 100°C without decomposition.

Before the analysis, a careful drying of the substances is carried out in vacuum at a temperature which is considerably lower than that of the melting point.

The identification proper results then from a complex of physical, physico-chemical, chemical, eventually microbiological reactions and observations. In completely new types of microbial conversions it is usually supplemented by experiments concerning partially synthetical preparation of the identified substance, whereby identity and correctness of the identification conclusions are best confirmed.

## Preliminary identification processes

**Colour and fluorescence reactions.** These reactions frequently indicate the presence of certain function groups in the molecule of the steroid e.g. $\Delta^4$-3-keto group, 17-keto group, α-ketol grouping, phenolic nucleus etc. There are relatively few reactions suitable for the identification of individual steroids, and thus sufficiently specific; here we have to quote e.g. the blue reaction with Lugol solution, given by a relatively small group of steroid compounds of various types (see p. 70). Some steroids also give important colour reactions with concentrated sulphuric acid e.g. cortexolone presents a bright red colour, which passes slowly (during 1 hour) into deep blue colour, cortisone and cortisol present a deep yellow colour, which in cortisone passes slowly into green and in cortisol into brown colour. Testosterone gives blue colour with the mixture of sulphuric and acetic acid. Most colour and fluorescence reactions are, however, unspecific and common to most steroids. Group colour reactions in pure isolated substances are often carried out as spot-reactions on a watch-glass or on filter paper, in steroid mixtures on paper chromatograms or in thin layers. Comprehensive lists and tables of colour and fluorescence reactions of steroids are presented by a series of authors (1035, 1081, 1138, 1141, 1161, 1183, 1184).

Reducing steroids (α-ketols) present a dark brown colour with an amnonia solution of $AgNO_3$ (Tollens' reaction) (1219) already at normal temperature. In the heat even steroids with a α-diol grouping (468, 494, 1092) react with this agent; this reaction was originally determined for detection of non-reducing sugars. Tollens' reaction for reducing steroids has recently been substituted by a more impressive colour reaction with tetrazolium salts, which change in an alkaline solution in the presence of reducing substances into red and blue formazanes (1122). Especially a reaction with blue tetrazolium is very sensitive. It was, however, found that this positive reaction is also given by some non-reducing steroids, e.g. $\Delta^4$-3-ketosteroids with a hydroxy group in position 6α and 7β (1131). Corticoids containing hydroxyl group in position 17α give the yellow Porter-Silber reaction with phenylhydrazine in alkaline solution with abs. maximum at 410 nm (1159, 1185).

Zimmermann reaction (*m*-dinitrobenzene in alkaline solution), which is very frequently used, presents a characteristic and very stable violet colour with 17-ketosteroids with abs. max. at 520 nm (1079). A less marked and less stable colour is found with this agent in keto groups in position 3 (blue-violet), 20 (pink) and 16 (brownish-orange), while in ketones in positions 6, 7, 11 and 12 the reaction is negative.

Considerably specific orange colour is given by conjugated 3-ketogroup after treatment with 2,4-dinitrophenylhydrazine, while the non-conjugated 3-ketone only shows yellow colour (1076). $\Delta^4$-3-Keto structure can also be well proved by colour reaction with hydrazide of isonicotinic acid (1191): $\Delta^4$-3-ketones give a yellow colour, which becomes brown and $\Delta^{1,4}$-3-ketosteroids give permanent yellow colour. Hüttenrauch (1094, 1095) has used a colour reaction with *p*-aminodimethylaniline for determination of conjugated unsaturated ketosteroids: $\Delta^4$-3-ketones nad $\Delta^1$-3-ketones are coloured yellow, while $\Delta^{1,4}$-3-ketones and $\Delta^{4,6}$-3-ketones become orange.

Diphenylamine in solution of acetic and sulphuric acid (Dische's reaction) gives with 11,20-diketo-21-ols a violet colour with abs. maximum at 530 nm, and with 11,17,21-triol-20-ones a green colour with abs. max. at 630–650 nm (1047).

Unsaturated $\Delta^5$-3$\beta$-ols can by proved by a orange or red colour reaction with $SbCl_3$, red-brown colour reaction with sulphuric acid in ethanol (1149), or by blue reaction with phosphomolybdenic acid.

A specific reaction for estrogens is the Kober reaction with sulphuric acid in the presence of phenol (also substituted): yellow colour, turning pink after dilution with water. An important blue reaction is given by phenolic steroids with a mixture of 1% solution $FeCl_3$ and 1% $K_3Fe(CN)$, the so called Turnbull blue reaction (1020).

The reaction with $FeCl_3$ (for diosphenols) is positive in steroids with hydroxy group in position 2 or 6$\beta$- characterized by marked violet colour (1072).

The acetoxy groups react with hydroxylamine yielding hydroxamic acids, which then give an intensive purple colour with $Fe^{\cdots}$ ions.

Some estrogens and corticoids give a marked yellow-green fluorescence by action of concentrated sulphuric acid (1165). This test can be also carried out as a spot-reaction on a watchglass. As to corticoids considerable fluorescence is given by corticosterone, epicorticosterone, cortisol and epicortisol, while cortexolone and pure cortisone do not produce this fluorescence. Estradiol, too, presents strong fluorescence with sulphuric acid. Differences in intesity of fluorescence reactions in various types of steroids for the purposes of quantitative analysis are illustrated in a survey by SWEAT (1198). On paper chromatograms the fluorescence reactions (in ultraviolet light) are most frequently carried out after detection with 15% phosphoric acid (1142) or with $SbCl_3$. Fluorescence spectra of various steroids in 85% phosphoric acid at various temperatures are described by KUBOTA and MIYAZAKI (1111). An increase in the intensity of fluorescence reactions in some steroids (e.g. in progesterone) can be obtained when the steroid is previously heated with an alcoholic hydroxide solution (1204).

A reaction with strong alkali is often used as a specific reaction for the $\Delta^4$-3-keto group, causing a typical yellow fluorescence (1034) with the exception of progesterone, some of its hydroxy derivatives (with OH in the positions: 6$\beta$, 11$\beta$, 14$\alpha$, 17$\alpha$, 18 and 20), and 11-ketoprogesterone which react only weakly (1068*a*, 1183). A conclusive colour reaction in cardioglycosides and cardenolides is the Legal reaction with natrium nitroprusside on an unsaturated 5-membered lactone ring (dark red colour) (1143).

**Paper or thin-layer chromatography.** Accoroing to the position of the metabolite on the chromatogram in various solvent systems with various methods of detection the character of conversion can be often judged. If there is little difference in the values $R_F$ between the initial steroid and the metabolite, it may be a matter of hydrogenation or dehydrogenation. When a considerably more polar substance is formed, the hydroxylation may be concluded upon, an essential decrease in polarity may mean that a microbial acetylation has occured.

In newly introduced hydroxy groups according to Savards' rule (1176) on the basis of $R_F$ values in Zaffaronis' systems the configuration ($\alpha$, $\beta$) may be concluded upon: steroids with an equatorial hydroxy group are usually more polar than the corresponding axial isomers. For the solving of configuration of hydroxyls in position 7 in progesterone this rule has not proved correct (259).

**Determination of melting point.** The melting point is usually determined by means of Kofler's block and the value found is corrected according to the thermometer used. The accuracy of the determination in substances with a higher melting point (over 200°C) often depends upon the course of heating, as some labile steroids begin to decompose already below the melting point. Hydrates and adducts with solvents often melt temporarily at a lower temperature, then crystallize again in the form of waterfree substances and melt once more at a higher temperature. Sometimes the same substance, crystalized from various solvents, presents different crystal forms, differing by the melting point, and therefore the solvent used for crystallization should always be mentioned.

For the identification and as a criterion of the substance purity the determination of the mixing melting point (mixture of metabolite and of standard) is still sometimes used, even though there are cases known, when a depression of the melting point did not occur, even in a mixture of nonidentical substances (mixture of cortisone and cortexolone).

**Optical rotation.** The determination of specific rotation is a very important identity mark of a steroid and together with the melting point is usually also a good criterion of the purity of the isolated substance. It is also the only analytical way of differentiation of optically active and racemic substances. The determination is carried out in microtubes of a content of 0·5 to 2·0 ml, 5 to 20 cm long, which are filled by help of a microfunnel drawn into a capillary reaching down to the bottom of the polarization tube. The measurement mostly proceeds in the light of a sodium-vapour lamp (5890 Å) at a laboratory temperature of 15 to 25°C, where there are no remarkable deviations. The solvents most frequently used are chloroform, acetone, alcohol or dioxane and in some cases also pyridine. As the value found for specific rotation changes with the kind of the solvent used, the solvent should always be precisely specified.

**Molecular rotation.** From the established specific rotation the molecular rotation $M_D$ is calculated according to the formula:

$$M_D = \frac{[\alpha]_D \,.\, \text{mol}\,.\, \text{weight}}{100}, \qquad [\alpha]_D = \text{specific rotation}\,.$$

As it is well known that there is a relation between optical rotation and the structure of the steroid the values of molecular rotation can be used for solving the constitution of an unknown metabolite. The method of molecular rotation differences is based on Freudenberg's shift-rule, derived from the principle of optical superposition. Differences of molecular rotations (usually designed $\Delta[M]_D$ in two similar steroid molecules in which the same substitution has been carried out, should be identical. Some function group in a molecule in the proximity of a newly introduced substituent can, however, bring about a so called vicinal effect, e.g. hydroxyls, keto groups or double bonds.

Identification by means of molecular rotation differences is, however, of importance, only when the character of the conversion is known, as it is a method absolutely empirical and neither designates the position of a substituent in a molecule, nor its configuration ($\alpha$, $\beta$). It can be considered a good identification method only in those cases, where the values of $\Delta[M]_D$ are sufficiently high, since the errors in measurement, too, must be taken into consideration. By means of the method of molecular rotation differences it was, for instance, possible to solve the configuration of the hydroxy groups in positions 1, 2, 6, 9, 10, 11, 16 and 20, while the values for the hydroxyls in position 7, 8, 12 and 15 are not convincing.

**Elementary analysis.** In isolated pure metabolites, especially in new so far not reported substances, after careful drying the elementary analysis is carried out, whose importance also consists in obtaining informations about the number of oxygen functions in the steroid molecule.

**Study of crystal structure.** Instead of laborious crystallographic investigations, the studies are nowadays more frequently carried out by means of X-rays. After passing through X-rays the powdered crystalline substance leaves records in the form of interference figures on a photographic plate according to the organization of the crystall lattice and the distance of the atoms. These interference diagrams are typical for every crystalline substance and can also serve as an identification criterion. The surveys of these diagrams are given by e.g. PARSONS (1153).

**Optical rotatory dispersion.** The optical rotation of an organic compound depends on the length of light used. The estimation of differences in the optical rotation dependent on the wave-length of ultraviolet radiation in the region 280 to 700 nm is carried out on a spectropolarimeter (1082, 1175). From the measured values a so called dispersion curve is constructed. This method is important especially for the localization of keto groups in the steroid molecule as well as for the kind of configuration of rings (*cis*, *trans*). The shape of the dispersion curve is also influenced by hydroxy groups and by acetoxy groups of the steroid molecule in the proximity of a keto group. For steroids this method was developed by DJERASSI (1055).

**Nuclear magnetic resonance.** The method belongs to the latest physical analytical processes for the study of complex organic compounds. It it of great importance in the characterization and in localization of substituents as well as for the configuration of individual rings. Tables of spectra in a solution of deuterochloroform (which least disturbs the adsorption proper of the measured substance) for steroids have been shown by SHOOLERY and ROGERS (1182). This method was also successfully used in solving the configuration of some microbiologically introduced hydroxyls for instance in positions 7, 8, 12 and 15 (465, 481, 481*a*), as it presents more convincing proofs than the method of molecular rotation differences.

## Chemical constitution proofs

For purpose of identification the isolated steroid metabolites are submitted to appropriate chemical reactions and so converted either into substances already known and described, or into such types of compounds which present important colour reactions or characteristic spectra in the ultraviolet and infrared region. The reactions should have an unambiguous course and should be suitable as to the stability of the analysed substance. Preliminary study of metabolites by means of paper chromatography, by help of colour reactions, spectra and elementary analysis usually represent a certain guide. Chemical constitutional proofs are either simple reactions of esterification, oxidation, reduction, dehydratation, isomerization etc., or multistep chemical processes where these operations are combined.

Changes, which occur in microbial transformations on the side-chain of 21-steroids (*oxidation, reduction, cleavage*), are studied by help of the well-known colour reactions (*Zimmermann* and *Porter-Silber reactions, tetrazolium salts* and others), according to infra-red spectra and on the basis of oxidation studies by help of periodic acid, natrium bismuthate, chromium trioxide etc. Changes on the steroid skeleton in substances with labile side-chain are mostly proved until after its cleavage by oxidative agents (53, 253, 312, 355, 405) or microbiologically; for instance in the hydroxy derivative of cortexone by means of the strain of *Penicillium lilacinum* (122). The resulting 17-ketosteroids, eventually etianic acids are essentially more stabile towards chemical agents so that it is already possible to use even more drastic chemical reactions for instance oxidation with chromium trioxide, the action of alkali and acids etc. Here no changes occur in substitutents on the nucleus except in the presence of a α-diol or α-ketol grouping. Identification processes in steroid hydroxy derivatives are given in a survey by SCHUBERT *et al.* (25). Chemical reactions of steroid compounds are reviewed in monographies e.g. FIESER and FIESER (1070, 1071), ČERNÝ *et al.* (1042), LETTRÉ *et al.* (1116), MOORE and HEFTMANN (1138). Chemical reactions in micro-scale for chromatographic identification processes are described by BUSH (1035).

**Esterification and saponification.** Esterification, on the one hand serves for the separation of mixtures of differently hydroxylated steroids (viz p. 91), on the other hand for the elucidation of their structures. Acetylation is usually carried out under so called "mild acetylation conditions" by an excess of acetanhydride in pyridine at room temperature within 4–20 hours. Under these conditions primary and secondary alcohols are acetylated besides the hydroxy group in position 11β which is strongly sterically hindered by the presence of the angular methyls. There is no acetylation in tertiary hydroxygroups except 17β-hydroxyl in 17-isosteroids of series $C_{21}$.

A non-esterified hydroxy group can be reliably proved by infra-red spectrum, hydroxyl in position 11β can be recognized besides tertiary hydroxyl groups, by its easy yielding to oxidation to a ketone. The hydroxyl group in position 15β under the mentioned acetylation conditions sometimes also presents difficulties in acetylation e.g. in 15β-hydroxyprogesterone (8) or in 7α, 15β-dihydroxyprogesterone (466*a*). The same group, on the contrary in 15β-hydroxy derivative of cortexolone is acetylated easily (187). Pseudoequatorial 15α-hydroxy group is mostly acetylated readily (52).

A primary hydroxyl group in position 21 can be proved directly by infrared spectrum even in the presence of other hydroxyls in the molecule. This hydroxy group can also be acetylated selectively by using 1 equivalent of acetanhydride in pyridine, or e.g. besides a simultaneously present 11α-hydroxy group also by means of an excess of acetanhydride in an acetic acid, or in a neutral solvent under catalytic action of some salts (e.g. of anhydrous potassium acetate (1114)). Hydroxyls in positions 3,16α- and 17β also acetylate easily: testosterone already yields acetate by the action of acetic acid under the catalysis of $H_2SO_4$ (121). Tertiary hydroxyl groups are acetylated only under more drastic reaction conditions e.g. at a higher temperature, or in the presence of acid catalysts (*p*-toluensulphonic acid, perchloric acid). As here dehydratations and the rise of D-Homosteroids has often been observed, this way of acetylation is usually not used in identification processes. Some labile acetoxy groups e.g. in positions 6 and 7, already split off in chromatography on neutral aluminium oxide and a 6-dehydro derivative is formed (299, 458).

**Saponification of esters.** Saponification of acetoxy groups is mostly carried out by means of alkalies ($NaHCO_3$, $Na_2CO_3$ and NaOH) in a methanol solution. Hydrolysis by means of diluted mineral acids proceeds considerably slower and there often arise undesirable by-products (1126). Esters of stabile steroids (estrogens, androgens) are saponified by means of alkaline hydroxides at normal temperature, or by heating with solutions of alkaline carbonates without the occurence of undesirable reactions. Compared to these, $C_{21}$-steroids (gestagens, corticoids) are more labile in solutions of alkaline hydroxides or acids and isomerization of the side-chain may happen. Greatest precaution is necessary when saponifying esters of corticoids. The 21-acetoxy group is saponified most frequently with $KHCO_3$ in diluted methanol according to REICHSTEIN (1099, 1169) or by means of sodium methylate solution in water-free methanol. In this case the saponification proceeds very quickly and is practically finished in several minutes (1093). As to corticoids, it is recommended to work in an inert atmosphere (under nitrogen) and only for as long a time as necessary, since in the presence of oxygen of after a longer time even under nitrogen, alkalies cause changes in the side-chain (oxidation, cleavage). Saponification of a simultaneously present $11\beta$-acetoxy group (e.g. in $11\beta$,21-diacetate or cortisol) cannot be carried out chemically and the only way of saponification of both acetoxy groups known so far is microbial method by means of the strain *Flavobacterium dehydrogenans* (84).

Saponification of labile acetoxy groups in positions 2 and 6 can be carried out by means of one equivalent KOH in methanol solution under nitrogen, otherwise dehydratations will occur (1031, 1164).

**Influence of strong acids and alkalies.** In the saponification of steroid esters by help of strong alkalies or acids sometimes occur undesirable side-reactions, which reduce the yield of isolated substances. The resulting substances are often very interesting and they have therefore been studied in detail and the laws, according to which the structure of the steroid metabolite may be concluded upon, have been determined. A considerable stability is shown by the hydroxy groups and ketones in positions 11 and 12 (57, 58), from tertiary ones hydroxyl in position 14α (57, 138).

In $\Delta^4$-3-ketosteroids hydroxylated in positions 1, 2, 6 and 7 there result, in the presence of alkalies (e.g. 0·1N-KOH in methanol) dehydratations giving rise to a double bond in positions 1, 2 or 6, 7. Dehydratation occurs especially at a higher temperature (60°C), in acetoxy groups already at normal temperature (258, 466*a*). The resulting conjugated systems show characteristic absorption in ultraviolet light at 235, resp. 285 nm (113, 258, 466*a*, 1132). The resulting dienone also shows strong absorption in infra-red region (466*a*).

If hydroxy groups in positions 1 and 6 are simultaneously present in the molecule of the steroid, only dehydratations on 6-hydroxy-$\Delta^{1,4}$-diene-3-one will occur under the action of alkalies (113, 481).

Strong acids (e.g. HCl in methanol solution) also cause dehydratation in the mentioned positions (213, 241, 320, 485). Hydroxyl in position 6 is stable under these conditions at 20°C, whereas hydroxyl in position 7, which is smoothly eliminated originating $\Delta^{4,6}$-diene-3-one (456). Acetoxy groups in positions 2β-(105, 109, 1048) and 6β- (369, 485) isomerize under the influence of HCl in water-free solvent into the corresponding 2α- and 6α-acetoxy derivatives.

6α- and 6β-acetoxy groups of $\Delta^4$-3-ketosteroids are subject to isomerization by the action of strong acids (e.g. HCl or $H_2SO_4$ in acetone) yielding saturated 5α-3,6-diketone **II** (138, 153, 473), which shows a characteristic absorption maximum.

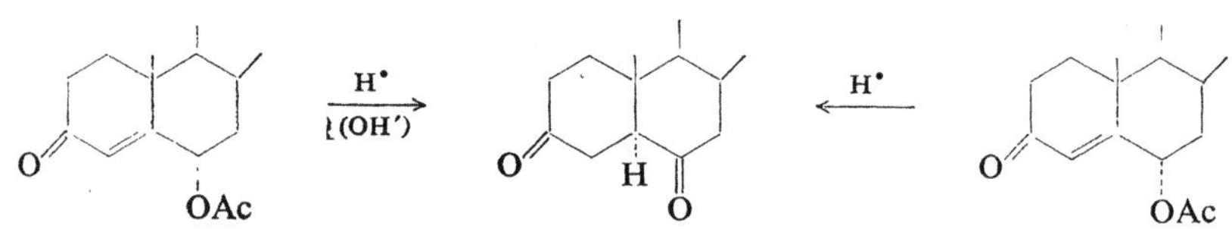


I II III

The same isomerization is also obtained in the 6β-acetoxy group by the influence of strong alkalies e.g. in saponification, while the 6α-acetoxy group gets saponified. These reactions are typical for hydroxyl in position 6 and they are not given by hydroxyl in position 7 (69). Compound **II** in alcoholic hydroxide solution further passes to $\Delta^4$-3,6-diketone (473, 1062); this substance has been determinated as one of the metabolites of dehydroepiandrosterone in microbial transformation by the strain *Bacillus pulvifaciens* (203).

The steroid β-diketones e.g. 16-ketoprogesterone in alcaline solution, change into enolform, showing a characteristic abs. maximum at 280 and 320 nm (373).

Acids and alkalies induce in some cases isomerization of the side-chain of $C_{21}$-steroids, giving use to biologically inactive 17-iso-steroids, also distinguished by a different specific rotation, which is substantially lower and sometimes negative, wherein the influence of some substituents, e.g. double bond in position 16–17, 15-keto group, 11α-hydroxy group and others makes itself felt (8, 154, 466*a*, 1053, 1054). In these substances the present 20-keto group is remarkable by a relatively small reactivity. By the effect of acids or alkalies 17-iso-steroids may often be converted again into more stable derivatives of the normal, natural series (472).

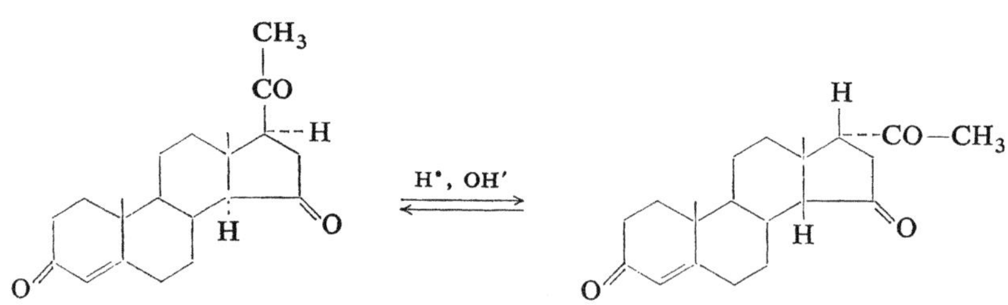


$[\alpha]_D$ + 204 ($CHCl_3$) $[\alpha]_D$ + 122 ($CHCl_3$)

In consequence of the effect of alkalies and acids the side-dihydroxyacetone chain of corticoids is subject to characteristic changes of similar type as reducing sugars i.e. to isomerizations (ketose ⇄ aldose), to conversions into substances of the saccharinic acids type and to cleavage. Upon the action of alkalies in presence of oxygen the side-chain cleaves into etianic acids **I** and formic acid (153, 1211). In prolonged action of alkalies even in the absence of oxygen (also in nitrogen atmosphere) isomerization occurs first to 20-hydroxy-21-aldehyde **II**, which then, on one hand splits into 17-keton **III**, on the other hand converts via the glyoxal form **IV** into 20-hydroxy-21-carboxylic-acid **V** (1216).

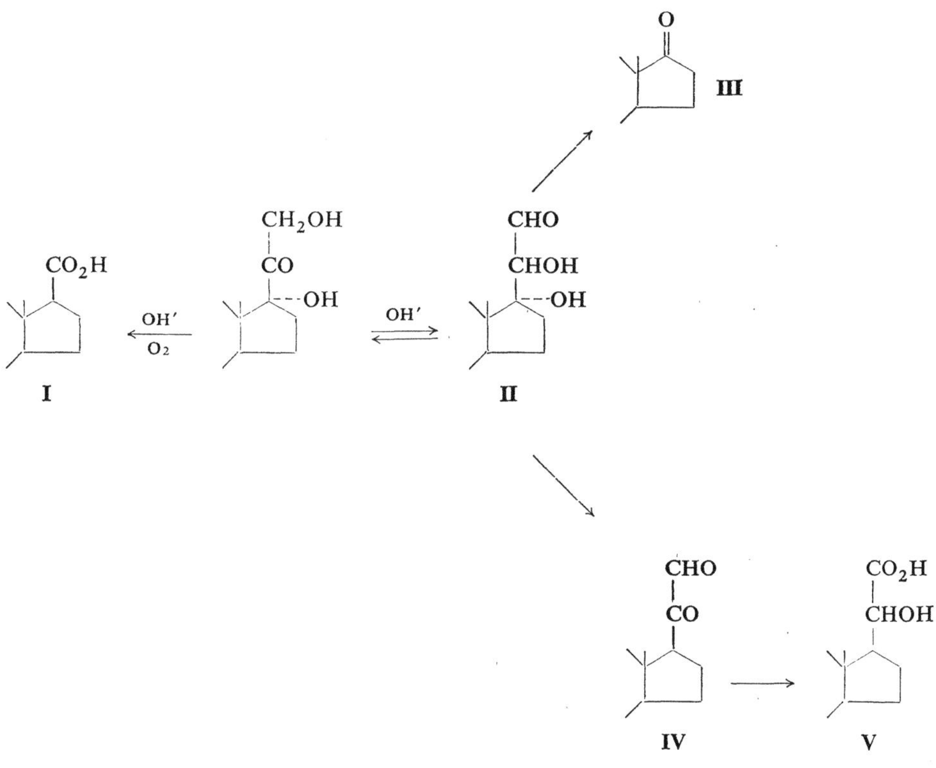


Similar conversions appear in some types of the side-chain of corticoids even upon the action of strong acids (e.g. HCl in absolute methanol). This acid-catalyzed rearrangement of the side-chain of corticoids in the presence of methanol has been studied by MATTOX (1127) who has found that dimethyl acetal of 20-keto-21-aldehyde (**II**) (see p. 102) is formed. The same substance results at similar reaction conditions also from 16α-hydroxy and 16-dehydro derivatives of cortexone (494, 1127).

Dimethoxyketone (**II**) results as undesired side-product in acid hydrolysis of cortisone acetate (1127). Similar conversions probably also occur in the Porter-Silber reaction with subsequent formation of 21-hydrazone (1021).

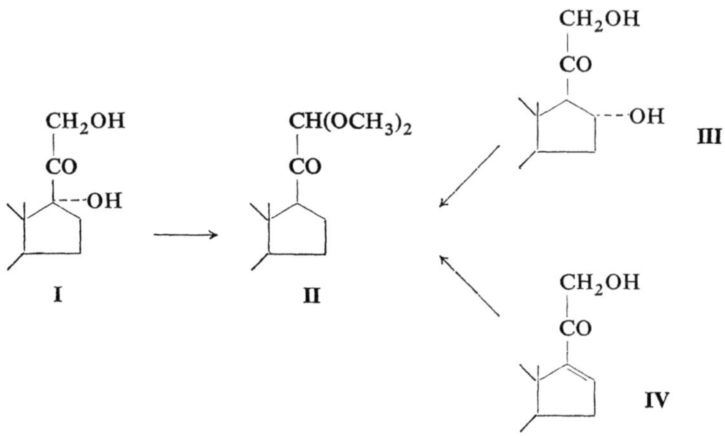

**Oxidation with chromium trioxide.** Oxidation with chromium trioxide in primary hydroxyl groups passes via aldehyde to carboxylic acids, secondary alcohols are oxidized into ketones and tertiary hydroxyl groups are not attacked. Oxidation is usually carried out in 90% acetic acid or in acetone in the presence of sulphuric acid (135, 213, 1027). The in acid medium unstable substances are oxidized by help of a complex compound of chromium trioxide and pyridine $CrO_3.2\ C_6H_5N$ according to Poose (1158) e.g. in steroids having hydroxyls in position 6 and 15 (53, 213, 472, 485).

The side-chain of corticoids after acetylation of hydroxyl in position 21 exhibits remarkable stability in oxidation with $CrO_3$. Nevertheless this chain is often removed by oxidation cleavage as in identification experiments concerning the position of hydroxyls. Oxidation with $CrO_3$ in acetic acid gives rise in steroid hydroxyderivatives at least partly to acid products, when the hydroxy group is in positions 1, 2, 16, 17, 18, 19 and 21 (456).

Ketones of androstane series (keto derivatives of androstenedione) are substances crystallizing well with remarkable physical constants. The abs. maximum of the $\Delta^4$-3-keto group (241 nm) in the presence of a further keto group in positions 1, 2, 6 and 7 shows characteristic shifts in alkaline solutions. Also 15-keto-androstenedione and 16-keto-progesterone ($\beta$-diketone) show characteristic changes by forming enol-form (1057).

Hydroxy groups in position 7$\alpha$ and 7$\beta$ cannot be directly oxidized with $CrO_3$ and hydrogenation of the $\Delta^4$ double bond or addition of halogens (25, 368) must be carried out previously.

In the oxidation with $CrO_3$ in $\alpha$-diol and $\alpha$-ketol groups on the nucleus the cleavage is effected in place between both oxygen functions and originates ketoacid or dicarboxylic acid. Oxidation e.g. of 14$\alpha$, 15$\beta$-dihydroxyprogesterone originates 14-keto-15-carboxylic acid well soluble in an alkaline aqueous solution (456, 468): similarly in 2-hydroxy-$\Delta^4$-3-ketones or in 16$\alpha$-hydroxy-17-ketones (8).

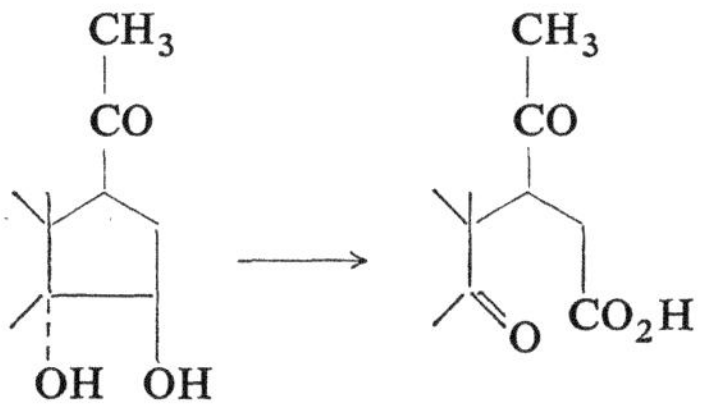


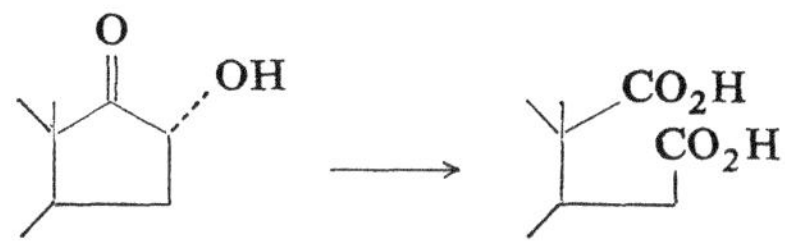


GRIMMER (1077) has described a method of determining the configuration of epimer hydroxy derivatives on the basis of the velocity, at which oxidation proceeds ($CrO_3$ in acetic acid). Here axial hydroxyls oxidize faster than the corresponding equatorial isomers. This rule was not, however, confirmed by a further study of the configuration of hydroxy derivatives in position 7 (172, 259).

Oxidation by chromium trioxide also helps to determine the configuration of the side-chain of $C_{21}$-steroids. Type **I**, when oxidized, does not change, type **II** is oxidized into type **I**, types **III**, **IV**, **VII** and **VIII** cleave oxidatively into 17-ketosteroid and types **V** and **VI** yield etianic acids.

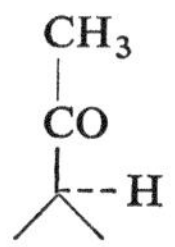


**I**

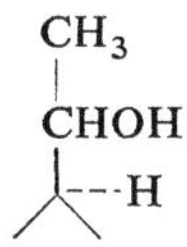


**II**

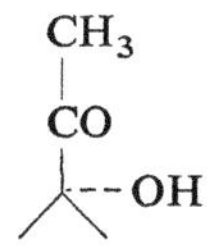


**III**

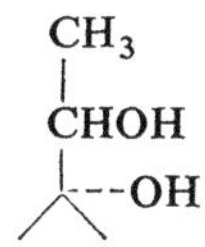


**IV**

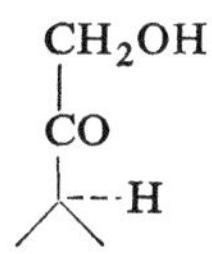


**V**

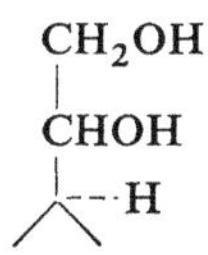


**VI**

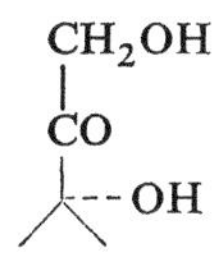


**VII**

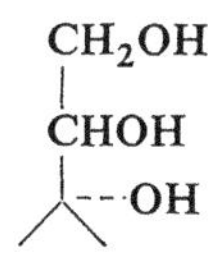


**VIII**

**Selective oxidation methods.** Selective oxidation in steroids with several hydroxyl groups in some cases can be carried out also by means of chromium trioxide, especially by modification with a pyridine complex. So e.g. in 6$\beta$,11$\beta$-dihydroxyandrostenedione the 11$\beta$-hydroxy group can be selectively oxidized under these conditions (485), or in epicortisol the oxidation of the hydroxy group in position 11 may be carried out even without previous acetylation in position 21; in this case the yields are unreliable regarding the lability of the side-chain (1209).

For oxidative cleavage of the $\alpha$-ketol or $\alpha$-diol grouping in the steroid side-chain or on the nucleus the periodic acid, natrium bismuthate and lead tetraacetate are often used. The most reliable results are obtained by help of the two first oxidation agents, in lead tetraacetate the results are usually not so good. Oxidation by help of $HIO_4$ is of great importance especially for analytical purposes in determining the contents of formaldehydogenous and acetaldehydogenous steroids, while the oxidation with $NaBiO_3$ is more frequently used for preparative purposes (1030).

When oxidized with periodic acid, the side-chains of types **I**, **II** and **III** do not change, types **IV** nad **VIII** are oxidized to 17-ketosteroid, wherein acetaldehyde is simultaneously released from type **IV** and formaldehyde from type **VIII**. Types **V**, **VI** and **VII** present on oxidation etianic acids and formaldehyde. Type **III** cannot be cleaved into 17-ketosteroid by oxidation until after reduction of the 20-keto group e.g. with $NaBH_4$.

Natrium bismuthate oxidizes the side-chains of type **IV**, **VII** and **VIII** into 17-ketone, type **V** into etianic acid and type **VI** into 20-aldehyde. Types **I**, **II** and **III** do not change. The oxidation is most frequently carried out in 50% acetic acid, but EDWARDS and KELLIE (420, 1061) recommend 75% ethanol with 4% of trichloroacetic acid as the oxidation medium, whereby the yields improve. Though $NaBiO_3$ mostly does not attack isolated hydroxyl groups simultaneously present in the nucleus, in some cases oxidation of hydroxyl in position 11$\beta$ (1071) has also been found.

From the quantity of periodic acid consumed for the oxidation, it may be concluded, whether the cleavage occured in one or more sites of the steroid molecule. For this purpose the quantity of not consumed periodic acid, determined in regular time intervals (within 24 hours) is established by means of titration with thiosulphate (indicator: KI and starch). Consumption of 1 equivalent of $HIO_4$ shows that oxidation occured in one site of the molecule only (456).

According to the rate, at which proceeds oxidation of the $\alpha$-diol group by $HIO_4$ or by lead tetraacetate the diol configuration (*cis, trans*) can be concluded upon: *cis*-$\alpha$-diols are oxidized under the same reaction conditions approximately twice as fast as the corresponding *trans*-$\alpha$-diols (468, 1117). The $\alpha$-glycol group in 14$\alpha$,15$\alpha$-dihydroprogesterone oxidized with periodic acid has proved relatively stable and did not undergo cleavage until when oxidized with chromium trioxide (456).

Lead tetracetate was used e.g. in oxidation cleavage of the $\alpha$-diol group in positions 14$\alpha$,15$\beta$-(468), 8$\beta$, 9$\alpha$-(369), 8$\beta$,14$\alpha$-, 16$\beta$,17$\beta$-(112) etc. When 16-keto testosterone ($\alpha$-ketol) is oxidized aldehyde-acid is formed.

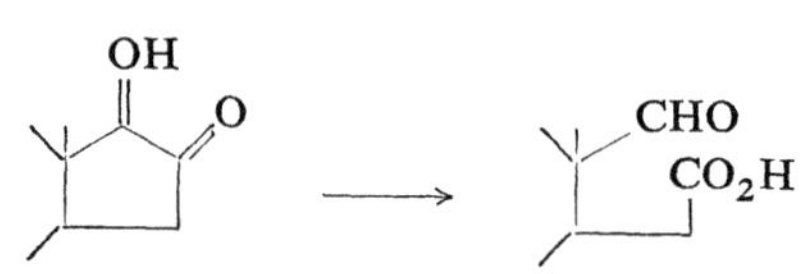


In chemical syntheses lead tetraacetate is usually also used for introduction of the hydroxy groups e.g. into position 2 (1033, 1174, 1194), 18 (1037*a*) and 21 (1062*a*).

Oxidation of the allylic hydroxy groups into ketones is rather often carried out by means of active manganese dioxide (1123). In steroids with double bond between carbons 4 and 5 hydroxyls in positions 3 and 6 can be oxidized (111, 1016).

In the presence of a double bond in positions 5, 6, hydroxyl can be oxidized in position 7 (e.g. in 7$\beta$, 11$\alpha$-dihydroxydiosgenine (179)). It has been found that equatorial allylic hydroxy groups are oxidized better at normal temperature, whereas in heat oxidation proceeds in both equatorial and in axial hydroxy groups. Oxidation of hydroxy group in position 17$\beta$ has been realized even in the presence of a further hydroxy group in position 11$\beta$, 12$\beta$ or 14$\alpha$ (341). The Oppenauer oxidation used in chemical synthesis nearly exclusively for the oxidation of the system $\Delta^5$-3$\beta$-hydroxy into $\Delta^4$-3-ketone ranges among further selective oxidation processes. Here we find cleavage of the side-chain and isomerization into D-Homosteroids in corticoids by the action of an alkaline medium. Dehydration may be seen in labile hydroxy groups on the nucleus: e.g. $\Delta^{4,6}$-androstadiene-3,17-dione results from oxidation of 7$\alpha$-hydroxy derivative of cortexone (108). Some strains of microorganisms carry out the same type of oxidation, which in this case is also called "*Oppenauer microbial oxidation*".

N-Bromacetamide and N-bromsuccinimide are very selective oxidation agents, which do not attack the $\alpha$-ketol grouping and therefore they are even used in corticoids for the oxidation of hydroxyl groups in positions: 3, 7, 11$\beta$, 12$\alpha$, 17$\beta$ and 20.

Another type of oxidation is dehydrogenation with selenium dioxide often used in steroids for introducing double bond into position 1, 2. In this way dehydro derivative, identical with substances prepared by microbial dehydrogenation, can be obtained. This reaction also serves to identify hydroxy group in position 19. By the action of alkalies or acids the resulting $\Delta^{1,4}$-3-keto-19-hydroxysteroid then easily passes into aromatic 3-hydroxy-1, 3, 5 (10)-estratriene with remarkable abs. maximum at 300 nm (312). In $\Delta^4$-3-ketosteroids a further double bond can be introduced into position 6, 7 by action of chloranil and so it is possible to carry out an identification of derivatives resulting e.g. from dehydration of 6-hydroxy- or 7-hydroxy-$\Delta^4$-3-ketosteroids.

**Reduction of keto groups.** Reduction processes are also sometimes brought about by microbial conversions. Especially reduction of keto groups in positions 3, 17 and 20 is not unfrequently seen and the resulting hydroxyl groups possess mostly configuration $\beta$. The same reduction can be carried out chemically for identification purposes, by catalytical hydrogenation ($PtO_2$) or by means of metal hydrides. Especially the method of reduction with $NaBH_4$ in a methanol solution at 0°C is very frequently used (420, 1146).

The configuration of the resulting hydroxyl group depends upon a number of factors, e.g. on steric conditions, the kind of hydride, pH etc. The keto group in position 11 is the most sterically hindered of all keto-groups so that in its presence keto groups in positions 3, 16, 17 and 20 (63, 112, 1146) can be selectively reduced. The reduction of the 11-keto group into alcohol with hydroxyl configuration $\beta$, is carried out by means of metal hydrides. A 11$\alpha$-hydroxy group is formed by reduction by means of alkaline metal in an alcoholic medium. In the presence of ketogroups in positions 3 or 20 they should be protected by preparation of ketals. One of the most reactive groups is the keto group in position 20, which is reduced preferentially before the 11-keto group and before conjugated $\Delta^4$-3-keto group (420, 1042). In the presence of a 17α-hydroxy group reduction proceeds nearly exclusively in position 20β (482, 1151). A very gentle method of selective reduction is the Meerwein-Ponndorf reduction, used e.g. to reduce a 17-keto group (reduction of androstenedione to testosterone) (1133).

The configuration of hydroxy groups in position 20 can be very well established from the values of the differences in molecular rotations in free hydroxy derivatives and after acetylation. In 20β-hydroxy group these values are positive, whereas in 20α-hydroxyl they are negative (1151).

**Hydrogenation of double bonds.** Microbial hydrogenation of double bonds in positions 1,2 and 4,5 is seen comparatively rarely and sooner occur as a side-reaction (295, 328, 331, 332). SHIRASAKA and TSURUTA (392) describe the strain *Alternaria batatícola,* which effects this reaction as a main transformation process in a certain number of $\Delta^4$-3-ketosteroids, at the same time there proceeds a reduction of the keto group.

The same reactions can be carried out chemically by catalytical hydrogenation in the presence of Pd or Pt on carriers. In the hydrogenation of double bond $\Delta^4$ mostly both isomers, 5α as well as 5β are formed, which then must be separated by fractional crystallization or by chromatography (369). The hydrogenation is mostly carried out in an alcoholic solution at normal pressure and temperature. In the presence of more double bonds sometimes even partial hydrogenation can be obtained, e.g. in $\Delta^{4,6}$-pregnadiene-15β-ol-3,20-dione the double bond in position 6,7 can be selectively hydrogenated (466*a*). Some substituents also decide about whether in hydrogenation a saturated 5α-or 5β-isomer is formed, as e.g. oxygen functions in positions 6,11 and 17 or pH value during the hydrogenation.

Hydrogenation of double bond $\Delta^4$ is also carried out, when its stability is endagered by some chemical reactions (dehydrogenations, oxidations etc.). This reduction is sometimes brought about by the action of zinc in acetic acid, e.g. $\Delta^4$-3,6-diketones are so conversed into saturated 5α-3,6-diketones. Identical substances also resulted in the hydrolysis of the 6β-acetoxy group.

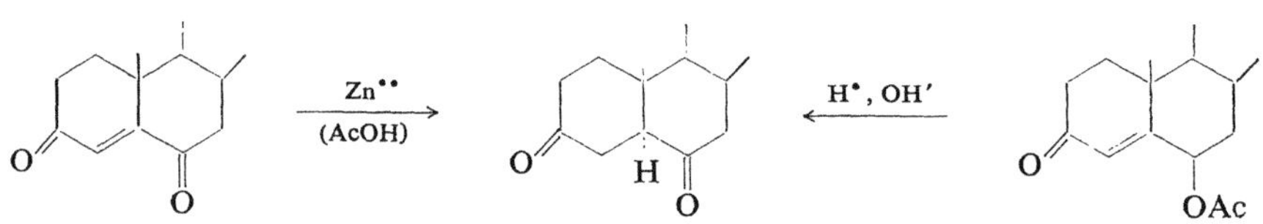


**Removal of hydroxy group under formation of double bond.** Procedure frequently used in characterization of a new hydroxy group is its removal under formation of a new double bond (dehydratation). In labile hydroxyls in positions 1, 2, 6 and 7 dehydratation is brought about by the action of diluted solutions of alkalies or strong acids, when a hydroxy group is in position 1 (1088) or 7 (299), it is sufficient already to filter the solution through a layer of aluminium oxide. The 11$\beta$-hydroxy group, contrary to the epimeric 11$\alpha$-hydroxy group is removed more easily by the action of strong acids, giving rise to a 9(11)-dehydro derivative.

Dehydratation of the other hydroxyl-bearing $\Delta^4$-3-ketosteroids is carried out by the action of strong acids and acid chlorides at a higher temperature. HCl or $H_2SO_4$ in chloroform solution, $POCl_3$, $SOCl_2$, solution of *p*-toluenesulphonic acid in benzene and others are used (70, 172, 328, 369, 413). Often also the corresponding tosyl- or mesyl-esters are prepared, whereupon the obtained *p*-toluenesulpho- or methyl-sulpho groups are cleaved off by the effect of organic bases. The position of the resulting double bond can be estimated by an infra-red spectrum. In microbially introduced hydroxyl groups this method has been used e.g. in the identification of hydroxyls in positions 9$\alpha$ (369, 413), 11$\alpha$ (374), 14$\alpha$ (70), 15 (172) and 16 (328). $\Delta^{4,14}$-Pregnadiene-3,20-dione can be relatively easily obtained by dehydratation of 15$\beta$-hydroxyprogesterone, while there is practically no dehydration in the 15$\alpha$-isomer (172).

**Reductive elimination of the hydroxy group with zinc in acetic acid.** Elimination of acetoxy groups or hydroxyl groups in positions 2 and 6 in $\Delta^4$-3-ketosteroids (186, 398, 402, 457, 458, 1194) is often brought about at normal temperature by the action of powdered zinc and diluted acetic acid. In these cases it is a matter of unsaturated systems of the following types:

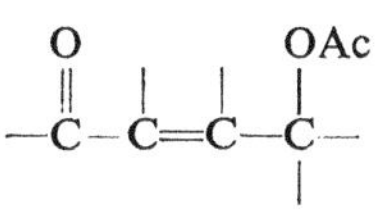

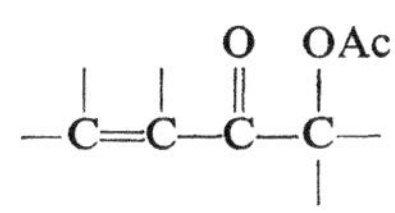

A reductive elimination proceeds, however, also in several saturated systems e.g. in 5-hydroxy-6-ketones, 11-hydroxy-12-ketones, 12-hydroxy-11-ketones and 17-hydroxy-20-ketones (312, 1173, 1207). Under these reduction, conditions there arise, however, in corticoids with a dihydroxyaceton side-chain also undesirable isomerizations to the 17-isosteroids or even a reduction of the 21-hydroxy group, especially if the reduction is carried out at higher temperature (1148).

**Preparation of acetonides.** The steroid *cis*-$\alpha$-diols e.g. 16$\beta$-hydroxytestosterone or triamcinolone, yield with acetone, under catalysis with strong acids (*p*-toluenesulphonic, hydrochloric, perchloric acids etc.) acetonides, which are isolated and identified by means of current physico-chemical processes (112, 134, 370, 420, 421, 1036, 1876). The 16,17-acetonide, prepared from 16$\beta$-hydroxytestosterone shows abs. maximum at 310 – 320 nm, whereas $\alpha$-diols with *trans*-configuration, e.g. 16$\alpha$-hydroxy-19-nor-testosterone do not yield any acetonides (134, 1036). In corticoids also 17,20-diols and 20,21-diols give acetonides. In such cases, where all present hydroxyl groups (e.g. in 16$\beta$-hydroxyprogesterone) are blocked by the formation of an acetonide, the prepared acetonide does not show absorption in the infra-red spectrum corresponding to free hydroxyl group. When there are two *cis*-α-diol groups in a steroid molecule, bis-acetonide is formed e.g. in the 20β-hydroxy derivative of triamcinolone (420).

**Dienon-phenol rearrangement.** This reaction discovered by INHOFFEN and HUANG-MINLON (1096) is usually used to identify the $\Delta^{1,4}$-3-keto groups. By the action of acetanhydride and of sulphuric acid the rearrangement into aromatic *p*-cresol **II** (473, 1060) proceeds as follows:

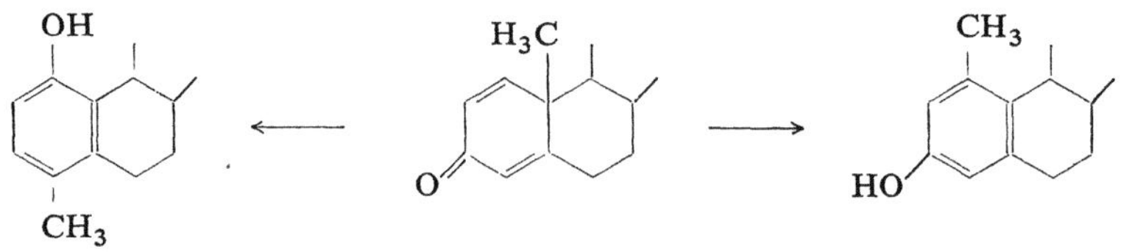


II I III

In aqueous medium on the other side (e.g. at the action of 36% HCl) 1-methyl-estrone **III** is formed from $\Delta^{1,4}$-androstadiene-3-17-dione (1059). TSUDA et al. (1208) have found that by the action of zinc in pyridine medium a rearrangement into phenol of type **II** with good yield also takes place. These phenolic steroids can be reliably proved by ultraviolet and infra-red spectra, by coupling with diazotized amines or by means of colour reactions.

## Identification of steroid metabolites by microbiological methods

For the completing of physical, physico-chemical and chemical identification procedures microbiological methods are also used, which in several cases can also be a decisive criterion in the solution of the configuration of a new steroid metabolite. Here are used such strains or microorganisms, eventually enzymes isolated from them, which carry out typical oxidation of reduction processes, especially hydroxylations into a certain position, dehydrogenation, cleavage of side-chain etc. In microbial introducing of a tertiary hydroxyl group we start from the experience that the introduced hydroxyl takes on the same configuration as the substituted hydrogen-atom i.e. 8β, 9α, 10β, 14α and 17α. No case of a configuration change is known in microbial hydroxylation (102, 182, 369, 1024, 1195).

Microbiological methods are sometimes also a part of chemical identification procedures. When establishing the structure of a new steroid metabolite it is sometimes necessary to remove a labile side-chain. In some metabolites of cortexone we find cleavage by the action of the strain *Penicillium lilacinum* giving rise to 17-ketosteroid (122). A derivative of etianic acid is always formed in the oxidation of this type of chain by chemical agents ($CrO_3$, $HIO_4$ or $NaBiO_3$).

Microbiological methods are mostly used in identification of unknown steroid hydroxy derivatives, especially of dihydroxysteroids. The position of hydroxyls can be preliminary concluded upon according to the configuration of simultaneously resulting mono-hydroxy derivatives, as it is known from experience that dihydroxysteroids formed by microbial hydroxylations have the hydroxyls in the same positions.

Microbial identification of dihydroxysteroid e.g. 6β,11α-dihydroxyprogesterone is effected by preparing dihydroxy derivatives from 6β-hydroxyprogesterone and from 11α-hydroxyprogesterone by help of the same microorganism and these products must be mutually identical with the unknown substance. It is possible to proceed in a way that 6β-hydroxyprogesterone is submitted to further hydroxylation by a microorganism, performing hydroxylation into position 11α exclusively (*Aspergillus niger, A. ochraceus* and others). 11α-Hydroxyprogesterone is further incubated with a strain, which hydroxylates into position 6β e.g. *Rhizopus arrhizus*. The metabotite, isolated from the adrenal glands, preliminarily identified as 6β-hydroxycorticosterone was proved by means of microbial hydroxylation of the 6β-hydroxy derivative of cortexone into position 11β by the strain *Curvularia lunata* and besides that also by hydroxylation of corticosterone into position 6β by *Trichothecium roseum* (1144). A considerable number of similar methods of microbial identification has already been described (25, 332, 365, 371, 396, 468).

Microbiological processes have been also used with advantage to characterise 19-nor-steroids. By help of the specific $\Delta^1$-dehydrogenase of the strain Nocardia restrictus the appropriate $\Delta^1$-dehydro derivative is formed, which easily passes into an estrone derivative. The phenolic character of the resulting steroid will be then proved by ultraviolet or infra-red spectra, or by the typical blue reaction with $FeCl_3$ and $K_3Fe(CN)_6$ or by coupling with diazotized sulphanilic acid (414).

The identification of the tertiary 9α-hydroxy group in androstenedione is also carried out microbiologically by transformation into phenolic 9,10-seco-3-hydroxy-1,3,5(10)-androstatrien-9,17-dione by help of various strains (110, 377, 378, 417). This conversion is not exhibited by steroids hydroxylated on carbon 10 or 14.

The microbiological method for the differentiation of steroids hydroxylated in positions 18 and 19 is also based on aromatization of ring A. By help of microorganisms, or of cell-free enzyme concentrates, capable to carry out $\Delta^1$-dehydrogenation, only the 19-hydroxysteroid can be converted into an estrone-derivative (414).

By means of the strain *Bacillus sphaericus,* effecting dehydrogenation only in position 1,2 it is possible to differentiate $\Delta^4$-3-ketones from the isomeric $\Delta^1$-3-ketones: only the former are subject to conversion into $\Delta^{1,4}$-3-ketones. A configuration proof of hydroxy groups in positions 1 and 2 can also be carried out by help of the enzyme system of the same microorganism. *B. sphaericus* effects 1-dehydrogenation only in the presence of hydroxyls in positions 2α and 1β, while there is no dehydrogenation in 1α- and 2β-hydroxysteroids (184, 258, 414).

# 8

# TRANSFORMATIONS OF STEROIDS BY VARIOUS STRAINS OF MICROORGANISMS

The types of microbial transformations of steroids have been dealt with from the chemical viewpoint in chapter 2. (see p. 18). This chapter deals with the basical transformation types, stating the microbe species possessing the corresponding enzyme system which catalyses the given reactions. Chapter 10. (see p. 129) gives reversely in the alphabetic order the used microbe and the type of catalyzed reaction, which proceeds on the steroid molecule, together with references.

## TYPE OF REACTION

### *A*) Oxidations

*a*) alcohol to ketone

*Absidia coerula, A. cylindrospora, A. glauca, A. orchidis, A. ramosa, A. regnieri, A. repens, A. turkestanica, Actinomyces albus, A. globisporus, A. species, A. viridochromogenes, Aerobacter aerogenes, Alcaligenes faecalis, Aspergillus alliaceus, A. amstelodami, A. atropurpureus, A. awamori, A. candiduds, A. chevalieri, A. clavatus, A. cinnamomeus, A. effusus, A. elegans, A. flavipes, A. flavus, A. fumigatus, A. luchuensis, A. melleus, A. nidulans, A. niger, A. ochraceus, A. oryzae, A. ostianus, A. parasiticus, A. penicilloides, A. rehmii, A. restrictus, A. sclerotiorum, A. sydowi, A. tamarii, A. terreus, A. terricola, A. ustus, A. variecolor, A. versicolor, Azotobacter species, Bacillus pulvifaciens, Bacterium cyclo-oxydans, B. species, Blakeslea trispora, Botrytis cinerea, Calonectria decora, Corynebacterium mediolanum, C. simplex, Cunninghamella albida, C. africana, C. blakesleeana, C. echinulata, C. elegans, C. homothallica, C. ramosa, C. verticilata, Eremothecium ashbyi, Escherichia coli, E. freundii, Flavobacterium androstenedionicum, F. carbonilicum, F. dehydrogenans, F. dehydrogenans* var. *hydrolyticum, F. maris, F. species, Helicostylum piriforme, Micrococcus dehydrogenans, Mucor parasiticus, Mycobacterium rhodochrous, M. species, M. tuberculosis, Nocardia species, Penicillium chrysogenum, Polyporus abietinus, Proactinomyces erythropolis, P. roseus, P. species, Pseudodiphteria bacilli, Pseudomonas specie, Rhisopus arrhizus, Streptomyces albus, S. aureofaciens, S. fradiae, S. gelaticus, S. griseus, S. rimosus, S. species, Thamnidium elegans, Wojnowicia graminis, Zygorhynchus heterogamus.*

*b*) hydroxylation

$C_1$

*Absidia orchidis, Haplosporella species, Mortiriella alphina, M. bainiery, M. kandelabrum, M. marburgensis, M. pusilla* var. *Isobellina, M. tuberosa, M. zonata, Penicillium species, Rhizoctonia ferrugena, R. solani, R. species, Streptomyces species, Xylaria species.*

$C_2$

*Bacillus sphaericus, Helminthosporium species, Penicillium species, Rhizoctonia ferrugena, R. species, Sclerotinia libertiana, S. species, Streptomyces roseochromogenes, S. species.*

$C_5$

*Absidia orchidis, Mucor parasiticus*

$C_6$

*Absidia glauca, A. hyalospora, A. ramosa, A. regnieri, A. turkestanica, Achromobacter Kashiwasakiensis, A. species, Aspergillus amstelodami, A. nidulans, A. ochraceus, A. saitoi, A. terreus, A. wentii, Bacillus cereus, B. pulvifaciens, Bacterium species, Blakeslea trispora, Botrytis cinerea, Circinella simplex, C. spinosa, Colletotrichum homoides, Coriolus versicolor, Corticium sasakii, Cunninghamella albida, C. africana, C. blakesleeana, C. echinulata, C. elegans, C. homothalica, C. ramosa, Curvularia geniculata, C. lunata, C. maculans, C. trifolii, Fusarium roseum, F. dimerum, F. avenaceum, F. avenaceum* var. *herbarum, F. semitectum, F. gibosum, F. gibosum* var. *acuminatum, F. graminearum, F. sambucinum, F. heterosporum, F. culmorum, F. macroceras, F. gigas, F. buxicola F. extansum, F. sphaeriae, F. cocophilum, F. nivale, F. aquaeductum, F. aquaeductum* var. *dimerum, F. microcrea, F. sporotrichiella, F. sporotrichiella* var. *poae, F. sarcochroum, F. oxysporum, F. oxysporum* var. *orthoceras, F. moniliforme, F. moniliforme* car. *subglutinans, F. moniliforme* var. *lactis, F. javanicum, F. metismoides, F. solani* var. *aduncisporum, F. solani* var. *argillaceum, F. solani* var. *coeruleum, F. solani* var. *redolens, Gibberella saubinetti, G. species, Gliocladium satenulatum, Gloesporium kaki, Glomerella lagenarium, Haplosporella species, Helicostylum piriforme, Helminthosporium species, Chaetocladium brefeldii, Chaetomium cochliodes, Ch. funicolum, Ch. sucineum, Lenzites abietina, Mucor adriaticus, M. adventitius, M. angulisporus, M. berolinensis, M. buntingii, M. circinelloides, M. corymbifer, M. dimorphosporus, M. dubius, M. erectus, M. genevensis, M. globosus, M. guilliermondii, M. hiemalis, M. humicola, M. humilis, M. hypochninus, M. javanicum, M. mandshuricus, M. microsporus, M. mucedo, M. murosum, M. parasiticus, M. piriformis, M. plumbens, M. pusillus, M. racemosus, M. ramannianus, M. rouxianus, M. rouxii, M. varians, Naematoloma sublateritium, Pellicularia filamentosa, Phycomyces blakesleanus, P. blakesleanus-theobromatus, P. nitens, Penicillium species, Polyporus orientalis, P. cinnabarinus, P. tulipiferus, Rhizoctonia solani, R. species, Rhizopus arrhizus, R. cambodjae, R. cohni, R. delemar, R. chinensis, R. japonicus, R. kazanensis, R. nidulans, R. nigricans, R. oryzae, R. pygmaeus, R. reflexus, R. species, R. shanghaiensis, R. tonkinensis, R. tritici, Scenedesmus species, Sclerotium hydrophilum, Streptomyces aureofaciens, S. diastochromogenes, S. fradiae, S. griseocarneus, S. rimosus, S. species, Syncephalas nodosa, Syncephalastrum racemosum, Thamnidium elegans, Trichothecium roseum, Xylaria species, Zygorhynchus heterogamus, Z. moelleri, Z. wuillemini.*

$C_7$

*Absidia orchidis, Absidia regnieri, A. species, Cephalosporium asperum, C. species, Cladosporium species, Coniothyrium helleborine, Corynebacterium simplex, Curvularia falcata, C. fallax, C. lunata, C. pallescens, C. species, Diplodia natalensis, D. tubericola, Fusidium species, Helicostylum piriforme, Helminthosporium sativum, H. species, Mucor griseo-cyanus, M. parasiticus, Penicillium species, Peziza species, Phycomyces blakesleanus, Rhizopus arrhizus, R. nigricans, R. species, Syncephalastrum racemosum, Xylaria species.*

$C_8$

*Corynespora melonis.*

$C_9$

*Absidia regnieri, Cercospora (Corynespora) melonis, Cunninghamella bertholetiae, C. blakesleeana, C. echinulata, C. elegans, C. verticillata, Curvularia lunata, Circinella species, Helicostylum piriforme, Mucor griseo-cyanus, M. parasiticus, Neurospora crassa, Nocarida restrictus, N. species, Streptomyces aureofaciens.*

$C_{10}$

*Curvularia lunata, Rhizopus nigricans.*

$C_{11}$

*Absidia coerula, A. cylindrospora, A. glauca, A. hyalospora, A. lichtheimi, A. orchidis, A. ramosa, A. regnieri, A. repens, A. species, A. spinosa, A. tieghemi, A. turkestanica, Aspergillus amstelodami, A. awamori, A. caespitosus, A. candidus, A. chevalieri, A. clavatus, A. conicus, A. effusus, A. echinulatus, A. fischeri, A. flavipes, A. flavus, A. fumigatus, A. giganteus, A. itaconicus, A. luchuensis, A. melleus, A. montevidensis, A. nidulans, A. niger, A. niveo-glaucus, A. niveus, A. ochraceus, A. oryzae, A. saitoi, A. species, A. sydowi, A. tamarii, A. terreus, A. usamii, A. ustus, A. variecolor, A. versicolor, A. wentii, Bacillus cereus, Bacillus cereus* var. *mycoides, Beauveria species, Blakeslea trispora, Botryodiplodia theobromae, Botrytis cinerea, B. peoniae, Calonectria decora, Circinella simplex, C. spinosa, Colletotrichum phomoides, C. pisi, Coniothyrium helleborine, C. species, Conocybe siligeneoides, Corticium microslerotia, C. procticola, C. sasakii, C. vagum, Coryneum cardinale, Cunninghamella albida, C. africana, C. bainieri, C. blakesleeana, C. echinulata, C. elegans, C. homothallica, C. ramosa, C. verticillata, Curvularia brachyspora, C. falcata, C. lunata, C. pallescens, C. species, Dactilium dendroides, Delacroixa coronata, Didymocladium ternatum, Dothichiza species, Eutorium chevalieri, Fusarium discolor, F. bulbigenum, F. culmorum, F. equiseti* var. *bullatum, F. heterosporum, F. oxysporum, F. reticulatum, F. redolens, F. solani, Gloesporium kaki, G. species, Glomerella lagenarium, Glomerella species, Gongronella urceolifera, Helicostylum piriforme, Helminthosporium species, Hormodendrum hordei, H. pallidum, Choanephora cucurbitarum, Mucor adriaticus, M. adventitius, M. angulisporus, M. dubuis, M. rectus, M. genevenzis, M. globosus, M. hiemalis, M. javanicus, M. mandshuricus, M. microspora, M. mucedo, M. plumbens, M. pusillus, M. racemosus, M. ramannianus, M. rouxianus, M. rouxii, Mycocladus hyalinus, Myrothecium roridum, Neurospora sitophila, Nigrospora oryzae, N. species, N. sphaerica, Pellicularia filamentosa, Penicillium atramentosum, P. corylophilum, P. ehrlichii, P. lilacinum, P. oxalicum, P. tardum, Pestalotia foedans, P. royenae, P. species, Phoma species, Phycomyces blakesleeanus-theobromatus, Polyporus orientalis, Pseudomonas boreopolis, P. species,*

*Psilocybe coeruslescens, P. mexicana, Pycnosporium species, Rhizopus arrhizus, R. cambodjae, R. cohnii, R. delemar, R. chinensis, R. japonicus, R. kazanensis, R. nidulans, R. nigricans, R. oryzae, R. reflexus, R. shanghaiensis, R. species, R. tonkinensis, R. tritici, Rhodoseptoria species, Scenedesmus species, Sclerotinia libertiana, Sclerotium hydrophilum, Spondylocladium species, Sporotrichum bombycinum, S. epigaeum, S. sulphurescens, Stachylidium bicolor, S. theobromae, Streptomyces fradiae, S. species, Stropharia cubensis, Syncephalastrum cinereum, S. racemosum, Syncephalas nodosa, Thamnidium elegans, Tieghemella hyalospora, T. orchidis, T. species, Trichoderma viridae, Trichothecium roseum, Zygorhynchus heterogamus, Z. moelleri.*

$C_{12}$

*Calonectria decora, Cercospora melonis, Colletotrichum phomoides, Coniothyrium helleborine, Fusarium lini, Gibberella saubinetti, Penicillium species, Thamnidium elegans.*

$C_{14}$

*Absidia regnieri, Achromobacter Kashiwasakiensis, A. species, Bacillus cereus, B. pulvifaciens, Bacterium species, Circinella species, Coriolus versicolor, Cunnighamella blakesleeana, C. elegans, Curvularia lunata, C. pallescens, C. species, Helicostylum piriforme, Helminthosporium sativum, H. species, Mucor corymbifer, M. griseo-cyanus, M. mucedo, M. parasiticus, M. species, Mycobacterium berlinensis, M. butyricum, M. friedmanii, M. lacticola, M. phlei, M. ranae, M. smegmatis, M. species, M. thamnopheos, M. tuberculosis, Naematoloma sublateritium, Pleospora gaeumanni, Polyporus cinnbarinus, P. tulipiferus, Poria cocos, Stachylidium bicolor, Syncephalastrum racemosum, Trichothecium roseum.*

$C_{15}$

*Absidia regnieri, Aspergillus species, Bacillus cereus, B. megatherium, Botrytis cinerea, Calonectria decora, Colletotrichum antirhini, Coriolus versicolor, Curvularia lunata, Diploidia tubericola, Fusarium avenaceum, F. avenaceum* var. *herbarum, F. aquaeductum, P. aquaeductum* var. *dimerum, F. buxicola, F. cocophilum, F. culmorum, F. expansum, G. gibosum, F. gibosum* var. *acuminatum, F. gigas, F. graminearum, F. heterosporum, F. javanicum, F. lini, F. lycopersici, F. macroceras, F. merismoides, F. microcrea, F. moniliforme, F. moniliforme* var. *subglutinans, F. moniliforme* var. *lactis, F. nivale, F. oxysporum, F. oxysporum* var. *orthoceras, F. roseum, F. sambucinum, F. sarcochroum, F. semitectum, F. solani, F. solani* var. *aduncisporum, F. solani* var. *argilaceum, F. solani* var. *coeruleum, F. solani* var. *redolens, F. species, F. sphaeria, F. sporotrichiella, F. sporotrichiella* var. *poae, Giberella baccata, G. saubinetti, G. species, Haplosporella species, Helminthosporium sativum, Hormoderndrum olivaceum, H. viride, Hypholoma species, Lenzites abietina, Nigrospora oryzae, N. species, N. sphaerica, Penicillium diversum, P. notatum, P. rugulosum, P. species, P. tardum, P. urticae, Phycomyces blakesleanus, Polyporus cinnabarinus, P. tulipiferus, Poria cocos, Scenedesmus species, Sclerotinia libertiana, Sclerotium hydrophilum, Spicaria simplicissima, S. species, Streptomyces aureus, Syncephalastrum racemosum, Xylaria species.*

$C_{16}$

*Actinomyces albidus, A. annulatus, A. coerulus, A. diastaticus, A. flaveolus, A. globosus, A. praecox, A. ruber, A. setonii, A. species, A. viridis, Corticium centrifugum, Didymella vodakii, Pestalotia funera, Streptomyces, achromogenes, S. annulatus, S. aureus, S. diastaticus, S. finicarius, S. flaveolus, S. griseus, S. griseoluteus, S. halstedii, S. chrysomallus, S. lavendulae, S. mediocidicus, S.*

*microflavus, S. olivaceus, S. parvus, S. phaeochromogenes, S. purpeochromogenes, S. roseochromogenes, S. rutgersensis, S. species, S. sulphureus, S. verne, S. viridis, Wojnowicia graminis.*

$C_{17}$

*Cucurbitaria laburni, Dactylium dendroides, Leptosphaeria maculans, Lophotrichus martinii, Melanospora parasitica, Sporodinia minima, Sporormia minima, Thielavia terricola, Trichoderma album, T. glaucum, T. konigi, T. lignorum, T. narcisi, T. nigrovirens, T. nunbergii, T. species, T. viridae, Trichothecium arrhenopum, T. candidum, T. plasmaparae, T. polybrochum, T. roseum, T. species.*

$C_{19}$

*Corticium microsclerotia, C. sasakii, C. vagum, Pelicularia filamentosa.*

$C_{21}$

*Aspergillus niger, Colletotrichum lindemuthianum, Hendersonia aberrans, H. acicola, H. herpotricha, H. fragmitis, H. rubi, Ophiobolus herpotrichus, O. species, Sclerotinia fructicola, Wojnowicia graminis.*

## *c*) epoxidation

*Cunninghamella blakesleeana, Curvularia lunata, C. species, Helicostylum piriforme, Mucor griseocyanus, M. parasiticus, Nocardia species, Phycomyces blakesleanus, Saccharomyces species.*

## *d*) dehydrogenation (new double bond formation)

*Alcaligenes species, Alternaria passiflorae, Arthrobacter simplex, A. species, Azotomonas fluorescens, Azotobacter agilis, A. indicus, Bacillus pulvifaciens, B. sphaericus, B. cyclo-oxydans, B. species, B. subtilis, Bacterium cyclo-oxydans, Calonectria decora, Colletotrichum atramentarium, C. Lindemuthianum, Corticium sasakii, Corynebacterium helvolum, C. hoagii, C. mediolanum, C. simplex, C. species, Cylindrocarpon radicicola, Didymella lycopersici, Fusarium bulbigenum, F. caucasicusm, F. javanicum, F. javanicum* var. *ensiforme, F. lateritium, F. lini, F. oxysporum, F. solani, F. zonatum, Gliocladium catenulatum, G. roseum, Gloesporium olivarum,Helminthosporium turcicum, Micromonospora chalcea, Mycobacterium berlinensis, M. butyricum, M. flavum, M. fortuitum, M. friedmanii, M. lacticola, M. phlei, M. ranae, M. rhodochrous, M. rubropertictum, M. smegmatis, M. species, M. thamnopheos, M. tuberculosis, Nocardia asteroides, N. aurantia, N. blakswelii, N. braziliensis, N. caviae, N. coeliaca, N. convulata, N. corallina, N. cuniculi, N. erythropolis, N. farcinica, N. formica, N. gardneri, N. globerula, N. keratolyticus, N. leishmanii, N. madurae, N. minima, N. opaca, N. polychromogenes, N. restrictus, N. rubra, N. salmonicolor, N. sylvodifera, N. species, Ophiobolus herpotrichus, O. heterosporus, O. miyabeanus, Proactinomyces species, Protaminobacter alboflavum, P. rubrum, Pseudomonas dacunhae, P. marcescens, P. olevorans, P. species, P. testosteroni, Pycnodothis species, Rhizoctonia solani, Sansevieria zeylanica, Sclerotinia fructicola, Septomyxa aesculi, S. affinis, S. corni, S. salicina, S. tulasnei, Serratia plymuthica, Sporotrichum epigacum, Stereum fasciatum, Streptomyces antibioticus, S. bobiliae, S. gelaticus, S. lavendulae, S. olivaceus, S. olivochromogenes, S. species.*

### *e*) aromatization of ring A

*Mycobacterium rhodochrous, M. smegmatis, Nocardia restrictus, Bacillus cyclo-oxydans, Pseudomonas testosteroni, Mycobacterium fortuitum, M. tuberculosis.*

### *f*) side chain ($C_{20}-C_{21}$) or ($C_{20}-C_{27}$) cleavage (and/or ring *D*-splitting)

*Arthrobacter species, Aspergillus alliaceus, A. chevalieri, A. fischeri, A. flavus, A. oryzae, A. penicilloides, A. ruber, A. sclerotiorum, A. terreus, A. terricola, A. ustus, Cephalosporium subverticillatum, Cladosporium resinae, Corynebacterium simplex, Cylindrocarpon radicicola, Fusarium caucasicum, F. javanicum* var. *ensiforme, F. lateritium, F. solani, Gliocladium catenulatum, Mucor mucedo, Mycobacterium flavum, M. fortuitum, M. smegmatis, M. tuberculosis, Nocardia restrictus, N. species, Penicillium adametzi, P. albidum, P. asperum, P. canescens, P. citreo-viride, P. citrinum, P. commune, P. corylophilum, P. cyaneofulvum, P. daleae, P. decunbens, P. chrysogenum, P. egyptaceum, P. frequetans, P. fuscum, P. godlewskii, P. implicatum, P. janthinellum, P. javanicum, P. jensenii, P. kapuscinskii, P. lilacinum, P. lividum, P. melearginum, P. melinii, P. miczynskii, P. multicolor, P. nalgiovensis, P. nigricans, P. notatum, P. ochro-chlorum, P. parvum, P. psittacinum, P. pisarium, P. pulvillorum, P. purpurescens, P. pusillum, P. raciborskii, P. raistrickii, P. restrictum, P. restriculosum, P. sclerotiorum, P. simplicissimum, P. soppi, P. species, P. spinulosum, P. steckii, P. subalteritium, P. terikowskii, P. thomii, Proactinomyces species, Pseudomonas species, Pycnodothis species, Rhizopus suinus, Septomyxa affinis, Sporotrichum epigaeum, Streptomyces lavendulae, S. species.*

### *g*) ring *B* – splitting

*Bacillus cyclo-oxydans, Mycobacterium fortuitum, M. rhodochrous, M. smegmatis, Nocardia restrictus, Pseudomonas testosteroni, Mycobacterium tuberculosis.*

## *B*) Reductions

### *h*) of keto group

*Actinomyces albus, Alternaria bataticola, Bacillus coli communis, B. putrificus, B. species, Bacterium coli, B. cyclo-oxydans, B. species, Candida pulcherrima, Catenabacterium catenaforme, Clostridium lentoputrescens, C. paraputrificum, C. tertium, Corticium centrifugum, Corynebacterium mediolanum, C. simplex, Curvularia lunata, Epicoccum oryzae, Mycobacterium flavum, M. lacticola, M. smegmatis, M. species, Penicillium lilacinum, P. thomii, Pseudomonas olevorans, P. species, Rhodotorula glutinis, R. longissima, Saccharomyces cerevisiae, S. pastorianus, S. species, Sclerotium hydrophilum, Sporotrichum gougeroti, Stereum fasciatum, Streptomyces albus, S. aureus, S. coelicolor, S. diastaticus, S. flavogriseus, S. griseus, S. hydrogenans, S. roseochromogenes, S. species, Trichomonas gallinae.*

### *i*) hydrogenation (double bond elimination)

*Bacillus putrificus, Clostridium paraputrificum, C. tertium, Giberella saubinetti, Sansevieria zeylanica, Streptomyces species.*

## *C*) Hydrolysis

### *j*) saponification of esters

*Actinomyces globisporus, A. viridochromogenes, Aspergillus oryzae, Bacillus megatherium, Curvularia lunata, Fusarium lini, Flavobacterium dehydrogenans, F. dehydrogenans* var. *hydrolyticum, Gibberella saubinetti,* and further species of genera *Actinomyces, Aspergillus, Cunninghamella, Penicillium, Rhizopus.*

### *k*) cleavage of glycoside bond

*Aspergillus niger, Penicillium cyclopium, P. chrysogenum.*

## *D*) Esterification

### *l*) formation of acetate

*Cladosporium resinae, Saccharomyces fragilis, Trichoderma glaucum.*

# 9

# APPLICATIONS FOR THE PREPARATION OF STEROID HORMONES AND THEIR ANALOGUES

The practical use of steroid hormones, their derivatives and analogues in medicine soon enforced the production on a larger scale of these substances, which are difficultly accessible and rather expensive. The isolation from animal material, current at the beginning of the research of these substances, has lost all importance for production in our days. The preparation of extracts of the adrenal glands, of corticoid concentrates on a certain limited scale and the isolation of estrone from horse urine have been still maintained, otherwise individual hormones of all four types are now produced only synthetically. *Partial syntheses* of steroid hormones have been described from the following sterols: cholesterol, ergosterol, sitosterols and stigmasterol, from bile acids, especially from hyodeoxycholic acid, further from sapogenins, especially diosgenin and hecogenin and *Solanum* glycoalkaloids, especially from solasodine. The question of suitable raw material for the partial syntheses of steroid hormones has been still an important problem from the economical point of view. Recently diosgenine and solasodine have been of the greatest importance.

*Total syntheses* of steroid hormones, whose importance had first been more theoretical than practical, start from relatively simple cyclic organic compounds and because of the stereochemical complexity of the steroid molecule meet with considerable difficulties owing to formation of involved racemic products, which can be separated only with difficulty. Recently these methods have become of practical interest especially for the preparation of estrogens and 19-norandrostane anabolic androgens.

L. Mamoli (14, 263–275) and A. Vercellone (489, 490) were the first, as early as in 1937, to have drawn attention to the possibility of using reactions catalyzed by *microbial* enzymes. They worked for partial syntheses of steroid hormones, especially at the synthesis of testosterone from dehydroepiandrosterone, which they oxidized by means of bacteria into androstenedione, which was then reduced into testosterone by yeast. Originally they worked with infected yeast and a direct conversion of dehydroepiandrosterone into testosterone was found without understanding the role of two microbes in this process. Wettstein (511) described a similar reduction of estrone into estradiol by means of yeast in the year 1939.

Further work was then aimed at the degradation of sterols, especially of cholesterol into androstene derivatives, as a possible way of avoiding chemical oxidative cleavage of the sidechain $C_{20}-C_{27}$ of sterol in the synthesis of steroid hormones (194, 232, 430, 475–480). In spite of giving very interesting theoretical results, these experiments have not found practical application seeing that the reaction course was not uniform. This problem was resolved by a number of further research workers who examined hundreds of different microbial strains without reaching practically utilizable results. The microbes, cleaving the sterol side-chain, always carried out also further degradation of the steroid ring system simultaneously (see p. 51).

The most interesting stimulus for the use of microbial enzymes in partial syntheses of steroid hormones was brought about by the findings about the *hydroxylation* in the steroid ring. The introduction of hydroxyl into a requested position is a rather difficult procedure from the chemical point of view, requiring several reaction steps and usually giving low yields. KRÁMLI and HORVÁTH (231) were first to effect the 7-hydroxylation of steroids in the year 1948. In 1952 PETERSON and MURRAY (331) published a fundamental work about the introduction of hydroxyl *into position 11* of progesterone, which considerably facilitated further partial syntheses of cortisone and similar corticoids. From this year on, dates a great interest in microbial conversions of steroids for the purpose of partial synthesis of steroid hormones (5, 8, 15, 18, 22, 23*a*, 24, 32), later also in modifications of heart glycosides and their aglycones (31).

Beside simple oxido-reductions of secondary alcoholic and ketonic groups, often with a stereospecific course, the following operations were carried out: hydroxylations specifically into many further positions of the steroid skeleton, introduction of double bond into position 1–2 or ring A and cleavage of the side-chain of pregnene steroids directly to androstene derivatives. Many hundreds of microbial species were used here (see pp. 111, 114, 115). However, a relatively small number of these operations can be taken into consideration for practical use where uniform course and high yields are required.

## Preparations of androgens

In the androstane series the interest was first of all focused on microbial preparation of male sexual hormone-testosterone (166, 264). Here the key intermediate is dehydroepiandrosterone, originally obtained by oxidation cleavage of the side-chain of sterols. By help of a number of microbes this steroid is converted oxidatively into $\Delta^4$-androstene-3,17-dione (41, 42, 126, 161, 266, 303, 304) with a yield up to over 80%. By further reduction by means of yeast over 80% of testosterone can be obtained (54, 75, 161, 264). An enzymatic system, oxidizing the structure 3-OH-$\Delta^5$ into the conjugated system 3-keto-$\Delta^4$ with simultaneous shift of the double bond, was found in a very great number of microbes (see p. 110). The species originally used were: *Corynebacterium mediolanum* (266, 272, 303), *Micrococcus dehydrogenans*

(41, 123), *Flavobacterium androstenedionicum* (126, 127, 304). Later on, as already mentioned (see p. 45), the individual enzymes induced in the species of *Pseudomonas testosteroni* were isolated (278, 448, 450).

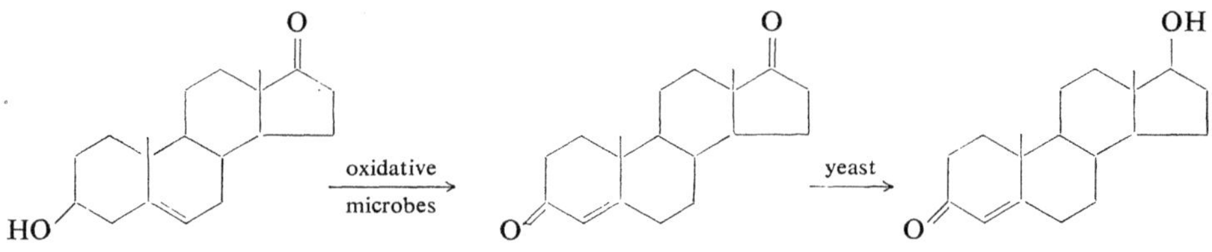


dehydroepiandrosterone androstenedione testosterone

The enzymatic system of the *Saccharomyces cerevisiae* converts specifically the 17-keto group into secondary alcohol with a configuration of hydroxyl 17$\beta$.

Similarly $\Delta^5$-androstene-3$\beta$, 17$\beta$-diol can also be converted by means of microbial enzymes, by a procedure analogous to the Oppenauer oxidation into testosterone (125, 162) as well as the corresponding 17$\alpha$-methyl derivative into methyltestosterone (274).

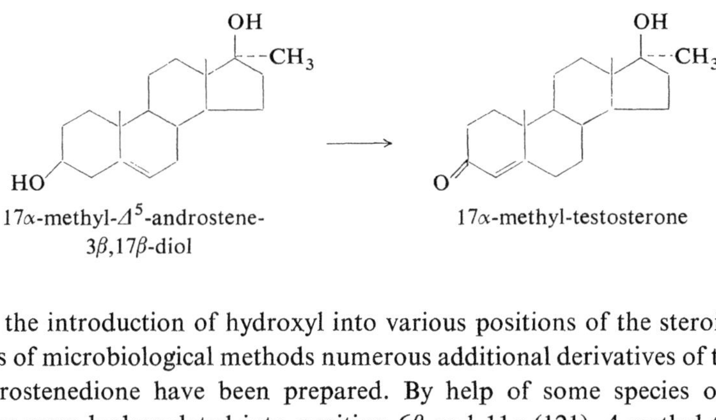

17$\alpha$-methyl-$\Delta^5$-androstene-3$\beta$,17$\beta$-diol 17$\alpha$-methyl-testosterone

By the introduction of hydroxyl into various positions of the steroid molecule by means of microbiological methods numerous additional derivatives of testosterone and androstenedione have been prepared. By help of some species of *Rhizopus*, androgens were hydroxylated into position 6$\beta$ and 11$\alpha$ (121), 4-methyl-testosterone has been hydroxylated into position 11$\alpha$ (213). The microbiological preparation of 11$\alpha$-hydroxy-androstenedione was described by Šebek (387). The 6$\beta$-hydroxylation of testosterone was carried out by help of the species *Wojnowicia graminis* by Herzog *et al.* (187*b*). 19-Nortestosterone was hydroxylated into position 6$\beta$ and 14$\alpha$ (138), into position 10 (135, 327) and 11 (135), further into positions 12 (136), 15 (137) and 16 (134). The 1$\beta$-hydroxylation (113) and 16-hydroxylation (112) of androstenedione has been described by Dodson.

A further series of microbiological procedures for the preparation of androgens starts from pregnene steroids, first of all from progesterone (166). Using *Fusarium solani* it has been possible for the first time to transform progesterone into androstenedione and androstadienedione (493). *Fusarium caucasicum* (518–521) and *F. lateritium* (78, 83) also posses enzymatic systems catalyzing a similar conversion. Representatives of the genera *Aspergillus* (72, 73), *Streptomyces* (336), *Cylindrocarpon* (336), *Cladosporium* (140) and *Penicillium* (80, 142, 160, 341, 388) carry out a direct cleavage of progesterone into testosterone. Using the species *Penicillium notatum* (160) the production of testosterone from progesterone with a yield of over 60% was made possible.

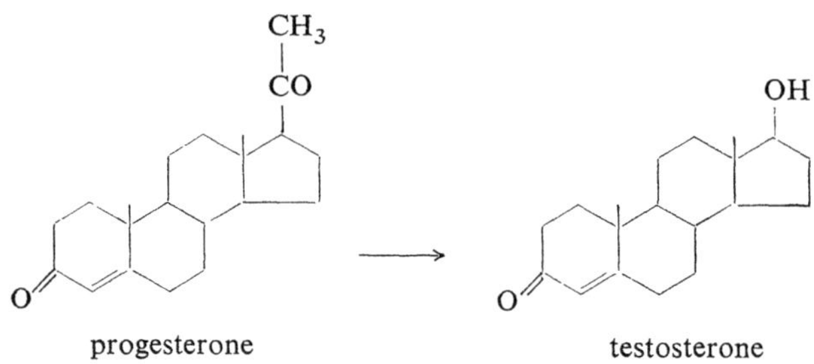


In an analogous way progesterone derivatives possessing an oxygen function in position 11, can be converted into corresponding substituted androstene derivatives (160).

The introduction of a further double bond in position 1–2 into ring A of androstane derivatives has been carried out using a number of microbes (83, 142, 495, 496, 518–521). This dehydrogenation is usually accompanied by side reactions such as the rise of testololactone or its 1-dehydro derivative.

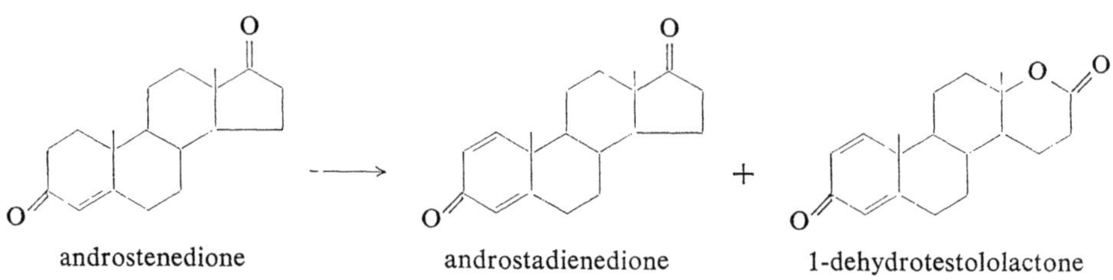


The formation of testololactone in the mentioned reactions catalyzed by the enzymatic system of *Fusarium causasicum* can be suppressed by the addition of some naphtalene derivatives (520) or other cyclic substances (521). The rise of testololactone from progesterone in a high yield has been observed in fermentation by help of *Aspergillus oryzae* (72, 223), whereby testosterone and androstenedione are formed as an intermediate, which are in oxido-reduction equilibrium.

## Preparations of estrogens

Analogous oxido-reductions on $C_{17}$ as in the system: androstenedione-testosterone, have also been observed in estrogens under the influence of microbial enzymes. By means of *Streptomycetes* estradiol is converted into estrone (188), whereas estrone was reduced into estradiol by means of yeasts (269, 344, 511) in an approximate yield of 50%.

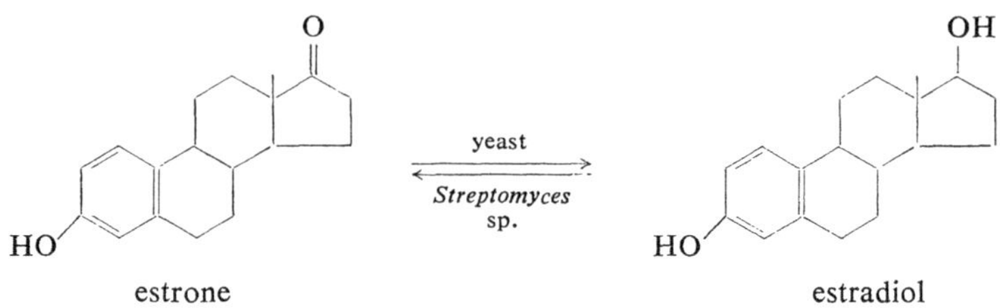


According to Inhoffen 1,4-androstadiene-3,17-dione already mentioned, was used for further synthesis of estrone by pyroreaction involving aromatization with cleavage of $C_{19}$. Most recently H. L. DRYEN Jr. *et al.* (*J. am. chem. Soc. 86*: 742, 1964) treated the 17-ethylene ketal of androstadienedione with an excess of the radical anion derived from Li and biphenyl in boiling tetrahydrofuran solution with resulting aromatization of the A- ring and expulsion of the angular methyl group as methyllithium. Yields of estrone were of up to 75%.

By help of the species *Corynebacterium simplex* (262) it was possible to carry out the aromatization of 19-nortestosterone also directly by a microbiological procedure.

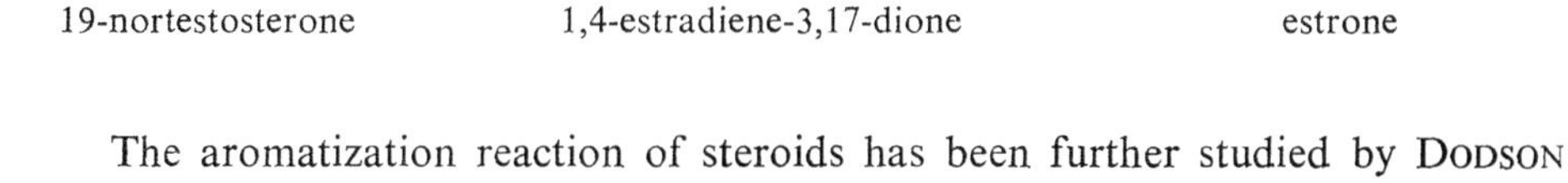


The aromatization reaction of steroids has been further studied by DODSON (106, 110) by help of the species *Pseudomonas* and *Arthrobacter*. 19-Hydroxyandrostenedione has been converted into estrone, while androstenedione yielded the 9,10-seco derivative with the aromatic ring A, similarly as has already been mentioned in the mechanism of cleavage of the ring system of steroids (see p. 52).

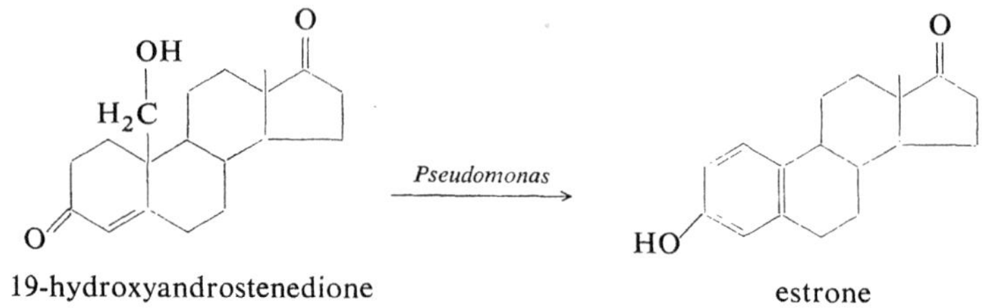


The sequence of reactions catalyzed by microbial enzymes, from progesterone to testosterone, with subsequent dehydrogenation of ring A, oxidation of $C_{19}$, its oxidative elimination and eventually enolic aromatization of ring A, leading up to estradiol, recalls the mechanism of steroid conversion in the biosynthesis of estrogens in the animal organism. An obvious difference consists, however, in the fact, that the animal cells first convert the progesterone into the 17α-hydroxy derivative, which is then cleaved by oxidation between the $C_{17}$ and $C_{20}$ carbons, whereas in microorganisms the cleavage of the sidechain is prevented by the presence of 17α-hydroxyl of the pregnene derivative.

CASAS-CAMPILLO *et al.* (86) have described microbial conversion of 19-nor-$\Delta^{4,7}$-androstadiene-3,17-dione directly into equiline, i.e. into estrogen with a further double bond in the ring B. By means of *Streptomycetes* it was possible to hydroxylate estrogens into position 16α (214), again similarly to the enzymatic systems of animal cells forming estriol. A certain equilibrium is again established between the 17-hydroxy- and 17-ketosteroid substituted in position 16.

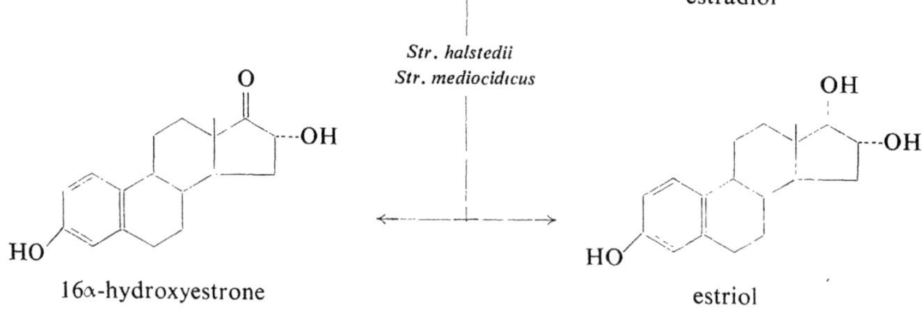

## Preparations of gestagens

Similarly to the androstene series the oxidations of the key intermediate, i.e. $\Delta^5$-pregnen-3$\beta$-ol-20-one into progesterone has been carried out in the preparation of *gestagens*. An enzymatic system catalyzing this dehydrogenation in position 3 and an isomerization, analogous to the chemical Oppenauer oxidation, has been found in a great number of microbes of all classes. MAMOLI (271) used *Corynebacterium mediolanum* for the conversion, D. PERLMAN (329) on the other hand, a certain number of *Actinomycetes* and moulds.

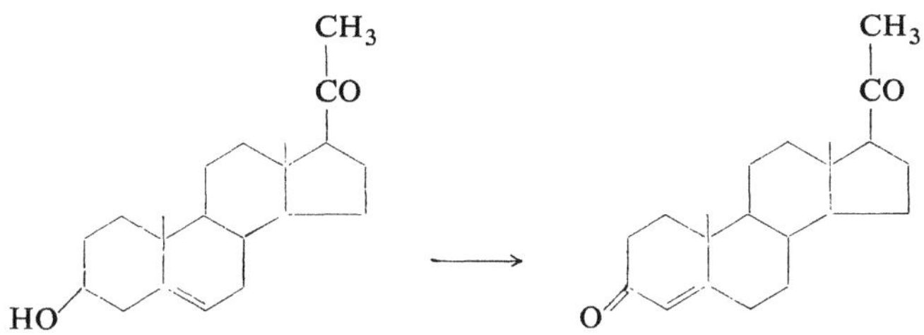


pregnenolone progesterone

The conversion of pregnenolone into progesterone was carried out with high yields with the species: *Streptomyces griseus, Str. fradiae, Str. aureofaciens, Str. rimosus, Phycomyces blakesleeanus, Aspergillus niger, Penicillium chrysogenum, Eremothecium ashbyi* and with the rust *Ustilago zeae.*

Being one of the most accessible steroids, progesterone further has been used for a wide range of hydroxylation processes catalyzed by the enzymatic systems of microorganisms *Aspergilli* (73, 115, 200), *Penicillia* (80, 115), *Basidiomycetes* (349, 383), *Absidia regnieri* (458), *B. cereus* and *B. megaterium* (257).

The following hydroxylations into positions:

*2* (391), *6* (65, 114, 78, 202, 296, 331, 378), *7* (368, 468, 472), *9* (366, 369, 466*a*), *11* (64, 65, 159, 331, 333, 486), *12* (242, 365, 368*a*), *14* (64, 202, 369, 371, 468, 472), *15* (68, 214*a*, 242, 466*a*, 472, 468), *16* (500), *17* (256, 296, 522), *18* (486), *21* (299*a*, 255, 513, 522) have been carried out.

Such derivatives can be used as intermediates for further syntheses of orally effective gestagens with the well-known anticonception effects, especially also in the 19-nor series of progesterone (57, 58, 87) and further as valuable intermediates for the synthesis of some corticoids. It is well known that just in this field the microbiological hydroxylations into position 11 was of importance decisive for the further development of the synthesis of numerous analogues of steroid hormones already known. As to the preparation proper of progesterone the microbial transformations of steroids can, however, hardly compete with purely chemical processes which do not offer any difficulties.

## Preparations of corticoids

The microbial conversions of steroids are, however, extensively used especially in partial syntheses of the *adrenal cortex hormones* and their analogues. First of all the conditions of biosynthesis of these hormones were studied directly in the adrenal tissue and a number of factors were found concerning optimum conditions in the incubation experiments with various steroid precursors, or in perfusion experiment. A number of oxido-reduction enzymes and of systems hydroxylating into position 11β, 17α and 21 of the steroid molecule, eventually in aldosterone also into position 18, participate in the biosynthesis of corticoids in the cells of adrenal cortex.

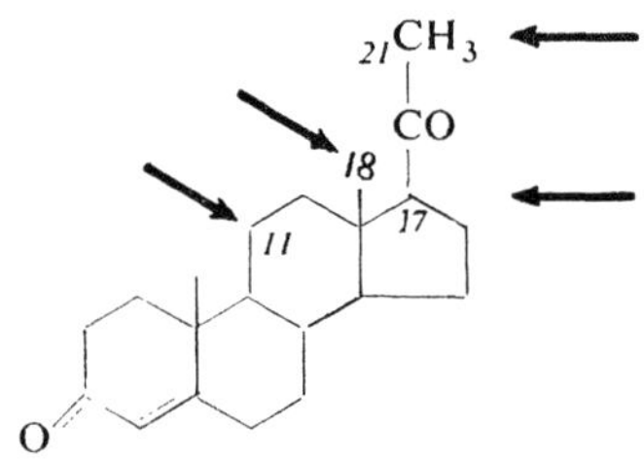


*Enzymatic oxidation of progesterone as the precursor of corticoids*

After chemical syntheses of corticoids had been carried out there followed investigations of the relations between the structure of hormones and their biological activity. These hormones can be divided according to their regulative effect on the body metabolism, into two groups: the *glucocorticoids* with antiinflammatory effect and *mineralocorticoids*. Cortisol and cortisone are the main representatives of the former, cortexone and aldosterone of the latter group. The ratio between both fundamental activities varies for the various synthetic analogues derived from these corticoids.

When in the year 1952 D. H. PETERSON and H. C. MURRAY (331) published the fundamental findings that some microbial strains similarly to adrenal cortex cells, can also hydroxylate steroid molecule *in position 11*, a new way opened for the preparation of glucocorticoids of the cortisone-cortisol type. By help of the strain *Rhizopus nigricans* it was possible to obtain from progesterone its 11α-hydroxy derivative with a yield of up to over 90%. This intermediate further served as a raw-material in the Mancera's ten-step synthesis of cortisone or in Hogg's "direct" synthesis.

A theoretically interesting process is the synthesis of cortisone from progesterone in 4 reaction steps, where 3 carried out by means of microbial enzymes.

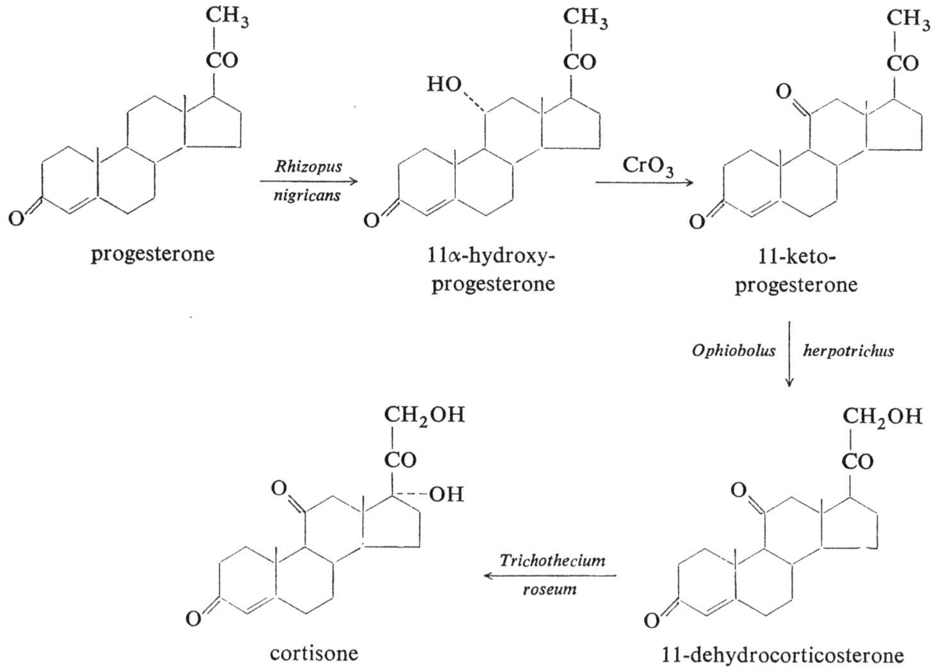


ZAFFARONI (522) studied the hydroxylation of progesterone and its oxygen derivatives into positions 21 by help of the strain *Aspergillus niger*. WIX *et al.* (513) also hydroxylated progesterone directly into 11α-epicorticosterone by help of *Aspergillus niger.*

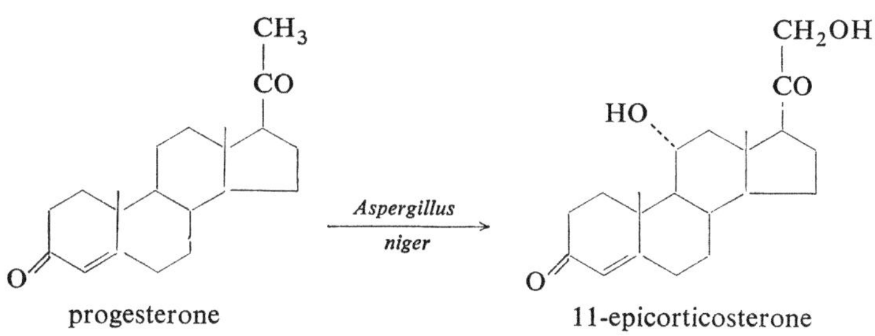


In the hydroxylation of progesterone into position 21 by means of the strain *Wojnowicia graminis* (255) MCALEER *et al.* have obtained yields of about 45% of cortexone. Progesterone has been hydroxylated into position 17α by means of the strain *Trichoderma viridae* (256). Also other species of the genus *Trichoderma* (*Tr. albanum, glaucum, konigi, lignorum, narcisi, nigrovirens, nunbergii*) were examined for the hydroxylation into position 17α of pregnene steroids. MEISTER *et al.* (297) used the species *Trichothecium roseum* for the conversion of cortexone into cortexolone.

Considerable interest further was concentrated in the hydroxylation of steroids into position 11*β*, namely for the synthesis of cortisol. The strains: *Streptomyces fradiae* (97), *Cunninghamella blakesleeana* (168), *Curvularia lunata* (406), *Absidia glauca* (251a), *Botrytis cinerea* (400), *Stachylidium bicolor* (403) were used for this purpose. Usually the starting steroid was cortexolone (steroid S-Reichstein), since the hydroxylation in progesterone usually proceeds into position 11*α* only. By the hydroxylation of cortexolone into position 11*β*, cortisol is then obtained directly.

In our experiments with 11*β*-hydroxylation of steroids we examined 64 different strains and found the required hydroxylation in 45 cases; in all cases representatives of the genus *Absidia* and *Cunninghamella* had been used, further in two strains of *Botrytis* and in the strain *Curvularia lunata*. Comparing the transformation of free cortexolone and its 21-acetate we find that cortisol is formed from the acetate twice as slow. First of all saponification of the ester takes place and only then the hydroxylation proper (79). The 11*β*-hydroxylation is usually accompanied by side-reactions, especially also by the formation of the 11*α*-epimer (165, 167). Using the species *Absidia orchidis* the metabolites of cortexolone were as follows: cortisol, 11*α*-epicortisol and in a small quantity also cortisone (167).

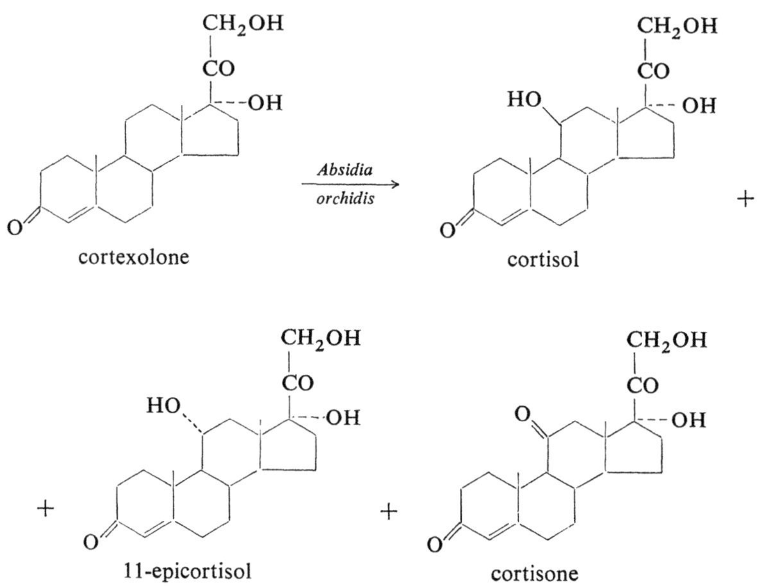


The 11*α*-hydroxylation proceeds generally with higher yields than the 11*β*-hydroxylation. Usually it was possible to obtain a mixture of both epimers in ratio approximately 1 : 1 from the starting cortexolone. For the production of cortisone we now hydroxylate cortexolone by the species *Beauveria bassiana* to epicortisol with 95% yield.

Analogously as in progesterone were attempted further 16α-hydroxylations of corticoids, namely by help of different *Streptomycetes* and *Actinomycetes.* Though in a number of estrogens the introduction of hydroxyl into position 16α lowers the activity of the steroid, the substitutent in position 16 in corticoids raises the effect. The chemical introduction of fluorine into position 9α, 16α or 6β brings about an increase in corticoid activity. The introduction of hydroxyl into position 16α frequently has been combined in order to obtain analogues having the fluorine atom in position 9α.

Further increase in the corticoid glucotropic activity with simultaneous limitation of mineralotropic effect was obtained by introducing another double bond into position 1–2 of the ring A in corticoids. Beside purely chemical ways, after bromination and dehydrobromination or by the action of $SeO_2$, microbial fermentation processes were used for the dehydrogenation of corticoids in position 1–2. In such a way it is possible to obtain in one single operation, prednisone from cortisone and prednisolone from cortisol. For such dehydrogenations a number of various species of microbes were used, such as: *Streptomycetes, Nocardia, Mycobacteria, Fusaria,* further the representatives of the genera *Corynebacterium, Arthrobacter,* the species *Didymella lycopersici, Septomyxa affinis, Bacillus subtilis, B. pulvifaciens, B. sphaericus, Protaminobacter alboflavum* etc. (see p. 114).

We have followed a similar dehydrogenation in 248 different strains of microbes with cortisone as the starting steroid (81). In all strains carrying out dehydrogenation in position 1–2, we have also found the ability of reducing the 20-keto group into the corresponding 20β-hydroxy derivative. There were considerable differences in the transformation rate among the individual strains, which had been examined. The most suitable qualities in this regard were shown by the strain *Mycobacterium flavum,* which under suitable conditions gave a yield of up to over 80% prednisone from cortisone (82). Fermentation with cortisone acetate as starting steroid (84) was also further examined and an ability of some microbes to carry out saponification of steroid ester, has been established. By means of *Bacillus repens* we now succeeded to obtain 80% of prednisone from cortisone, without simultaneous formation of 20β-hydroxy derivative.

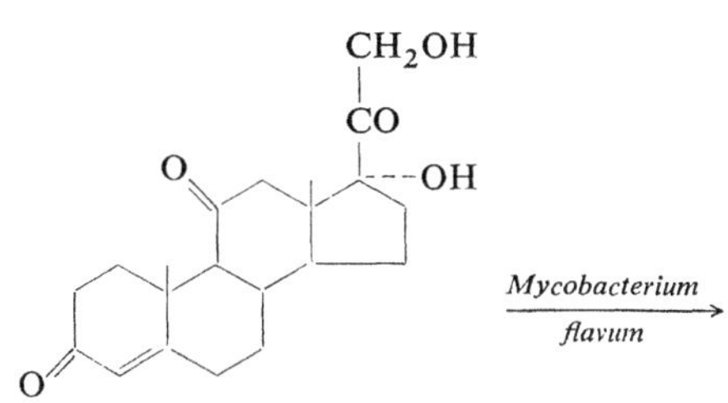


cortisone

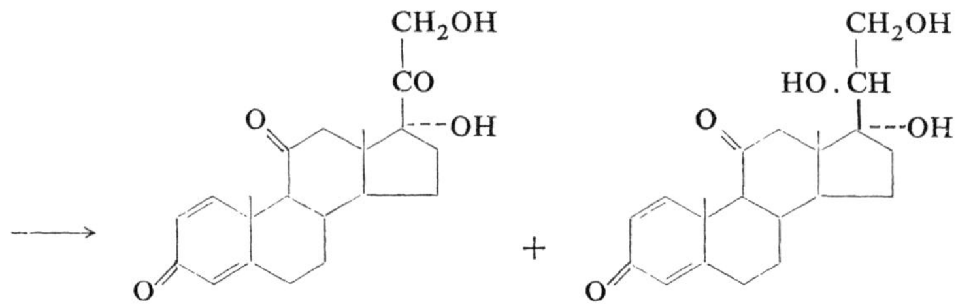


prednisone 20-hydroxy derivative

Kondo and Masuo (226) have described conversion of cortisol into prednisolone in high concentrations in the fermentation medium by the use of *Arthrobacter simplex.*

Fermentation processes in the corticoids series also were used in the synthesis of aldosterone. Vischer *et al.* (497) have proved that by help of enzymatic systems of microbes the d,1-steroids can be separated in a way that the *d-form,* i.e. the component corresponding to the natural steroid, *is oxidizable* while the 1-form from the racemate remains unchanged. For microbiological cleavage of racemates the following cultures have been used: *Proactinomyces erythropolis, P. aquosus, P. restrictus, P. roseus,* the species *Mycobacterium* and *Pseudomonas, Micrococcus dehydrogenans, Alcaligenes faecalis,* yeasts etc. For the cleavage of racemic intermediates of the total synthesis of aldosterone, especially the species *Ophiobolus herpotrichus* and *Didymella lycopersici* have been used by Vischer *and al.*

As follows from a concise review of some results, the microbial conversions of steroids have been utilized in synthetical preparation of steroid hormones, especially in stereospecific reductions of keto groups into the required alcohol derivative with a configuration of hydroxyl $\alpha$ or $\beta$, in the introduction of the oxygen function into position 11 of ring C, for the formation of a new double bond conjugated in the positions 1–2 of ring A, or cleavage of the side-chain of pregnene steroids on $C_{17}$. The other operations are rather of theoretical importance for the obtaining of differently substituted derivatives and analogues of steroid hormones, otherwise chemically accessible only with difficulty. This is how they are also utilized in preparative chemistry.

The decisive question concerning the use of microbial conversions of steroids for the production of steroid hormones and their analogues, remains the *economical side* of these processes. It should not be forgotten that the conversions proceed under relatively low concentrations and the whole operation is considerably expensive in exploitation both regarding the fermentation proper and the isolation part of the procedure. Economically microbial transformations can be compared in practice with purely chemical processes only in such cases, where the fermentation technique substitutes a synthesis of several steps and proceeds uniformly and with high yields.

# 10

# LIST OF MICROORGANISMS PERFORMING TRANSFORMATIONS OF STEROIDS

In this chapter the microorganisms are listed according to species in alphabetic order, which have been used for various types steroid transformations, the following *abbreviations* have been used to designate the individual reaction types:

| | | |
|---|---|---|
| Oxidation (alcohol to ketone) | O: | (position) OH — CO |
| Hydroxylation | Hxl: | (position) |
| Epoxidation | Ep: | (position) |
| Dehydrogenation (introduction of new double bond) | D: | (position) |
| Aromatization of ring A | Ar: | |
| Side chain cleavage (and/or ring D splitting) | Cl: | (position) |
| Ring B splitting | Sp: | B |
| Reduction of keto group | R: | (position) |
| Hydrogenation (double bond elimination) | Hdg: | (position) |
| Saponification of esters | S: | (position) |
| Cleavage of glycoside bond | Cl - Gl | |
| Esterification | E: | (position) |
| Isomerization (double bond shift from 5—6 to 4—5) | I: | |

| | | |
|---|---|---|
| *Absidia coerula* | Hxl: | 11α-OH, 11β-OH (79,129,130) |
| | O: | 11-OH → 11-CO (79, 129) |
| *Absidia cylindrospora* | Hxl: | 11α-OH, 11β-OH (129, 130) |
| | O: | 11-OH → 11-CO (129,130) |
| *Absidia glauca* | Hxl: | 11α-OH, 11β-OH (531, 360, 79, 130), 6β-OH (531) |
| | O: | 11-OH → 11-CO (79, 130) |
| *Absidia hyalospora* | Hxl: | 11α-OH, 11β-OH, 6β-OH (129) |
| *Absidia Lichtheimii* | Hxl: | 11α-OH, 11β-OH (129) |
| *Absidia orchidis* | Hxl: | 1β-OH, 5β-OH, 7β-OH (318, 320, 207, 321); 11α-OH, 11β-OH (165, 167, 79, 129, 130) |
| | O: | 11-OH → 11-CO (79, 129) |

| | | |
|---|---|---|
| *Absidia ramosa* | Hxl: | 11α-OH, 11β-OH, 6β-OH (129) |
| | O: | 11-OH → 11-CO (129) |
| *Absidia regnieri* | Hxl: | 14α-OH, 6β-OH, 7α-OH, 15β-OH (396, 399, 458) |
| *Absidia repens* | Hxl: | 11α-OH, 11β-OH (79, 129, 130) |
| *Absidia species* | Hxl: | 7α-OH, 11α-OH (999) |
| *Absidia spinosa* | Hxl: | 11α-OH, 11β-OH (129) |
| *Absidia Tieghemii* | Hxl: | 11α-OH, 11β-OH (129) |
| *Absidia turkestanica* | Hxl: | 11α-OH, 11β-OH, 6β-OH (129) |
| | O: | 11-OH → 11-CO (129) |
| *Actinomyces albidus* | Hxl: | 16α-OH (500, 501) |
| *Actinomyces albus* | O: | 17β-OH → 17-CO (510) |
| *Actinomyces annulatus* | Hxl: | 16α-OH (500) |
| *Actinomyces coerulus* | Hxl: | 16α-OH (501) |
| *Actinomyces diastaticus* | Hxl: | 16α-OH (501) |
| *Actinomyces flaveolus* | Hxl: | 16α-OH (501) |
| *Actinomyces globisporus* | O: | 3β-OH → 3-CO (161) |
| | I: | (161) |
| *Actinomyces globosus* | Hxl: | 16α-OH (500, 501) |
| *Actinomyces praecox* | Hxl: | 16α-OH (501) |
| *Actinomyces ruber* | Hxl: | 16α-OH (501) |
| *Actinomyces setonii* | Hxl: | 16α-OH (500, 501) |
| *Actinomyces species* | Hxl: | 16α-OH (328, 566, 664) |
| | O: | 3α-OH, 12α-OH → 3,12-di-CO (176, 177) |
| *Actinomyces viridochromogenes* | O: | 3β-OH → 3-CO (161) |
| | I: | (161) |
| *Actinomyces viridis* | Hxl: | 16α-OH (501) |
| *Aerobacter aerogenes* | O: | 3β-OH → 3-CO (279) |
| *Achromobacter species* | Hxl: | 6β-OH, 14α-OH (202) |
| *Achromobacter kashiwasakiensis* | Hxl: | 6β-OH, 14α-OH (467) |
| *Alcaligenes faecalis* | O: | 3α-OH, 7α-OH, 12α-OH → 3,7,12-tri-CO (182, 198, 359) |
| *Alcaligenes species* | D: | 1—2 (436) |
| | R: | 20-CO → 20β-OH (436) |
| *Alternaria bataticola* | R: | 3-CO → 3β-OH (392, 393) |
| | Hdg: | 4—5 (393) |
| *Alternaria passiflorae* | D: | 1—2, 4—5 (708) |
| *Arthrobacter simplex* | D: | 1—2 (226, 230, 618, 487) |
| *Aspergillus alliaceus* | O: | 3β-OH → 3-CO (73) |
| | Cl: | 17—20 (73) |
| *Aspergillus amstelodami* | Hxl: | 11α-OH (115, 73, 675); 15β-OH (73); 6β-OH (675) |
| | O: | 3β-OH → 3-CO (73) |
| *Aspergillus atropurpureus* | Hxl: | 15β-OH (73) |
| | O: | 3β-OH → 3-CO (73) |
| *Aspergillus awamori* | Hxl: | 11α-OH (73, 200) |
| | O: | 3β-OH → 3-CO (73) |
| | Cl-Gl: | (235) |
| *Aspergillus caespitosus* | Hxl: | 11α-OH (115) |

| | | |
|---|---|---|
| *Aspergillus candidus* | Hxl: | 11α-OH (115); 15β-OH (73) |
| | O: | 3β-OH → 3-CO (73) |
| *Aspergillus carbonarius* | Hxl: | 15β-OH (115) |
| *Aspergillus chevalieri* | Hxl: | 11α-OH (115, 72, 666) |
| | Cl: | 17—20 (73) |
| | Cl–Gl: | (235) |
| *Aspergillus cinnamomeus* | Hxl: | 15β-OH (73) |
| | O: | 3β-OH → 3-CO (73) |
| *Aspergillus clavatus* | Hxl: | 11α-OH (73, 115, 534) |
| | O: | 3β-OH → 3-CO (73) |
| *Aspergillus conicus* | Hxl: | 11α-OH (115) |
| *Aspergillus effusus* | Hxl: | 11α-OH (73) |
| | O: | 3β-OH → 3-CO (73) |
| *Aspergillus echinulatus* | Hxl: | 11α-OH (115) |
| *Aspergillus elegans* | Hxl: | 15β-OH (73) |
| | O: | 3β-OH → 3-CO (73) |
| *Aspergillus fischeri* | Hxl: | 11α-OH (115, 534, 802) |
| | Cl: | 17—20 (73) |
| *Aspergillus flavipes* | Hxl: | 11α-OH (115) |
| | Cl: | 17—20 (142, 73) |
| | O: | 3β-OH → 3-CO (73) |
| *Aspergillus flavus* | Hxl: | 11α-OH (115, 534, 802) |
| | Cl: | 17—20 (73, 334) |
| | O: | 3β-OH → 3-CO (73) |
| | Cl–Gl: | (235) |
| *Aspergillus fumigatus* | Hxl: | 11α-OH (115); 11β-OH (533); 15β-OH (73) |
| | O: | 3β-OH → 3-CO (73) |
| *Aspergillus giganteus* | Hxl: | 15β-OH (115, 612); 11α-OH (612) |
| *Aspergillus glaucus* | Cl–Gl: | (235) |
| *Aspergillus gracilis* | Hxl: | 15β-OH (115) |
| *Aspergillus itaconicus* | Hxl: | 11α-OH (534, 802) |
| *Aspergillus luchuensis* | Hxl: | 11α-OH (73, 115, 534, 802) |
| *Aspergillus melleus* | Hxl: | 11α-OH (73) |
| *Aspergillus montevidensis* | Hxl: | 11α-OH (115) |
| *Aspergillus nidulans* | Hxl: | 6β-OH (8, 463*a*, 675); 11α-OH (115, 534, 73, 463*a*, 675, 850, 830, 994) |
| *Aspergillus niger* | Hxl: | 11α-OH (531, 141, 115, 73, 513, 534, 516, 463*a*); 21-OH (522, 513, 516, 770, 853); 6β-OH (8); |
| | O: | 3β-OH → 3-CO (329); 11-OH → 11-CO (670) |
| | Cl–Gl: | (235, 583) |
| *Aspergillus niveo-glaucus* | Hxl: | 11α-OH (115) |
| *Aspergillus niveus* | Hxl: | 11α-OH (115) |
| *Aspergillus ochraceus* | Hxl: | 11α-OH (114, 115, 210, 73, 504, 463*a*, 591, 533, 534, 852, 841, 834, 831, 802, 932); 6β-OH (114, 463*a*) |
| *Aspergillus oryzae* | Hxl: | 11α-OH (115); |
| | Cl: | 17—20 (73, 72, 223); |
| | S: | 16, 17, 20, 21 (313*a*) |

| | | |
|---|---|---|
| *Aspergillus ostianus* | Hxl: | 15β-OH (73) |
| *Aspergillus parasiticus* | Hxl: | 15β-OH (73, 115) |
| *Aspergillus penicilloides* | C: | 17—20 (73) |
| *Aspergillus phoenicis* | Cl–Gl: | (235) |
| *Aspergillus rehmii* | Hxl: | 15β-OH (73) |
| *Aspergillus restrictus* | Hxl: | 15β-OH (73) |
| *Aspergillus ruber* | Cl: | 17—20 (73) |
| *Aspergillus saitoi* | Hxl: | 6β-OH, 11α-OH (200) |
| *Aspergillus sclerotiorum* | Cl : | 17—20 (73) |
| *Aspergillus species* | Hxl: | 11α-OH (534, 509, 675); 21-OH (509); 6β-OH (675) |
| *Aspergillus sydowi* | Hxl: | 11α-OH, 15β-OH, (73) |
| *Apergillus tamarii* | Hxl: | 11α-OH (73) |
| | Cl–Gl: | (235) |
| *Aspergillus terreus* | Hxl: | 11α-OH, 6β-OH (115) |
| | Cl: | 17—20 (73) |
| | Cl–Gl: | (350) |
| *Aspergillus terricola* | Cl: | 17—20 (73) |
| *Aspergillus usamii* | Hxl: | 11α-OH (200) |
| *Aspergillus ustus* | Hxl: | 11α-OH (73, 534, 802) |
| *Aspergillus variecolor* | Hxl: | 11α-OH (73) |
| *Aspergillus versicolor* | Hxl: | 11α-OH, 15β-OH (73) |
| *Aspergillus wentii* | Hxl: | 11α-OH (534, 463*a*); 6β-OH (463*a*) |
| | Cl–Gl: | (235) |
| *Azotobacter agilis* | D: | 1—2 (624) |
| *Azotobacter indicus* | D: | 1—2 (624) |
| *Azotobacter species* | O: | 3β-OH → 3-CO (194, 231) |
| | D: | 7—8 (194) |
| *Azotomonas fluorescens* | D: | 1—2 (624) |
| *Bacillus cereus* | Hxl: | 14α-OH (8, 393); 6β-OH (435, 393); 11α-OH (435, 257, 393) |
| *Bacillus coli communis* | O: | 3α,12α-di-OH → 3,12-di-CO (143, 261) |
| *Bacillus cyclo-oxydans* | D: | 1—2 (148, 618, 594) |
| | Ar: | (417) |
| | Sp: | B (417) |
| *Bacillus fusiformis* | D: | 1—2 (712) |
| *Bacillus macerans* | Cl–Gl: | (358*a*) |
| *Bacillus megatherium* | Hxl: | 15β-OH (435, 257, 610, 609); 15α-OH (187); |
| | S: | 21 (437, 427) |
| *Bacillus pulvifaciens* | D: | 1—2 (203, 473, 354) |
| *Bacillus putrificus* | O: | 3β-OH → 3-CO (273); |
| | Hdg: | 4—5 (273) |
| | R: | 3-CO → 3α-OH (273) |
| *Bacillus species* | D: | 1—2 (393) |
| | R: | 20-CO → 20β-OH (488) |
| *Bacillus sphaericus* | D: | 1—2 (433, 153, 431, 225, 758, 753, 730, 729, 724, 722, 712, 619, 851, 846, 832, 845, 145) |
| | O: | 3β-OH → 3-CO (729) |
| *Bacillus subtilis* | D: | 1—2 (250) |
| *Bacterium coli* | R: | 7-CO → 7-OH (143) |

| | | |
|---|---|---|
| *Bacterium cyclo-oxydans* | D: | 1—2 (594, 754, 701, 836, 931) |
| | R: | 20-CO → 20β-OH (420, 147) |
| | O: | 20β-OH → 20-CO (604) |
| *Bacterium species* | O: | 3β-OH → 3-CO (274, 267, 268, 271) |
| | Hxl: | 6β-OH, 14α-OH (202) |
| | Hdg: | 5—6 (424) |
| *Beauveria species* | Hxl: | 11α-OH (981) |
| *Blakeslea trispora* | Hxl: | 11α-OH, 11β-OH, 6β-OH (129) |
| | O: | 11-OH → 11-CO (129) |
| *Botryodiplodia theobromae* | Hxl: | 11α-OH (766, 865) |
| *Botrytis cinerea* | Hxl: | 11β-OH (944, 79, 393, 689, 585, 820); 11α-OH (79); 6β-OH (400, 393); 15β-OH (400); 10β-OH (135) |
| | O: | 11-OH → 11-CO (79) |
| *Botrytis peoniae* | Hxl: | 11β-OH (944) |
| *Calonectria decora* | Hxl: | 12β-OH (183, 242, 365, 368*a*, 370, 373, 998); 15β-OH (365, 368*a*, 370, 998) 15α-OH (242, 373); 11β-OH (365, 368*a*, 370); 11α-OH (374) |
| | D: | 1—2 (495, 708) |
| | D: | 4—5 (708) |
| | O: | 3β-OH → 3-CO (317) |
| *Candida pulcherrima* | R: | 20-CO → 20β-OH (444) |
| *Catenabacterium catenaforme* | R: | 3-CO → 3β-OH (310) |
| *Cephalosporium asperum* | Hxl: | 7α-OH (970) |
| *Cephalosporium species* | Hxl: | 7α-OH (854) |
| *Cephalosporium subverticillatum* | Cl: | 17—20 (56) |
| *Carcospora melonis* | Hxl: | 8β-OH (465) |
| *Chaetocladium Brefedii* | Hxl: | 6β-OH (129) |
| *Chaetomium cochliodes* | Hxl: | 6β-OH (914) |
| *Chaetomium funicolum* | Hxl: | 6β-OH (716, 914) |
| *Chaetomium sucineum* | Hxl: | 6β-OH (914) |
| *Choanephora cucurbitarium* | Hxl: | 11α-OH, 11β-OH, 6β-OH (129) |
| *Circinella simplex* | Hxl: | 11α-OH, 11β-OH, 6β-OH (129) |
| *Circinella species* | Hxl: | 9α-OH (367, 1001); 14α-OH (367, 369) |
| *Circinella spinosa* | Hxl: | 11α-OH, 6β-OH (129) |
| *Cladosporium resinae* | Cl: | 17—20 (140) |
| | E: | 17α-OH (140) |
| *Cladosporium species* | Hxl: | 7α-OH (259) |
| *Clostridium lentoputrescens* | R: | 17-CO → 17β-OH (273) |
| *Clostridium paraputrificum* | Hdg: | 4—5 (381, 382) |
| | R: | 3-CO → 3α-OH (382) |
| *Clostridium terium* | Hdg: | 4—5 (382) |
| | R: | 3-CO → 3α-OH (382) |
| *Colletotrichum antirhini* | Hxl: | 15α-OH (8, 574) |
| *Colletotrichum atramentarium* | D: | 1—2 (952) |
| *Colletotrichum lindemuthianum* | D: | 1—2 (821) |
| *Colletotrichum phomoides* | Hxl: | 11β-OH (669, 586); 6β-OH, 12α-OH (austr. pat. 220.072) |
| *Colletotrichum pisi* | Hxl: | 11β-OH (669, 586) |

| | | |
|---|---|---|
| *Coniothyrium helleborine* | Hxl: | 11β-OH (669, 341, 721, 771, 587, 819, 963); 12β-OH (341, 771, 819, 934); 6β-OH (341); 7α-OH (590) |
| *Coniothyrium species* | Hxl: | 21-OH (723, 839); 11β-OH (723); 11α-OH (8) |
| *Coriolus versicolor* | Hxl: | 6β-OH (619); 14α-OH, 15-OH (619) |
| *Corticium centrifugum* | R: | 17-CO → 17β-OH (112) |
| | Hxl: | 16β-OH (112) |
| | O: | 16-OH → 16-CO (112) |
| *Corticium microsclerotia* | Hxl: | 11β-OH (737, 817); 19-OH (615, 858, 933) |
| *Corticium praticola* | Hxl: | 11β-OH (737, 817, 933) |
| *Corticium sasakii* | Hxl: | 11β-OH (171, 155, 737, 817, 933); 11α-OH (171, 155); 19-OH (156) |
| *Corticium vagum* | Hxl: | 11β-OH (737, 817, 933); 11α-OH (737); 19-OH (737) |
| *Corynebacterium helvolum* | D: | 1—2 (272) |
| *Corynebacterium hoagii* | D: | 1—2 (618) |
| *Corynebacterium mediolanum* | D: | 1—2 (272) |
| | O: | 3β-OH → 3-CO (36, 267, 271, 275, 490, 765, 629, 985) |
| *Corynebacterium simplex* | D: | 1—2 (313, 436, 341, 226, 758, 726, 717, 707, 692, 691, 584, 180, 215) |
| | R: | 20-CO → 20β-OH (147, 420) |
| | O: | 3β-OH → 3-CO (529); 20β-OH → 20-CO (604) |
| *Corynebacterium species* | D: | 1—2 (752) |
| *Coryneum cardinale* | Hxl: | 11α-OH (732) |
| *Cucurbitaria laburni* | Hxl: | 17α-OH (895) |
| *Cunninghamella albida* | Hxl: | 11α-OH, 11β-OH, 6β-OH (129) |
| *Cunninghamella africana* | Hxl: | 11α-OH, 11β-OH (129) |
| | O: | 11-OH → 11-CO (129) |
| *Cunninghamella bertholetiae* | Hxl: | 9α-OH (751) |
| *Cunninghamella blakesleeana* | Hxl: | 11β-OH (168, 277, 325, 531, 48, 225, 765, 79, 129, 985, 487);11α-OH (79, 129); 6β-OH (79, 129, 371); 14α-OH (371); 9α-OH (751) |
| | Ep: | 9—11 (55); 14α, 15α (690); |
| *Cunninghamella echinulata* | Hxl: | 11α-OH (79, 129, 592); 11β-OH (79, 129); 9α-OH (751); 6β-OH (79, 129) |
| | O: | 11-OH → 11-CO (79, 129) |
| *Cunninghamella elegans* | Hxl: | 6β-OH (371, 79, 129); 14α-OH (371); 9α-OH (751); 11α-OH, 11β-OH (129, 79) |
| | O: | 11-OH → 11-CO (79, 129) |
| *Cunninghamella homothallica* | Hxl: | 11α-OH, 11β-OH, 6β-OH (129) |
| | O: | 11-OH → 11-CO (129) |
| *Cunninghamella ramosa* | Hxl: | 11α-OH, 11β-OH (79, 129); 6β-OH (79); |
| | O: | 11-OH → 11-CO (79, 129) |
| *Cunninghamella verticillata* | Hxl: | 11α-OH, 11β-OH (129); 9α-OH (751) |
| | O: | 11-OH → 11-CO (129) |
| *Curvularia brachyspora* | Hxl: | 11β-OH (696, 703) |
| *Curvularia falcata* | Hxl: | 11β-OH (577); 7α-OH (299) |
| *Curvularia fallax* | Hxl: | 7α-OH (299) |

| | | |
|---|---|---|
| *Curvularia geniculata* | Hxl: | 11β-OH (227) |
| *Curvularia lunata* | Hxl: | 11β-OH (535, 38, 361, 406, 351, 523, 117, 884, 525, 57, 225, 79, 758, 728, 718, 696, 600, 577, 838, 527, 70, 487, 909); 14α-OH (38, 361, 371, 227, 718, 702, 582, 838, 70, 487); 7α-OH (299, 227, 582, 838); 6β-OH (371, 227): 9α-OH (227); 21-OH (351); 11α-OH (57); 15α-OH (70); 10β-OH (135) |
| | Ep: | 11β-9α (55, 240); 14α-15α (690, 815, 407) |
| | R: | 20-CO → 20β-OH (150) |
| | O: | 11-OH → 11-CO (79) |
| *Curvularia maculans* | Hxl: | 11β-OH (227) |
| *Curvularia pallescens* | Hxl: | 11β-OH (535, 696, 679); 8β-OH (432, 34); 14α-OH (738, 736, 734); 7α-OH (299) |
| *Curvularia species* | Hxl: | 7α-OH (299, 368); 11β-OH (713, 759) |
| | Ep: | 9α-11α (240) |
| *Curvularia trifolii* | Hxl: | 11β-OH (227) |
| *Cylindrocarpon radicicola* | D: | 1—2 (142, 336, 618, 596) |
| | Cl | 17—20 (142, 336, 596) |
| *Dactylium dendroides* | Hxl: | 11α-OH (116, 769, 607, 822); 17α-OH (116, 769, 607, 822) |
| *Delacroixa coronata* | Hxl: | 11α-OH (735) |
| *Didymella lycopersici* | D: | 1—2 (496, 497, 708); 4—5 (708) |
| *Didymella vodakii* | Hxl: | 16α-OH (842) |
| *Didymocladium ternatum* | Hxl: | 11α-OH (766, 865) |
| *Diplodia natalensis* | Hxl: | 7α-OH (612) |
| *Diplodia tubericola* | Hxl: | 7β-OH (466, 46) |
| *Dothichiza species* | Hxl: | 11β-OH (686) |
| *Epicoccum oryzae* | R: | 20-CO → 20β-OH (26) |
| *Eremothecium ashbyi* | O: | 3β-OH → 3-CO (329) |
| *Escherichia coli* | R: | 7-CO → 3α-OH (143) |
| *Escherichia freundii* | O: | 3α-OH → 3-CO (174) |
| *Eurotium chevalieri* | Hxl: | 11α-OH (666) |
| *Flavobacterium androstendionicum* | O: | 3β-OH → 3-CO (126, 127) |
| *Flavobacterium carbonilicum* | O: | 3β-OH → 3-CO (302) |
| *Flavobacterium dehydrogenans* | O: | 3β-OH → 3-CO (42, 41, 125, 123, 953); 17β-OH → 17-CO (124); |
| | R: | 20-CO → 20α-OH (324) |
| | S: | 11β, 21 (92) |
| *Flavobacterium dehydrogenans* var. *hydrolyticum* | S: | 3β (694) |
| *Flavobacterium maris* | O: | 3β-OH → 3-CO (44) |
| *Flavobacterium species* | O: | 3β-OH → 3-CO (40*a*) |
| | Cl–Gl: | (358*c*) |
| *Fusarium avenaceum* | Hxl: | 6β-OH, 15α-OH (78) |
| *Fusarium avenaceum* var. *herbarum* | Hxl: | 6β-OH, 15α-OH (78) |
| *Fusarium aquaeductum* | Hcl: | 6β-OH, 15α-OH (78) |
| *Fusarium aqueaductum* var. *dimerum* | Hxl: | 6β-OH, 15α-OH (78) |

| | | |
|---|---|---|
| *Fusarium bulbigenum* | D: | 1—2 (606) |
| | Hxl: | 11α-OH (958) |
| *Fusarium buxicola* | Hxl: | 6β-OH, 15α-OH (78) |
| *Fusarium caucasicum* | D: | 1—2 (493, 78, 840, 825, 824, 520) |
| | Cl: | 17—20 (493, 78, 840, 825) |
| *Fusarium cocophilum* | Hxl: | 6β-OH, 15α-OH (78) |
| *Fusarium culmorum* | Hxl: | 15α-OH (214*a*, 78, 997); 6β-OH (78); 11α-OH (605, 958) |
| *Fusarium dimerum* | Hxl: | 6β-OH (224) |
| *Fusarium discolor* | Hxl: | 11α-OH (605) |
| *Fusarium equiseti* var. *bullatum* | Hxl: | 11α-OH (605, 958) |
| *Fusarium expansum* | Hxl: | 6β-OH, 15α-OH (78) |
| *Fusarium gibosum* | Hxl: | 6β-OH, 15α-OH (78) |
| *Fusarium gibosum* var. *acuminatum* | Hxl: | 6β-OH, 15α-OH (78) |
| *Fusarium gigas* | Hxl: | 6β-OH, 15α-OH (78) |
| *Fusarium graminearum* | Hxl: | 6β-OH, 15α-OH (78) |
| *Fusarium heterosporum* | Hxl: | 6β-OH, 15α-OH (78); 11α-OH (958) |
| *Fusarium javanicum* | Hxl: | 6β-OH, 15α-OH (78) |
| | D: | 1—2 (618) |
| *Fusarium javanicum* var. *ensiforme* | D: | 1—2 (596) |
| | Cl: | 17—20 (596) |
| *Fusarium lateritium* | D: | 1—2 (78, 83) |
| | Cl: | 17—20 (78, 83) |
| *Fusarium lini* | Hxl: | 12β-OH (453, 154, 508, 154*a*, 453*a*); 15α-OH (453, 154, 456); |
| | D: | 1—2 (606) |
| | S: | 3β (454) |
| *Fusarium lycopersici* | Hxl: | 15α-OH (214*a*, 997); 6β-OH (393) |
| *Fusarium macroceras* | Hxl: | 6β-OH, 15α-OH (78) |
| *Fusarium merismoides* | Hxl: | 6β-OH, 15α-OH (78) |
| *Fusarium microcrea* | Hxl: | 6β-OH, 15α-OH (78) |
| *Fusarium monoliforme* | Hxl: | 6β-OH, 15α-OH (78) |
| *Fusarium monoliforme* var. *subglutinans* | Hxl: | 6β-OH, 15α-OH (78) |
| *Fusarium monoliforme* var. *lactis* | Hxl: | 6β-OH, 15α-OH (78) |
| *Fusarium nivale* | Hxl: | 6β-OH, 15α-OH (78) |
| *Fusarium oxysporum* | Hxl: | 6β-OH, 15α-OH (78); 11α-OH (958) |
| | D: | 1—2 (606) |
| *Fusarium oxysporum* var. *orthoceras* | Hxl: | 6β-OH, 15α-OH (78) |
| *Fusarium redolens* | Hxl: | 11α-OH (605, 958) |
| *Fusarium reticulatum* | Hxl: | 11α-OH (958) |
| *Fusarium sambucinum* | Hxl: | 6β-OH, 15α-OH (78); 11α-OH (958) |
| *Fusarium sarcochroun* | Hxl: | 6β-OH, 15α-OH (78) |
| *Fusarium semitectum* | Hxl: | 6β-OH, 15α-OH (78) |
| *Fusarium solani* | Hxl: | 15α-OH (214*a*, 997); 11α-OH (766, 865) |

| | | |
|---|---|---|
| | D: | 1—2 (493, 495, 311, 439, 224, 78, 758, 618, 812, 880) |
| | Cl: | 17—20 (493, 311, 439, 224, 812, 880) |
| | R: | 20-CO → 20β-OH (439) |
| *Fusarium solani* var. *aduncisporum* | Hxl: | 6β-OH, 15α-OH (78) |
| *Fusarium solani* var. *argilaceum* | Hxl: | 6β-OH, 15α-OH (78) |
| *Fusarium solani* var. *coeruleum* | Hxl: | 6β-OH, 15α-OH (78) |
| *Fusarium solani* var. *redolens* | Hxl: | 6β-OH, 15α-OH (78) |
| *Fusarium species* | Hxl: | 6β-OH (224) |
| | O: | 3β-OH → 3-CO (209*b*) |
| *Fusarium sphaeria* | Hxl: | 6β-OH, 15α-OH (78) |
| *Fusarium sporotrichiella* | Hxl: | 6β-OH, 15α-OH (78) |
| *Fusarium sporotrichiella* var. *poae* | Hxl: | 6β-OH, 15α-OH (78) |
| *Fusarium zonatum* | D: | 1—2 (606) |
| *Fusidium species* | Hxl: | 7α-OH (108) |
| *Gibberella baccata* | Hxl: | 15α-OH (299, 485, 909) |
| *Gibberella fujikuroi* | O: | 3β-OH → 3-CO (209*b*) |
| *Gibberella saubinetti* | Hxl: | 6β-OH, 15α-OH (485, 405, 393); 12β-OH (326) |
| | S: | 3β-(326) |
| *Gibberella species* | Hxl: | 15α-OH (485, 299); 6β-OH (485) |
| *Gliocladium catenulatum* | Hxl: | 6β-OH (567) |
| | Cl: | 17—20 (334, 567) |
| | D: | 1—2 (618) |
| *Gliocladium roseum* | D: | 1—2 (404, 393) |
| *Gloesporium kaki* | Hxl: | 11α-OH (401, 393); 6β-OH (393) |
| *Glomerella lagenarium* | Hxl: | 11α-OH (401, 393); 6β-OH (393) |
| *Glomerella species* | Hxl: | 11α-OH (981) |
| *Gongronella urceolifera* | Hxl: | 11β-OH (746) |
| *Haplosporela species* | Hxl: | 1β-OH,15β-OH, 7β-OH, 6β-OH (113) |
| *Helicostylum piriforme* | Hxl: | 14α-OH (543, 646, 531, 122, 57, 736, 641, 570, 546, 719); 9α-OH (169, 751); 11α-OH (531, 179, 355); 7α-OH (355, 179); 8-OH (564, 719, 640, 663); |
| | Ep: | 14α, 15α (55, 690); |
| | O: | 11β-OH → 11-CO (169); 7β-OH → 7-CO (179) |
| *Helmintosporium sativum* | Hxl: | 15α-OH (469, 472, 172); 15β-OH (472, 172); 7β-OH (472); 14α-OH, 7α-OH (472) |
| *Helmintosporium species* | Hxl: | 7α-OH (259, 472); 14α-OH (220, 472); 7β-OH, 15α-OH, 15β-OH (472) |
| *Helmintosporium turcicum* | D: | 1—2 (404, 393) |
| *Hendersonia aberrans* | Hxl: | 21-OH (677) |
| *Hendersonia acicola* | Hxl: | 21-OH (677) |
| *Hendersonia herpotricha* | Hxl: | 21-OH (677) |

| | | |
|---|---|---|
| *Hendersonia phragmitis* | Hxl: | 21-OH (677) |
| *Hendersonia rubi* | Hxl: | 21-OH (677) |
| *Hormodendrum hordei* | Hxl: | 11α-OH (766, 865) |
| *Hormodendrum olivaceum* | Hxl: | 15α-OH (52) |
| *Hormodendrum pallidum* | Hxl: | 11α-OH (766, 865) |
| *Hormodendrum viride* | Hxl: | 15α-OH (53) |
| *Hypholoma species* | Hxl: | 15α-OH (610) |
| *Lenzites abietina* | Hxl: | 15β-OH (299, 909) |
| *Leptosphaeria maculans* | Hxl: | 17α-OH (895, 703) |
| *Lichtheimia corymbifera* | Hxl: | 11α-OH, 11β-OH (130) |
| *Lichtheimia ramosa* | Hxl: | 11α-OH, 11β-OH (130) |
| *Lophotrichus martinii* | Hxl: | 17α-OH (895) |
| *Melanospora parasitica* | Hxl: | 17α-OH (895) |
| *Micrococcus dehydrogenans* | O: | 3β-OH → 3-CO (41, 123, 125, 124) |
| *Micromonospora chalcea* | D: | 1—2 (714) |
| *Mortierella alphina* | Hxl: | 1-OH (614) |
| *Mortierella bainieri* | Hxl: | 1-OH (614) |
| *Mortierella candelabrum* | Hxl: | 1-OH (614) |
| *Mortierella marburgensis* | Hxl: | 1-OH (614) |
| *Mortierella polycephala* | Hxl: | 1-OH (614) |
| *Mortierella pusilla* | Hxl: | 1-OH (614) |
| var. *isobellina* | | |
| *Mortierella tuberosa* | Hxl: | 1-OH (614) |
| *Mortierella zonata* | Hxl: | 1-OH (614) |
| *Mucor adriaticus* | Hxl: | 11α-OH, 6β-OH (129) |
| *Mucor adventitius* | Hxl: | 11α-OH, 6β-OH (531) |
| *Mucor angulisporus* | Hxl: | 11α-OH, 6β-OH (129) |
| *Mucor berolinensis* | Hxl: | 6β-OH (129) |
| *Mucor buntingii* | Hxl: | 6β-OH (129) |
| *Mucor circinelloides* | Hxl: | 6β-OH (129) |
| *Mucor corymbifer* | Hxl: | 6β-OH, 14α-OH (64) |
| *Mucor dimorphosporus* | Hxl: | 6β-OH (129) |
| *Mucor dubius* | Hxl: | 11α-OH, 6β-OH (531) |
| *Mucor erectus* | Hxl: | 11α-OH, 6β-OH (129) |
| *Mucor genevensis* | Hxl: | 11α-OH, 6β-OH (531, 129); 11β-OH (129) |
| *Mucor globosus* | Hxl: | 11α-OH, 11β-OH, 6β-OH (129) |
| *Mucor griseo-cyanus* | Hxl: | 14α-OH (538, 543, 369, 736, 570); 9α-OH (122) |
| | Ep: | 14α-15α (55) |
| | Hxl: | 7α-OH, 14α-OH (69) |
| *Mucor Guilliermondii* | Hxl: | 6β-OH (129) |
| *Mucor hiemalis* | Hxl: | 11α-OH (531, 834); 6β-OH (531, 129) |
| *Mucor humicola* | Hxl: | 6β-OH (129) |
| *Mucor humilis* | Hxl: | 6β-OH (129) |
| *Mucor hypochninus* | Hxl: | 6β-OH (129) |
| *Mucor javanicus* | Hxl: | 11α-OH, 6β-OH (531, 129) |
| *Mucor mandshuricus* | Hxl: | 6β-OH, 11α-OH (129) |
| *Mucor microsporus* | Hxl: | 11α-OH, 6β-OH (531) |
| *Mucor mucedo* | Hxl: | 11α-OH, 6β-OH (531, 129); 14α-OH (371) |
| | Cl: | 17—20 (142) |

| | | |
|---|---|---|
| *Mucor murorum* | Hxl: | 6β-OH (129) |
| *Mucor parasiticus* | Hxl: | 11α-OH, 6β-OH (531); 14α-OH (543, 736, 570, 122); 9α-OH (122); 5β-OH (205); 7β-OH (319) |
| | Ep: | 14α, 15α (55, 690, 815) |
| | O: | 3β-OH → 3-CO (319) |
| *Mucor piriformis* | Hxl: | 6β-OH (129) |
| *Mucor plumbens* | Hxl: | 11α-OH, 6β-OH (531, 129) |
| *Mucor pusillus* | Hxl: | 6β-OH, 11α-OH, 11β-OH (129) |
| *Mucor racemosus* | Hxl: | 11α-OH, 6β-OH (531, 129) |
| *Mucor Ramannianus* | Hxl: | 6β-OH, 11α-OH (129) |
| *Mucor Rouxianus* | Hxl: | 11α-OH (129, 531); 6β-OH (531) |
| *Mucor varians* | Hxl: | 6β-OH (129) |
| *Mycobacterium berlinensis* | Hxl: | 14α-OH (910, 810) |
| | D: | 1—2 (910, 739, 810); 4—5 (739) |
| *Mycobacterium butyricum* | Hxl: | 14α-OH (810) |
| | D: | 1—2 (739, 618, 810); 4—5 (739) |
| *Mycobacterium flavum* | D: | 1—2 (81, 82) |
| | R: | 20-CO → 20β-OH (81, 82) |
| | Cl: | 17—20 (82) |
| *Mycobacterium fortuitum* | D: | 1—2 (739, 736); 4—5 (739) |
| | Ar: | (376) |
| | Sp: | B (376) |
| *Mycobacterium friedmannii* | D: | 1—2 (379, 810); 4—5 (739) |
| | Hxl: | 14α-OH (810) |
| *Mycobacterium lacticola* | D: | 1—2 (436, 910, 462*a*, 844, 810, 845) |
| | Hxl: | 14α-OH (910) (436) |
| | R: | 20-CO → 20β-OH (436) |
| *Mycobacterium phlei* | D: | 1—2 (600, 742, 739, 720, 700, 687, 844, 832, 827, 810, 845); 4—5 (739) |
| | Hxl: | 14α-OH (810) |
| *Mycobacterium ranae* | D: | 1—2 (810) |
| | Hxl: | 14α-OH (810) |
| *Mycobacterium rhodochrous* | D: | 1—2 (462*a*) |
| | O: | 3β-OH → 3-CO (477); 20β-OH → 20-CO (604) |
| | Cl: | 20—27 (479) |
| | Ar: | (417) |
| | Sp: | B (417) |
| *Mycobacterium rubropertinctum* | D: | 1—2 (462*a*) |
| *Mycobacterium smegmatis* | D: | 1—2 (910, 376, 462*a*, 472, 739, 700, 687, 603, 844, 598, 845); |
| | Ar: | (375, 378, 377, 376) |
| | Hxl: | 14α-OH (910) |
| | Sp: | B (375, 378, 377, 376) |
| *Mycobacterium species* | D: | 1—2 (713, 234, 233, 910, 753, 722) |
| | Hxl: | 14α-OH (910) |
| | O: | 3β-OH → 3-CO (511) |
| | R: | 20-CO → 20β-OH (427) |

| | | |
|---|---|---|
| *Mycobacterium thamnopheos* | D: | 1–2 (910, 810) |
| | Hxl: | 14α-OH (910, 810) |
| *Mycobacterium tuberculosis* | D: | 1–2 (844, 810, 845, 376) |
| | Ar: | (376) |
| | Sp: | B (376) |
| | Hxl: | 14α-OH (810) |
| *Mycocladus hyalinus* | Hxl: | 11α-OH (129, 130); 11β-OH (130) |
| *Myrothecium roridum* | Hxl: | 11α-OH (766, 865) |
| *Naematoloma sublateritium* | Hxl: | 14α-OH (371, 1000); 6β-OH (371) |
| *Neurospora crassa* | Hxl: | 9α-OH (432) |
| *Neurospora sitophila* | Hxl: | 11α-OH (555, 802) |
| *Nigrospora oryzae* | Hxl: | 11α-OH, 15β-OH (595) |
| *Nigrospora species* | Hxl: | 11α-OH, 15β-OH (595) |
| *Nigrospora sphaerica* | Hxl: | 11α-OH, 15β-OH (595) |
| | O: | 3β-OH → 3-CO (317) |
| *Nocardia asteroides* | D: | 1–2 (462*a*, 706, 707) |
| *Nocardia aurantia* | D: | 1–2 (462*a*) |
| *Nocardia blackwelii* | D | 1–2 (434, 462*a*, 706, 816, 832) |
| *Nocardia braziliensis* | D: | 1–2 (462*a*) |
| *Nocardia caviae* | D: | 1–2 (707) |
| *Nocardia coeliaca* | D: | 1–2 (462*a*) |
| *Nocardia convoluta* | D: | 1–2 (706, 707) |
| *Nocardia corallina* | D: | 1–2 (462*a*, 710, 706, 707) |
| | Hxl: | 9α-OH (747) |
| | Cl: | 17–20 (747) |
| *Nocardia cuniculi* | D: | 1–2 (707) |
| *Nocardia erythropolis* | D: | 1–2 (462*a*, 826) |
| *Nocardia farcinica* | D: | 1–2 (826) |
| *Nocardia formica* | D: | 1–2 (706, 832) |
| *Nocardia gardneri* | D: | 1–2 (462*a*, 710, 707) |
| *Nocardia globerula* | D: | 1–2 (688, 462*a*, 706, 832, 829, 707) |
| *Nocardia leishmanii* | D: | 1–2 (706, 832, 707) |
| *Nocardia madurae* | D: | 1–2 (462*a*) |
| *Nocardia minima* | D: | 1–2 (706) |
| *Nocardia opaca* | D: | 1–2 (688, 462*a*, 829) |
| *Nocardia polychromogenes* | D: | 1–2 (707) |
| *Nocardia restrictus* | D: | 1–2 (410, 413, 414) |
| | Ar: | (413) |
| | Sp: | B (413) |
| | Hxl: | 9α-OH (413) |
| *Nocardia rubra* | D: | 1–2 (462*a*) |
| *Nocardia salmonicolor* | D: | 1–2 (462*a*) |
| *Nocardia species* | D: | 1–2 (107, 940, 753, 752, 722, 846) |
| | Ar: | (107) |
| | Sp: | B (107, 111) |
| | Hxl: | 9α-OH (111, 107) (358) |
| | Cl: | 17–20 (111) |
| | O: | 3α-OH → 3-CO (158) |
| *Nocardia sylvodifera* | D: | 1–2 (707) |
| *Nocardia transvalensis* | D: | 1–2 (826) |

| | | |
|---|---|---|
| *Ophiobolus herpotrichus* | Hxl: | 21-OH (299*a*, 497, 770, 580, 853, 703, 915, 909) |
| | D: | 1—2 (393, 404, 708); 4—5 (708) |
| *Ophiobolus Miyabeanus* | D: | 1—2 (708); 4—5 (708) |
| *Ophiobolus species* | Hxl: | 21-OH (509) |
| *Penicillium adametzi* | Cl: | 17-20 (334, 80) |
| *Penicillium albidum* | Cl: | 17—20 (80) |
| *Penicillium asperum* | Cl: | 17—20 (80) |
| *Penicillium atramentosum* | Hxl: | 11α-OH (80) |
| *Penicillium canescens* | Cl: | 17—20 (80) |
| *Penicillium citreo-viride* | Cl: | 17—20 (80) |
| *Penicillium citrinum* | Cl: | 17—20 (72, 80, 160) |
| *Penicillium commune* | Cl: | 17—20 (80) |
| *Penicillium corylophilum* | Cl: | 17—20 (80) |
| | Hxl: | 11α-OH (115) |
| *Penicillium cyaneofulvum* | Cl: | 17—20 (80) |
| *Penicillium cyclopium* | Cl-Gl: | (235, 583) |
| *Penicillium daleae* | Cl: | 17—20 (80) |
| *Penicillium decumbens* | Cl: | 17—20 (163, 160, 80) |
| *Penicillium diversum* | Hxl: | 15β-OH (80) |
| *Penicillium egyptaceum* | Cl: | 17—20 (80) |
| *Penicillium ehrlichii* | Hxl: | 11α-OH (766, 865) |
| *Penicillium frequentans* | Cl: | 17—20 (80) |
| *Penicillium fuscum* | Cl: | 17—20 (80) |
| *Penicillium godlewskii* | Cl: | 17—20 (80) |
| *Penicillium chrysogenum* | Cl: | 17—20 (142, 72, 80) |
| | O: | 3β-OH → 3-CO (329) |
| | Hxl: | 11α-OH (533) |
| | Cl-Gl: | (235) (350, 583) |
| *Penicillium implicatum* | Cl: | 17—20 (80) |
| *Penicillium janthinellum* | Cl: | 17—20 (80) |
| *Penicillium javanicum* | Cl: | 17—20 (80) |
| *Penicillium jensenii* | Cl: | 17—20 (80) |
| *Penicillium kapuscinskii* | Cl: | 17—20 (80) |
| *Penicillium lilacinum* | Cl: | 17—20 (334, 383, 80, 576) |
| | R: | 20-CO → 20β-OH (388, 576) |
| | Hxl: | 11α-OH (115) |
| *Penicillium lividum* | Cl: | 17—20 (80) |
| *Penicillium melearginum* | Cl: | 17—20 (80) |
| *Penicillium melinii* | Cl: | 17—20 (80) |
| *Penicillium miczynskii* | Cl: | 17—20 (80) |
| *Penicillium multicolor* | Cl: | 17—20 (80) |
| *Penicillium nalgiovensis* | Cl: | 17—20 (80) |
| *Penicillium nigricans* | Cl: | 17—20 (80) |
| *Penicillium notatum* | Cl: | 17—20 (163, 160, 80) |
| | Hxl: | 15α-OH (68) |
| *Penicillium ochro-chlorum* | Cl: | 17—20 (80) |
| *Penicillium oxalinum* | Hxl: | 11α-OH (80) |
| *Penicillium parvum* | Cl: | 17—20 (80) |
| *Penicillium piscarium* | Cl: | 17—20 (80) |

| | | |
|---|---|---|
| *Penicillium psittacinum* | Cl: | 17—20 (80) |
| *Penicillium pulvillorum* | Cl: | 17—20 (80) |
| *Penicillium purpurescens* | Cl: | 17—20 (80) |
| *Penicillium pusillum* | Cl: | 17—20 (80) |
| *Penicillium raciborskii* | Cl: | 17—20 (80) |
| *Penicillium raistrickii* | Cl: | 17—20 (80) |
| *Penicillium restrictum* | Cl: | 17—20 (80) |
| *Penicillium restriculosum* | Cl: | 17—20 (80) |
| *Penicillium rugulosum* | Hxl: | 15β-OH (80) |
| *Penicillium sclerotiorum* | Cl: | 17—20 (80) |
| *Penicillium simplicissimum* | Cl: | 17—20 (80) |
| *Penicillium soppi* | Cl: | 17—20 (80) |
| *Penicillium species* | Hxl: | 15α-OH (574, 481*a*, 115); 1α-OH (105, 109, 481*a*); 2β-OH (105, 109); 6β-OH (481*a*); 12β-OH, 15β-OH (481*a*) |
| *Penicillium spinulosum* | Cl: | 17—20 (80) |
| *Penicillium steckii* | Cl: | 17—20 (72, 80) |
| *Penicillium subalteritium* | Cl: | 17—20 (80) |
| *Penicillium tardum* | Hxl: | 11α-OH (115); 15β-OH (80) |
| *Penicillium terlikowski* | Cl: | 17—20 (80) |
| *Penicillium thomii* | Cl: | 17—20 (80) |
| | Hxl: | 19-OH (418) |
| *Penicillium urticae* | Hxl: | 15-OH (684, 683) |
| *Pestalotia foedans* | Hxl: | 11α-OH (568) |
| *Pestalotia funera* | Hxl: | 16α-OH (843) |
| *Pestalotia royenae* | Hxl: | 11α-OH (568) |
| *Pestalotia species* | Hxl: | 11α-OH (759) |
| *Peziza species* | Hxl: | 7α-OH (299, 909) |
| *Phoma species* | Hxl: | 11β-OH (981) |
| *Phycomyces blakesleanus* | Hxl: | 6β-OH (570); 7α-OH (574); 15β-OH, 7β-OH (574) |
| | Ep: | 14α, 15α (815) |
| *Phycomyces blakesleanus — theobromatus* | Hxl: | 6β-OH, 11α-OH (531) |
| *Phycomyces nitens* | Hxl: | 6β-OH (129) |
| *Pleospora gaeumanni* | Hxl: | 14α-OH (738, 736, 734, 898, 909) |
| *Polyporus cinnabarinus* | Hxl: | 6β-OH, 14α-OH, 15β-OH (619) |
| *Polyporus tulipiferus* | Hxl: | 6β-OH, 14α-OH, 15β-OH (619) |
| *Poria cocos* | Hxl: | 6β-OH, 14α-OH, 15β-OH (619) |
| *Protaminobacter alboflavum* | D: | 1—2 (695) (579) |
| *Protaminobacter rubrum* | D: | 1—2 (695) |
| *Proactinomyces erythropolis* | O: | 3β-OH → 3-CO (477, 478) |
| *Proactinomyces roseus* | Hxl: | 7α-OH (231, 232) |
| | O: | 3β-OH → 3-CO (231, 232) |
| *Proactinomyces species* | O: | 3β-OH → 3-CO (478) (479) |
| | Cl: | 17—20 (479) |
| *Protoabsidia blakesleana* | Hxl: | 11α-OH (130) |
| *Pseudodiphteria bacilli* | O: | 17β-OH → 17-CO (526); 3β-OH → 3-CO (526) |
| *Pseudomonas boreopolis* | Hxl: | 11β-OH (744) |

| | | |
|---|---|---|
| *Pseudomonas dacunhae* | D: | 1—2 (393) |
| *Pseudomonas fluorescens* | R: | 20-CO → 20β-OH (744) |
| *Pseudomonas marcescens* | D: | 1—2 (744) |
| *Pseudomonas olevorans* | D: | 1—2 (744) |
| | R: | 20-CO → 20β-OH (744) |
| *Pseudomonas species* | Hxl: | 11β-OH (307, 482, 447) |
| | R: | 20-CO → 20β-OH (307, 484, 483, 482); |
| | D: | 1—2 (484, 483, 447) |
| | O: | 3β-OH → 3-CO (448, 927); 17β-OH → → 17-CO (448, 927) |
| *Pseudomonas testosteroni* | D: | 1—2 (247) (248, 249) |
| | O: | 3α-OH → 3-CO (278); 3β-OH → 3-CO (278) |
| | Ar: | (417) |
| | Sp: | B (417) |
| *Pycnodothis species* | D: | 1—2 (26, 674) |
| | O: | 3β-OH → 3-CO (26) |
| *Pycnosporium species* | Hxl: | 11β-OH (26, 679) |
| *Rhizoctonia ferrugena* | Hxl: | 1-OH (151, 258); 2β-OH (151) |
| *Rhizoctonia solani* | Hxl: | 1-OH, 6β-OH (356) |
| *Rhizoctonia species* | Hxl: | 6β-OH (356, 121); 1-OH, 2β-OH (356) |
| *Rhizopus arrhizus* | Hxl: | 11α-OH (645, 540, 559, 298, 121, 531, 332, 331, 296, 644, 648, 671, 562, 545); 6β-OH (645, 544, 298, 121, 531, 332, 331, 296, 169, 129, 642, 644, 671, 563, 545, 909, 570); 7α-OH (559, 648); 7β-OH (204, 319); 22-OH (298) |
| *Rhizopus cambodjae* | Hxl: | 6β-OH, 11α-OH (65, 129) |
| *Rhizopus cohni* | Hxl: | 11α-OH, 6β-OH (531, 129); 11β-OH (129) |
| | O: | 11-OH → 11-CO (129) |
| *Rhizopus delemar* | Hxl: | 6β-OH (570, 531); 11α-OH (531) |
| *Rhizopus chinensis* | Hxl: | 6β-OH, 11α-OH (531, 129) |
| *Rhizopus japonicus* | Hxl: | 11α-OH, 6β-OH (531) |
| *Rhizopus kazanensis* | Hxl: | 6β-OH, 11α-OH (531) |
| *Rhizopus nidulans* | Hxl: | 6β-OH, 11α-OH (8) |
| *Rhizopus nigricans* | Hxl: | 11α-OH (541, 531, 536, 540, 537, 542, 121, 119, 120, 551, 159, 74, 331, 332, 333, 335, 102, 295, 296, 298, 463, 250, 681, 213, 51, 77, 76, 345, 767, 57, 241, 129, 743, 728, 723*a*, 643, 648, 656, 658, 665, 667, 668, 565, 558, 553, 552, 541, 50, 909); 6β-OH (542, 121, 331, 333, 332, 335, 295, 298, 296, 129, 578, 557, 550, 544, 570); 10-OH (327); 11β-OH (129, 57); 7β-OH (213, 241) |
| *Rhizopus oryzae* | Hxl: | 11α-OH, 6β-OH (531) |
| *Rhizopus pygmacus* | Hxl: | 6β-OH (129) |
| *Rhizopus reflexus* | Hxl: | 11α-OH (531, 549, 121); 6β-OH (531, 121, 571, 556, 570) |
| *Rhizopus shanghaiensis* | Hxl: | 6β-OH, 11α-OH (531) |
| *Rhizopus species* | Hxl: | 6β-OH (121); 7α-OH, 7β-OH (108) |
| *Rhizopus suinus* | Cl: | 17—20 (812) |

| | | |
|---|---|---|
| *Rhizopus tonkinensis* | Hxl: | 6β-OH (129, 531); 11α-OH (531) |
| *Rhizopus tritici* | Hxl: | 6β-OH, 11α-OH (531) |
| *Rhodoseptoria species* | Hxl: | 11β-OH (685, 26) |
| *Rhodotorula glutinis* | R: | 20-CO → 20α-OH (444) |
| *Rhodotorula longissima* | R: | 20-CO → 20β-OH (85) |
| *Saccharomyces cerevisiae* | R: | 17-CO → 17β-OH (497, 511, 270, 75, 266, 265, 264, 268, 263, 489, 344, 61*a*, 528, 83, 699, 901); 3-CO → 3β-OH (63, 676, 877); 3-CO → 3α-OH (128) |
| | Hdg: | 1—2 (61*a*) |
| *Saccharomyces fragilis* | R: | 17-CO → 17β-OH (260) |
| | E: | 17β-OH (260) |
| *Saccharomyces pastorianus* | Hxl: | 11β-OH (533) |
| *Sansevieria zeylanica* | D: | 1—2 (613) |
| | Hdg: | 1—2 (613); 4—5 (613) |
| *Scenedesmus species* | Hxl: | 11α-OH, 6β-OH, 15β-OH (253) |
| *Sclerotinia fructicola* | Hxl: | 21-OH (580) |
| *Sclerotinia libertiana* | Hxl: | 2β-OH (394, 457, 398, 391, 393); 15β-OH (394, 398, 393); 11α-OH, 11β-OH (393) |
| *Sclerotinia species* | Hxl: | 2β-OH (390) |
| *Sclerotium hydrophilum* | Hxl: | 11α-OH, 6β-OH (402, 393); 15β-OH (402) |
| *Septomyxa aesculi* | D: | 1—2 (837) |
| *Septomyxa affinis* | D: | 1—2 (386) (742*a*, 337, 757, 756, 755, 754, 750, 749, 711, 618, 96, 216, 837) |
| | Ar: | (740) |
| | Cl: | 17—20 (711) |
| *Septomyxa corni* | D: | 1—2 (837) |
| *Septomyxa salicina* | D: | 1—2 (837) |
| *Septomyxa tulasnei* | D: | 1—2 (837) |
| *Serratia plymuthicum* | D: | 1—2 (744) |
| *Spicaria simplicissima* | Hxl: | 15β-OH (52) |
| *Spicaria species* | Hxl: | 15β-OH (53) |
| *Spondylocladium species* | Hxl: | 11β-OH (809) |
| *Sporodinia minima* | Hxl: | 17α-OH (593) |
| *Sporormia minima* | Hxl: | 17α-OH (731, 847) |
| *Sporotrichum bombycinum* | Hxl: | 11α-OH (945) |
| *Sporotrichum epigacum* | Hxl: | 11α-OH (945) |
| | D: | 1—2 (618) |
| *Sporotrichum gougeroti* | R: | 20-CO → 20β-OH (444) |
| *Sporotrichum sulphurescens* | Hxl: | 11α-OH (745, 811) |
| *Stachylidium bicolor* | Hxl: | 14α-OH (403); 11α-OH (597); 11β-OH (403, 393, 597) |
| *Stachylidium theobromae* | Hxl: | 11β-OH, 11α-OH (715, 597) |
| *Stereum fasciatum* | D: | 1—2 (443) |
| | R: | 20-CO → 20β-OH (443) |
| *Streptomyces achromogenes* | Hxl: | 16α-OH (501) |
| *Streptomyces albus* | O | 17β-OH → 17-CO (510) |
| | R: | 20-CO → 20β-OH (228); 17-CO → 17β-OH (927) |
| *Streptomyces annulatus* | Hxl: | 16α-OH (501) |

| | | |
|---|---|---|
| *Streptomyces antibioticus* | D: | 1—2 (606) |
| *Streptomyces aureofaciens* | Hxl: | 6β-OH, 9α-OH (8) |
| | O: | 3β-OH → 3-CO (329, 601) |
| *Streptomyces aureus* | R: | 3-CO → 3β-OH (229) |
| | Hxl: | 16α-OH (500, 501); 15α-OH (574) |
| *Streptomyces bobiliae* | D: | 1—2 (606) |
| *Streptomyces coelicolor* | R: | 20-CO → 20β-OH (901) |
| *Streptomyces diastaticus* | R: | 20-CO → 20β-OH (228) |
| | Hxl: | 16α-OH (501) |
| *Streptomyces diastochromogenes* | Hxl: | 6β-OH (501) |
| *Streptomyces fimicarius* | Hxl: | 16α-OH (500, 501) |
| *Streptomyces flaveolus* | Hxl: | 16α-OH (500) |
| *Streptomyces flavogriseus* | R: | 20-CO → 20β-OH (228) |
| *Streptomyces fradiae* | Hxl: | 11β-OH (97, 98); 11α-OH (502); 6β-OH (501, 502) |
| | O | 3β-OH → 3-CO (329, 601, 927) |
| *Streptomyces gelaticus* | D: | 7—8 (194) |
| | O: | 3α-OH → 3-CO (176); 12α-OH → 12-CO (176) |
| *Streptomyces griseocarneus* | Hxl: | 6β-OH (501) |
| *Streptomyces griseoluteus* | Hxl: | 16α-OH (501) |
| *Streptomyces griseus* | Hxl: | 16α-OH (500, 501) |
| | O: | 3β-OH → 3-CO (329, 601) |
| | R: | 20-CO → 20β-OH (85); 3-CO → 3β-OH (974) (499) |
| | Hdg: | 4—5 (974) |
| *Streptomyces halstedii* | Hxl: | 16α-OH (214) |
| *Streptomyces hydrogenans* | R: | 20-CO → 20β-OH (251, 195, 197, 196, 308) |
| *Streptomyces chrysomallus* | Hxl: | 16α-OH (500, 501) |
| *Streptomyces lavendulae* | D: | 1—2 (8, 142, 336, 618, 588, 575) |
| | Cl: | 17—20 (8, 142, 336, 575) |
| | R: | 20-CO → 20β-OH (901) |
| *Streptomyces lipmanii* | Hxl: | 16α-OH (500, 501) |
| *Streptomyces mediocidicus* | Hxl: | 16α-OH (214) |
| *Streptomyces microflavus* | Hxl: | 16α-OH (500, 501) |
| *Streptomyces olivaceus* | Hxl: | 16α-OH (500, 501) |
| | D: | 1—2 (606) |
| *Streptomyces olivochromogenes* | D: | 1—2 (606) |
| *Streptomyces parvus* | Hxl: | 16α-OH (500, 501) |
| *Streptomyces phaeochromogenes* | Hxl: | 16α-OH (501) |
| *Streptomyces purpeochromogenses* | Hxl: | 16α-OH (500, 501) |
| *Streptomyces rimosus* | Hxl: | 6β-OH (501) |
| | O: | 3β-OH → 3-CO (329, 601) |
| *Streptomyces roseochromogenes* | Hxl: | 16α-OH (8, 461, 148, 149, 500, 501, 763, 761, 760, 855, 843, 833, 816); 2β-OH (149, 421) |
| | R: | 20-CO → 20β-OH (422) |
| *Streptomyces rutgersensis* | Hxl: | 16α-OH (501) |
| *Streptomyces species* | Hxl: | 2β-OH (186); 1-OH (258); 16α-OH (494, 330, 763, 764, 760, 761, 843, 833); 11β-OH (533); 11α-OH (395, 393); 6β-OH (395, 393) |

| | | |
|---|---|---|
| | O: | 3β-OH → 3-CO (329) |
| | R: | 3-CO → 3β-OH (616, 849) |
| | Hdg: | 4—5 (152) |
| *Streptomyces sulphureus* | Hxl: | 16α-OH (501) |
| *Streptomyces verne* | Hxl: | 16α-OH (501) |
| *Streptomyces viridis* | Hxl: | 16α-OH (500) |
| *Syncephalastrum cinereum* | Hxl: | 11α-OH (129) |
| *Syncephalastrum racemosum* | Hxl: | 7α-OH, 15β-OH, 14α-OH (468), 11α-OH, 11β-OH, 6β-OH (129) |
| *Syncephalas nodosa* | Hxl: | 11α-OH, 6β-OH (531) |
| *Thamnidium elegans* | Hxl: | 11α-OH, 11β-OH (129) |
| | O: | 11-OH → 11-CO (129) |
| *Thielavia terricola* | Hxl: | 17α-OH (895) |
| *Tieghemella cylindrospora* | Hxl: | 11α-OH, 11β-OH (130) |
| *Tieghemella hyalospora* | Hxl: | 11α-OH (426, 130); 11β-OH (130) |
| *Thieghemella species* | Hxl: | 11α-OH (1002, 1006) |
| *Tieghemella spinosa* | Hxl: | 11α-OH, 11β-OH (130) |
| *Trichoderma albanum* | Hxl: | 17α-OH (673) |
| *Trichoderma glaucum* | Hxl: | 17α-OH (673) |
| | E: | 21 (192) |
| *Trichoderma konigi* | Hxl: | 17α-OH (673) |
| *Trichoderma lignorum* | Hxl: | 17α-OH (673) |
| *Trichoderma narcisi* | Hxl: | 17α-OH (673) |
| *Trichoderma nigrovirens* | Hxl: | 17α-OH (673) |
| *Trichoderma nunbergii* | Hxl: | 17α-OH (673) |
| *Trichoderma species* | Hxl: | 17αOH (673) |
| *Trichoderma viridae* | Hxl: | 17α-OH (276); 11α-OH (515) |
| *Trichomonas gallinae* | R: | (384) |
| *Trichothecium arrhenopum* | Hxl: | 17α-OH (908) |
| *Trichothecium candidum* | Hxl: | 17α-OH (908) |
| *Trichothecium plasmoparae* | Hxl: | 17α-OH (908) |
| *Trichothecium polybrochum* | Hxl: | 17α-OH (908) |
| *Trichothecium roseum* | Hxl: | 11β-OH (577); 11α-OH (297, 463*a*, 669); 17α-OH (297, 299, 703, 908, 909, 602); 6β-OH (731, 464, 463*a*); 14α-OH (731) |
| | O: | 11-OH → 11-CO (299) |
| *Wojnowicia graminis* | Hxl: | 21-OH (255, 770, 853, 834, 909, 677); 16α-OH, 6β-OH (187*b*) |
| | O: | 16-OH → 16-CO (187*b*) |
| *Xylaria species* | Hxl: | 1β-OH, 15β-OH, 7β-OH, 6β-OH (113) |
| *Zygorhynchus heterogamus* | Hxl: | 11α-OH, 6β-OH (531); 11β-OH, 6β-OH (129) |
| | O: | 11-OH → 11-CO (129) |
| *Zygorhynchus moelleri* | Hxl: | 11α-OH, 6β-OH (531, 129); 11β-OH (129) |
| *Zygorhynchus Vuillemini* | Hxl: | 6β-OH (129) |

# 11

# LIST OF STEROID METABOLITES

This chapter summarises the most important metabolites obtained by microbial transformations of the individual steroids. The principal steroids (often under *generic names*) are arranged in *alphabetical order*, with the *corresponding derivatives given under the parent compound*. The type of transformation indicates the reaction(s) giving rise to the metabolite. Here the ***abbreviations*** have been used already explained on *page 129*. When the type of transformation (column 2.) is too complicated, only the starting steroid has been given. The reference numbers correspond to the literature data compiled in chapter 12.

Rotation in solvent is indicated (An, acetone: C, chloroform: Di, dioxane: E, ethanol: M, methanol: Py, pyridine).

| *Metabolite* | Type of transformation (*see p. 129*) | m. p. °C | $[\alpha]_D$ | Reference (*see p. 181*) |
|---|---|---|---|---|
| 21-Acetamido-9α-fluoro-$\Delta^{1,4}$-pregnadiene-11β, 17α-diol-3,20-dione | D: 1—2 | 252—255 | +106,9 (M) | 1193 |
| 21-Acetamido-9α-fluoro-$\Delta^{4}$-pregnene-11β, 17α-diol-3,20-dione | (9α-Fluorocortisol) | 263—264 | +132 (M) | 1193 |
| Aldosterone | Hxl: 21-OH | 164—169 | +161 (C) | 497 |
| 1-dehydro- | D: 1—2 | 182—185 | | 497 |
| 14α-hydroxy- | Hxl: 14α-OH | | | 736, 898 |
| $\Delta^{1,4}$-Androstadiene-3,17-dione | D: 1—2 | 140—141 | +115 (C) | 83, 247, 248, 433, 579, 598*b*, 674, 739 |

| *Metabolite* | Type of transformation (*see p. 129*) | m. p. °C | $[\alpha]_D$ | Reference (*see p. 181*) |
|---|---|---|---|---|
| $\Delta^{1,4}$-Androstadiene-3,17-dione | D: 1—2, 4—5 | | | 248, 708 |
| | (Testosterone) | | | 248 |
| | (Dehydroepiandrosterone) | | | 203, 729 |
| | ($\Delta^{5}$-Androstene-3β-, 17β-diol) | | | 93 |
| | ($\Delta^{16}$-5α-Pregnene-3,20-dione) | | | 825, 977 |
| | ($\Delta^{5,16}$-Pregnadiene-3-ol-20-one) | | | 825, 977 |
| | (Progesterone) | | | 142, 311, 336, 439, 493, 575, 598*e*, 674 |
| | ($\Delta^{5}$-Pregnene-3β-ol-20-one) | | | 493 |
| | (5α-Pregnane-3,20-dione) | | | 493 |
| 2α-hydroxy- | D: 1—2 | 148—150 | + 67 (C) | 153, 184 |
| 11α-hydroxy- | Hxl: 11α-OH | 212—214 | + 86,5 (C) | 598*d*, 945 |
| 11β-hydroxy- | D: 1—2 | 176—179 | | 93, 624*f* |
| 12β-hydroxy- | D: 1—2 | 164—166 | | 341 |
| 14α-hydroxy- | (Dehydroepiandrosterone) | | | 203 |
| 5α-Androstane-3β, 17α-diol | (5α-Androstane-3,17-dione) | 214 | | 273 |
| 5α-Androstane-3β, 17β-diol | (5α-Androstane-3,17-dione) | | | 273, 489 |
| 5β-Androstane-3α, 17β-diol | (Testosterone) | 229—230 | + 24 (E) | 273, 382 |
| | ($\Delta^{4}$-Androstene-3,17-dione) | | | 273 |
| 5α-Androstane-1α, 3β-diol-17-one | ($\Delta^{4}$-Androstene-3,17-dione) | 202—203,5 | + 90 (C) | 105, 109 |
| 5β-Androstane-3α, 11α-diol-17-one | (5β-Pregnane-3α, 11α, 17α-triol-20-one) | | | 576 |
| 5β-Androstane-11α, 17α-diol-3-one | Hxl: 11α-OH | | | 598*d* |
| 5β-Androstane-11α, 17β-diol-3-one | Hxl: 11α-OH | | | 598*d* |

| *Metabolite* | Type of transformation (*see p. 129*) | m. p. °C | $[\alpha]_D$ | Reference (*see p. 181*) |
|---|---|---|---|---|
| 5β-Androstane-3,17-dione | Hdg: 1—2<br>Hdg: 1—2, 4—5 | 131—132 | +113 (E) | 273<br>381 |
| 5α-Androstan-1α-ol-3,17-dione | ($\Delta^4$-Androstene-3,17-dione) | 204—206 | +114 (C) | 109 |
| 5α-Androstan-11α-ol-2,17-dione | Hxl: 11α-OH | | | 598*d* |
| 5β-Androstan-11α-ol-3,17-dione | Hxl: 11α-OH | | | 598*d* |
| 5α-Androstan-17β-ol-3,6-dione | Hdg: 4—5 +<br>O: 6-OH → 6-CO | 233—234 | — 9 (E) | 121, 554 |
| 5α-Androstan-3α-ol-17-one | R: 3-CO → 3α-OH | 184—185 | + 87,8 (Di) | 270, 449*a* |
| 5α-Androstan-3β-ol-17-one | R: 3-CO → 3β-OH | 177—178 | + 82 (M) | 270, 273 |
| 5β-Androstan-3α-ol-17-one | ($\Delta^{1,4}$-Androstadiene-3,17-dione) | 143—144 | +111 (E) | 381 |
| 5α-Androstan-11α-ol-3-one | Hxl: 11α-OH | | | 598*d* |
| 5β-Androstan-17β-ol-3-one | (Testosterone) | 140 | + 33 (E) | 273 |
| | ($\Delta^4$-Androstene-3,17-dione) | | | 273 |
| 5α-Androstane-3β, 11α, 17-triol | Hxl: 11α-OH | | | 598*d* |
| 5β-Androstane-3α, 11α, 17-triol | Hxl: 11α-OH | | | 598*d* |
| 5β-Androstane-3β, 11α, 17-triol | Hxl: 11α-OH | | | 598*d* |
| 5α-Androstane-3,6,17-trione | (Dehydroepiandrosterone) | 191,5—202 | + 71 (C) | 203 |
| 5β-Androstane-3,6,17-trione | (5β-Pregnane-3,6,20-trione) | | | 576 |
| 5α-Androstane-3,11,17-trione | (5α-Pregnane-3,11,20-trione) | 180—181 | +155 (C) | 576 |
| $\Delta^{1,4,6}$-Androstatriene-3,17-dione | D: 1—2 | 164,5—166 | + 82,6 (C) | 591*d* |
| $\Delta^5$-Androstene-3β, 17β-diol | R: 17-CO → 17β-OH | 184 | — 63 (C) | 263, 269, 273, 385 |
| $\Delta^5$-Androstene-1α, 3β-diol-17-one | Hxl: 1α-OH | 275—277,5 | + 10,6 (C) | 105 |
| $\Delta^5$-Androstene-3β, 7α-diol-17-one | Hxl: 7α-OH | 181,5—183,5 | — 70,7 (C) | 108 |
| $\Delta^5$-Androstene-3β, 7β-diol-17-one | Hxl: 7β-OH | 214—216 | + 67,5 (C) | 108 |

| *Metabolite* | Type of transformation (*see p. 129*) | m. p. °C | $[\alpha]_D$ | Reference (*see* p. *181*) |
|---|---|---|---|---|
| $\Delta^5$-Androstene-3β,11α-diol-17-one | Hxl: 11α-OH | | | 962 |
| $\Delta^1$-5α-Androstene-3,17-dione | D: 1—2 | 140—142 | +128 (C) | 184 |
| $\Delta^1$-5β-Androstene-3,17-dione | Hdg: 4—5 | 168—169 | +222 (E) | 381 |
| $\Delta^4$-Androstene-3,17-dione | O: 17β-OH → 17-CO | 175—176,5 | +194 (C) | 385, 389*a*, 448, 478 |
| | O: 3β-OH → → 3-CO + I | | | 41, 93, 161, 203, 271, 302, 478, 528, 729, 589*a* |
| | O: 3β,17β-diOH → 3,17-diCO + I | | | 42, 93, 125, 126 478, 490, 528, 589*a* |
| | (Progesterone | | | 334, 576 |
| | (Cortexone) | | | 576 |
| 1β,6β-dihydroxy- | Hxl: 1β,6β-diOH | 237—239 | + 68 (C) | 113 |
| 6β-,11β-dihydroxy- | Hxl: 6β-OH | 275—284 | +132(C-M) | 485 |
| 11β,15α-dihydroxy- | Hxl: 15α-OH | 221—222 | +222 (C) | 485 |
| | Hxl: 11β-OH | | | 485 |
| 1α-hydroxy- | Hxl: 1α-OH | 218—221 | +184 (C) | 105, 109, 591*a* |
| 1β-hydroxy | Hxl: 1β-OH | 155—156,5 | +126 (C) | 113 |
| 2β-hydroxy- | Hxl: 2β-OH | 143—145 | + 36,8 (C) | 105, 109, 591*a* |
| 6β-hydroxy- | Hxl: 6β-OH | 191—194 | +107 (C) | 113, 121, 485 |
| | (Testosterone) | | | 187*b* |
| | (Dehydropiandrosterone) | | | 203 |
| 7α-hydroxy- | Hxl: 7α-OH | 255—256,5 | +164 (C) | 481*a*, 612 |
| 7β-hydroxy- | Hxl: 7β-OH | 221—225 | +178 (C) | 110, 113, 481*a* |
| 9α-hydroxy- | Hxl: 9α-OH | 223,5—224,5 | +182,8 (C) | 107, 111, 747 |

| *Metabolite* | Type of transformation (*see p. 129*) | m. p. °C | $[\alpha]_D$ | Reference (*see p. 181*) |
|---|---|---|---|---|
| $\Delta^4$-Androstene-3,17-dione | | | | |
| 11α-hydroxy- | Hxl: 11α-OH | 226—227 | +162 (C) | 121, 537 |
| | ($\Delta^4$-Pregnene-11α-ol-3,20-dione) | | | 72 |
| | ($\Delta^4$-Pregnene-11α,17α-diol-3,20-dione) | | | 576 |
| | (11-Epicortisol) | | | 576 |
| 11β-hydroxy- | O: 17β-OH → 17-CO | 200—201 | +208 (C) | 385, 589*a* |
| 12β-hydroxy- | Hxl: 12β-OH | | | 341 |
| 14α-hydroxy- | (Dehydroepiandrosterone) | 262—265 | +171 (C) | 203, 576 |
| 15α-hydroxy- | Hxl: 15α-OH | 190—198 | +217 (C) | 154, 456, 485 |
| 15β-hydroxy- | Hxl: 15β-OH | 192—197 | +136 (C) | 113 |
| 16α-hydroxy | Hxl: 16α-OH | 185—187 | +194 (C) | 8 |
| 19-hydroxy- | Hxl: 19-OH | 165—168 | +182 (C) | 615 |
| $\Delta^5$-Androsten-3β-ol-7,17-dione | (Dehydroepiandrosterone) | 243—244,5 | — 82,8 (C) | 108 |
| $\Delta^1$-5α-Androsten-17β-ol-3-one | R: 17-CO → 17β-OH | 158—159 | — 42,3 (E) | 61*b* |
| $\Delta^5$-Androstene-3β,11α,17β-triol | Hxl: 11α-OH | | | 598*d* |
| $\Delta^1$-5α-Androstene-3,11,17-trione | D: 1—2 | 244,5—246,5 | +188 (C) | 291 |
| $\Delta^1$-5β-Androstene-3,11,17-trione | D: 1—2 | 174,5—177,5 | +205 (C) | 291 |
| Adrenosterone | O: 17β-OH → 17-CO | 220—223 | +284 (C) | 385 |
| | (11-Ketoprogesterone) | | | 72, 576 |
| | (Cortisone) | | | 576 |
| Bufalin | | | | |
| 12β-hydroxy- | Hxl: 12β-OH | 240—246 | — 16 (M) | 455 |
| Cholanic acid | | | | |
| 3α,12α-dihydroxy-7-keto- | (Cholic acid) | | | 359 |
| 3α-hydroxy-7,12-diketo- | (Cholic acid) | | | 359 |
| 7ξ-hydroxy-3,12-diketo- | (Cholic acid) | | | 143 |
| 3,7,12-triketo- | (Cholic acid) | 236—237 | | 359, 479 |
| $\Delta^{1,4}$-Cholestadien-3-one | D: 1—2 | 112 | + 28 (C) | 636 |
| $\Delta^{5,7}$-Cholestadien-3β-ol | D: 7—8 | | | 194 |
| $\Delta^5$-Cholestene-3β,7β-diol | Hxl: 7β-OH | 178 | + 7 (C) | 231 |

| *Metabolite* | Type of transformation (*see p. 129*) | m. p. °C | $[\alpha]_D$ | Reference (*see p. 181*) |
|---|---|---|---|---|
| $\Delta^4$-Cholesten-3-one | O: 3β-OH → → 3-CO + I | 81—82 | + 89 (C) | 39, 194, 231, 232, 415, 430, 478, 479, 528, |
| Coprostan-3-one | O: 3β-OH → 3-CO | 61 | + 36 (C) | 478 |
| Cortexolone | Hxl: 17α-OH | 210—212 | +124 (Di) | 299*a*, 593, 847 |
| | Hxl: 21-OH | | | 299*a*, 677, 723, 591*e* |
| | ($\Delta^5$-Pregnene-3β,17α, 21-triol-20-one 21-acetate or 3,21-diacetate) | | | 93, 694, 810*b*, 953 |
| 1-dehydro | D: 1—2 | 246—250 | + 76 (C) | 187*a*, 307, 313, 404, 433, 483, 495, 591*d*, 598*b*, 634, 810*b* |
| | D: 1—2, 4—5 | | | 434 |
| | ($\Delta^5$-Pregnene-3β,17α, 21-triol-20-one 3,21-diacetate) | | | 93, 810*b* 951 |
| 6β-hydroxy- | D: 1—2 | 229—231 | | 988 |
| 7α-hydroxy- | Hxl: 7α-OH | | | 612 |
| 11α-hydroxy- | Hxl: 11α-OH | 246—247 | + 73 (M) | 460*a*, 735, 958 |
| 12β-hydroxy- | D: 1—2 | 232—234 | | 341, 988 |
| 14α-hydroxy- | D: 1—2 | 230—231 | +132 (C) | 598, 988 |
| 16α-hydroxy- | D: 1—2 | | | 634 |
| 16α-methyl- | D: 1—2 | 209—212 | + 46 (Di) | 988 |
| 6β,14α-dihydroxy- | Hxl: 6β,14α-diOH | | | 227 |
| 7α,14α-dihydroxy- | Hxl: 7α,14α-diOH | 238—240 | + 47,7 (C) | 227, 582 |
| 1ξ-hydroxy- | Hxl: 1ξ-OH | 193—207 | + 89 (Di) | 151 |
| 2β-hydroxy- | Hxl: 2β-OH | 225,5—228 | — 58 (Di) | 151, 186, 457 |

| *Metabolite* | Type of transformation (*see p. 129*) | m. p. °C | $[\alpha]_D$ | Reference (*see p. 181*) |
|---|---|---|---|---|
| Cortexolone | | | | |
| 6β-hydroxy- | Hxl: 6β-OH | 234—236 | + 53 (E) | 202, 254, 332, 395, 402, 405, 435, 531, 544, 638, 642, 637*a*, 968 |
| | Hxl: 17α-OH | | | 297 |
| 7α-hydroxy- | Hxl: 7α-OH | 209—211 | + 94 (M) | 612, 854, 970 |
| 7β-hydroxy- | Hxl: 7β-OH | 246—248 (7,21-diacetate) | +107 (C) | 46*a*, 481*a* |
| 8β-hydroxy- | Hxl: 8β-OH | 225—227 (248—252 | +115 (?) +107 (C)) | 122, 531, 640, 654, 660 |
| 9α-hydroxy- | Hxl: 9α-OH | | | 227, 747 |
| 12β-hydroxy- | Hxl: 12β-OH | 179—181 | | 341, 771 819 |
| 14α-hydroxy | Hxl: 14α-OH | 234—237 | +155 (M) | 122, 227, 399, 406, 531, 546, 646 |
| 15α-hydroxy- | Hxl: 15α-OH | 227—230 | +146 (M) | 52, 53, 373, 405, 469, 997 |
| 15β-hydroxy- | Hxl: 15β-OH | 253—255 | +103 (E) | 52, 53, 373, 141, 187, 625*a*, 638, 675 |
| 19-hydroxy- | Hxl: 19-OH | 235—241 | +119 (Di) | 156, 312, 615, 635*a* |
| 16α-methyl-15α-hydroxy- | Hxl: 15α-OH | 224—226 | +133 (Di) | 70 |
| 16-methylene- | (16-Methylene-$\Delta^5$-pregnene-3β,17α,21-triol-20-one 3,21-diacetate) | 205—206 | + 46,6 (C) | 985*a* |

| *Metabolite* | Type of transformation (*see p. 129*) | m. p. °C | $[\alpha]_D$ | Reference (*see p. 181*) |
|---|---|---|---|---|
| Cortexone | Hxl: 21-OH | 142—143 | +178 (E) | 255, 299*a*, 522, 529, 580, 591*e*, 617, 670, 677, 723, 921 |
| | O: 3β-OH → 3-CO + + I | | | 275 |
| 1-dehydro- | D: 1—2 | 189—195 | +120 (C) | 313, 404, 433, 495, 634 |
| | Hxl: 21-OH | | | 580 |
| 16-α-hydroxy- | D: 1—2 | | | 634 |
| 16-dehydro- | Hxl: 21-OH | 152—154 | +146 (C) | 617 |
| 2β,15β-dihydroxy- | Hxl: 2β,15β-diOH | 210—219 | − 47 (M) | 398 |
| 7α,14α-dihydroxy- | Hxl: 7α,14α-diOH | 245—248 | +146 (Di) | 69 |
| 14α,15α-dihydroxy- | Hxl: 15α-OH | 208—222 | +172 (C) | 456 |
| 6β-hydroxy- | Hxl: 6β-OH | 198—202 | + 97 (C) | 299*a*, 531 |
| | Hxl: 6β-OH + S: 21-OAc | | | 119, 402 |
| | Hxl: 21-OH | | | 191*e* |
| 7α-hydroxy- | Hxl: 7α-OH | 216—225 | +144 (C) | 259, 299, 300, 854, 970 |
| 7β-hydroxy- | Hxl: 7β-OH | 178—181,5 | +151 (C) | 259 |
| 8β(9α?)-hydroxy- | Hxl: 8β(9α?)-OH | 182,5—184 | +167 (C) | 122, 432, 747, 1001 |
| 14α-hydroxy- | Hxl: 14α-OH | 175—176 | +190 (C) | 122, 276, 403, 456, 569, 570, 1000 |
| 15α-hydroxy- | Hxl: 15α-OH | 215—220 | +217 (C) | 25, 68, 154, 299, 405, 456, 595 |
| 15β-hydroxy- | Hxl: 15β-OH | 216—222 | +212 (C) | 299, 400, 683, 998 |
| 16α-hydroxy- | Hxl: 16α-OH | 203—205 | +114,5 (E) | 494, 623 |
| 19-hydroxy- | Hxl: 19-OH | 165—168 | +180 (C+E) | 615 |

| *Metabolite* | Type of transformation (*see p. 129*) | m. p. °C | $[\alpha]_D$ | Reference (*see p. 181*) |
|---|---|---|---|---|
| Corticosterone | Hxl: 21-OH | 177—179 | +220 (E) | 522, 591*e*, 670, 677, 723 |
| | Hxl: 11β-OH | | | 276, 403, 406, 615 |
| 1-dehydro- | D: 1—2 | 227,5—230,5 | +173 (M) | 187*a*, 203, 313, 404, 433, 495, 700 |
| 9α-chloro- | D: 1—2 | | | 951 |
| 6α-fluoro- | D: 1—2 | | | 758*a* |
| 12α-fluoro- | D: 1—2 | | | 730 |
| 17α-methyl- | D: 1—2 | | | 691 |
| 6β,15α-dihydroxy- | Hxl: 6β,15α-diOH | | | 405 |
| 2β-hydroxy- | Hxl: 2β-OH | 183—186 | — 7 (M) | 394, 398 |
| 14α-hydroxy- | Hxl: 11β-OH | 206—215 | +180 (M) | 396, 399 |
| 15β-hydroxy- | Hxl: 15β-OH | 230—235 | +180 (Py) | 394, 398, 400, |
| 16α-hydroxy- | Hxl: 16α-OH | | | 623 |
| 18-hydroxy- | Hxl: 11β-OH | 165—168 | | 942 |
| 19-hydroxy- | Hxl: 11β-OH | 163—164 | +210 (C) | 48 |
| 17α-methyl- | Hxl: 11β-OH | | | 691 |
| Cortisol | Hxl: 11β-OH | 218—220 | +165 (E) | 97, 98, 117, 168, 277, 325, 360, 403, 406, 535, 561, 585, 587, 597, 629, 638, 678, 679, 685, 686, 689, 746, 809, 820, 937*a*, 966,1012 |
| | Hxl: 17α-OH | | | 299*a*, 593, 731, 847 |
| | Hxl: 21-OH | | | 677, 723 |
| 6-dehydro- | Hxl: 11β-OH | 239—241 | +177 (Di) | 696 |
| 9α,16α-dihydroxy- | Hxl: 16α-OH | | | 623 |

| *Metabolite* | Type of transformation (*see p. 129*) | m. p. °C | $[\alpha]_D$ | Reference (*see p. 181*) |
|---|---|---|---|---|
| Cortisol | | | | |
| 2β-16α-dihydroxy-9α-fluoro- | Hxl: 2β,16α-diOH | 220–223 | – 26,5 (M) | 421, 149 |
| 6α,16α-dimethyl- | Hxl: 11β-OH | 154–158 (21-acetate) | +115 (C) | 362 |
| 9α-fluoro- | R: 21-al → 21-OH | 260–262 | +139 (E) | 422 |
| 6α-fluoro-16α-methyl- | Hxl: 11β-OH | 216,5–218,5 | | 487 |
| 7α-hydroxy- | Hxl: 7α-OH | | | 612 |
| 14α-hydroxy- | Hxl: 14α-OH | 241–242 | +182,6 (E) | 535, 537, 582 |
| 15β-hydroxy- | Hxl: 11β,15β-diOH | | | 625*a* |
| 19-hydroxy- | Hxl: 19-OH | | | 635*b* |
| 1-hydroxy-9α-fluoro- | Hxl: 1-OH | 247–250 | | 258, 621, |
| 2β-hydroxy-9α-fluoro- | Hxl: 2β-OH | 245–249 | + 1,5 (M) | 149, 421 |
| 6β-hydroxy-9α-fluoro- | Hxl: 6β-OH | | | 621 |
| 7α-hydroxy-9α-fluoro- | Hxl: 7α-OH | | | 612 |
| 14α-hydroxy-9α-fluoro- | Hxl: 14α-OH | | | 702 |
| 16α-hydroxy-9α-fluoro- | Hxl: 16α-OH | 257–260 | + 91 (Py) | 421, 461, 623,630*a* |
| 16α-hydroxy-6α-fluoro- | Hxl: 16α-OH | 234–236 | + 95 (Di) | 1134 |
| 16α-hydroxy-6α,9α-difluoro- | Hxl: 16α-OH | 247–251 | + 58 (Di) | 760, 976 |
| 16α-methyl- | Hxl: 11β-OH | 220–222 | +110 (Di) | 70, 638 |
| 16β-methyl- | Hxl: 11β-OH | | | 638 |
| 16α-methyl-14α-hydroxy- | Hxl: 11β,14α-diOH | 221–226 | +145 (Di) | 70 |
| 16-methylene- | Hxl: 11β-OH | 232–233 | + 69,2 (Di) | 985*a* |
| Cortisone | (Cortexolone) | 217–221 | +211 (M) | 97, 157, 165, 168, 406 |
| | (Corticosterone) | | | 297, 299*a* |
| | Hxl: 17α-OH | | | 297, 299*a* |
| | S: 21-OAc | | | 84 |
| 7α-hydroxy- | Hxl: 7α-OH | 278–283 (21-acetate) | +151,8 (Py) | 612 |
| 15β-hydroxy- | Hxl: 15β-OH | | | 625*a* |
| 3,5-Cycloandrostan-17β-ol-6-one | R: 17-CO → → 17β-OH | 182–183 | | 61*c* |
| 3,5-Cyclopregnane-6β,11α-diol-20-one | Hxl: 11α-OH | 211,5–217 | + 92 (C) | 238*a*, 505 |
| (20 S)-18,20-Cyclo-$\Delta^4$-pregnene-11α,20-diol-3-one | Hxl: 11α-OH | 205–208 | +113 (Di) | 506 |

| *Metabolite* | Type of transformation (*see p. 129*) | m. p. °C | $[\alpha]_D$ | Reference (*see p. 181*) |
|---|---|---|---|---|
| 7-Dehydrocholesterol | D: 7—8 | 143 | −114 (C) | 194 |
| 11-Dehydrocorticosterone | Hxl: 21-OH | 177—179,5 | +258 (E) | 299*a*, 522,591*e* |
| 6β-hydroxy- | (Corticosterone) | 195—200 | +146,5 (M) | 402 |
| 14α-hydroxy- | (Corticosterone) | 205—215 | + 69 (M) | 396, 399 |
| 15β-hydroxy- | (Corticosterone) | 195—200 | +190 (M) | 402 |
| Dexamethasone | D: 1—2<br>Hxl: 21-OH | 268—271 | | 487, 976<br>676*a* |
| 11-Desoxycorticosterone, see Cortexone | | | | |
| 16-Diazo-$\Delta^4$-androstene-3,17-dione | (16-Diazo-$\Delta^5$-androstene-3β-ol-17-one) | 217—218 | −112 (C) | 342 |
| 5,6-Dichloro-5β-pregnane-17α,21-diol-3,20-dione | Hxl: 21-OH | | | 677 |
| 5,6-Dichloro-5β-pregnane-3β,21-diol-3,20-dione | Hxl: 21-OH | | | 677 |
| 5,6-Dichloro-5β-pregnan-21-ol-3,20-dione | Hxl: 21-OH | | | 677 |
| 6α,9α-Difluoro-$\Delta^{1,4}$-pregnadiene-11β,16α,17α-triol-3,20-dione | Hxl: 16α-OH | | | 761 |
| 6α,9α-Difluoro-$\Delta^4$-pregnene-11β,16α,17α-triol-3,20-dione | Hxl: 16α-OH | | | 761 |
| Diginatigenin | Hxl: 12β-OH<br>Hxl: 12β-OH +<br>S: 16β-OAc | 155—156 | + 34 (M) | 453*a*<br>326 |
| Digitoxigenin | R: 3-CO → 3β-OH | 248—250 | + 19 (M) | 319 |
| 16-dehydro- | D: 16—17 | | | 464 |
| 1β,7β-dihydroxy- | Hxl: 1β,7β-diOH | 244—254 | + 19,1 (M) | 207, 320 |
| 5β,7β-dihydroxy- | Hxl: 5β,7β-diOH | 249—250 | + 49,1 (M) | 207, 320 |
| 1β-hydroxy- | Hxl: 1β-OH | 276—282 | + 3,2 (M) | 318–320 |
| 6β-hydroxy- | Hxl: 6β-OH | 268—275 | + 37 (M) | 464 |
| 7β-hydroxy- | Hxl: 7β-OH | 266—275 | + 37,3 (M) | 204, 206, 209*a*, 320, 321 |
| Digitoxigenone | O: 3β-OH → 3-CO<br>O: 3α-OH → 3-CO | 198—204 | + 33 (C) | 209*a*<br>322 |
| 7β-hydroxy- | Hxl: 7β-OH<br>Hxl: 7β-OH +<br>O: 3β-OH → 3-CO | 266—273 | + 55,8 (M) | 319<br>319 |
| Digoxigenin | Hxl: 12β-OH | 220—222 | + 22,5 (M) | 209*a*, 317, 326, 454 |

| *Metabolite* | Type of transformation (*see p. 129*) | m. p. °C | $[\alpha]_D$ | Reference (*see p. 181*) |
|---|---|---|---|---|
| Digoxigenin | | | | |
| 3-dehydro- | Hxl: 12β-OH | 247—252 | + 40 (M) | 323, 454 |
| | Hxl: 12β-OH + O: 3α-OH → 3-CO | | | 322 |
| | Hxl: 12β-OH + O: 3β-OH → 3-CO | | | 317 |
| Diosgenin | | | | |
| 7β,11α-dihydroxy- | Hxl: 7β,11α-diOH | 266—270 | + 23 (E) | 179 |
| 11β-hydroxy-7-keto- | (Diosgenin) | 221—225 | —137,1 (E) | 179 |
| 11-Epicorticosterone, see Δ⁴-Pregnene-11α, 21-diol-3,20-dione | | | | |
| 11-Epicortisol | Hxl: 11α-OH | 216—219 | +112 (E) | 130, 141, 155, 165, 167, 332, 360, 399, 401, 402, 426, 435, 531, 592, 615, 638, 669, 675, 732, 735, 937*a*, 981, 1002 |
| | Hxl: 21-OH | | | 677 |
| | Hxl: 11α,17α-diOH | | | 297 |
| 6α-chloro- | Hxl: 11α-OH | | | 624*a* |
| 1-dehydro- | Hxl: 11α-OH | 246—247 | + 73 (M) | 187*a*, 460*a*, 753, 958 |
| | D: 1—2 | | | 810*b* |
| 16β-ethyl- | Hxl: 11α-OH | | | 638 |
| 16α-methyl- | Hxl: 11α-OH | | | 638 |
| 16β-methyl- | Hxl: 11α-OH | | | 638 |
| 6α-fluoro- | Hxl: 11αOH | 201—203 | +100 (E) | 624*a* |
| 6α-fluoro-16α-methyl- | Hxl: 11α-OH | 214—217 | | 631 |
| 16β-methyl- | Hxl: 11α-OH | | | 625, 638 |
| 16-methylene- | Hxl: 11α-OH | 199—201 | + 41,7 (Di) | 985*a* |
| 3-Epidigitoxigenin | R: 3-CO → 3α-OH | 274—282 | + 27 (M) | 319, 326, 454 |
| 7β-hydroxy- | Hxl: 7β-OH + R: 3-CO → 3α-OH | 255—259 | + 47,9 (M) | 319 |
| 3-Epidigoxigenin | Hxl: 12β-OH + R: 3-CO → 3α-OH | 249—262 | + 27 (M) | 326, 454 |

| *Metabolite* | Type of transformation (*see p. 129*) | m. p. °C | $[\alpha]_D$ | Reference (*see p. 181*) |
|---|---|---|---|---|
| Epitestosterone | O: 3β-OH → → C-3O + I | 215—218 | + 68 (E) | 162 |
| Equilin | (19-Nor-$\Delta^{4,7}$-androstadiene-3,17-dione) | 238—240 | +308 (Di) | 86 |
| Estradiol-17β | R: 17-CO → → 17β-OH | 177—179,5 | + 83 (E) | 269, 385, 497, 511, 810*a* |
| $\Delta^1$-Estrene-3,17-dione | (5α-Estrane-17β-ol-3-one) | 188—189 | +221 (C) | 248, 249 |
| Estriol | Hxl: 16α-OH + R: 17-CO → 17β-OH | 281—283 | + 56 (Di) | 214, 623 |
| Estrone | O: 17β-OH → 17-CO | 260—262 | +154 (Di) | 42, 385, 479, 510, 589*a* |
| | ($\Delta^4$-Androstene-19-ol-3,20-dione) | | | 106, 110 |
| | (19-Nortestosterone) | | | 248, 249 |
| 16α-hydroxy- | Hxl: 16α-OH | 214—216 | +174,8 (M) | 214, 623 |
| 2α-methyl- | (2α-Methyl-19-nortestosterone) | 243—244 | +167 (C) | 337, 740 |
| 4-methyl- | (4-Methyl-19-nortestosterone) | 244—247 | +147 (C) | 337, 741 |
| 17α-Ethyl-testosterone | | | | |
| 1-dehydro- | D: 1—2 | | | 638*c* |
| 11β-hydroxy- | D: 1—2 | | | 638*c* |
| 11-keto- | D: 1—2 | | | 638*c* |
| 17α-Ethinyl-$\Delta^{1,4}$-androstadien-17β-ol-3-one | D: 1—2 | 228—233 | − 17 (C) | 708 |
| 12α-Fluoro-$\Delta^{1,4}$-androstadiene-11β,17β-diol-3-one | D: 1—2 | | | 347 |
| 9α-Fluoro-17α-ethynyl-$\Delta^{1,4}$-androstadiene-11β,17β-diol-3-one | D: 1—2 | | | 624*f* |
| 9α-Fluoro-17aβ-hydroxymethyl-$\Delta^{1,4}$-D-homo-androstadiene-11β,16α,17aα-triol-3,17-dione | D: 1—2 | 254—255 | + 45,6 (M) | 148 |
| 9α-Fluoro-6α-methyl-$\Delta^{1,4}$-pregnadiene-11β,16α,17α-triol-3,20-dione | Hxl: 16α-OH | | | 763*a* |
| 9α-Fluoro-16α-methyl-$\Delta^{1,4}$-pregnadiene-11β,17α,21-triol-3,20-dione, see Dexamethasone | | | | |

| *Metabolite* | Type of transformation (*see p. 129*) | m. p. °C | $[\alpha]_D$ | Reference (*see p. 181*) |
|---|---|---|---|---|
| 6α-Fluoro-16α-methyl-$\Delta^{4}$-pregnene-14α,17α,21-triol-3,20-dione | Hxl: 14α-OH | 247—251 | +124 (Di) | 487 |
| 9α-Fluoro-6α-methyl-$\Delta^{4}$-pregnene-11β,16α,17α-triol-3,20-dione | Hxl: 16α-OH | | | 763*a* |
| 6α-Fluoro-$\Delta^{1,4}$-pregnadiene-11α,17α-diol-3,20-dione | Hxl: 11α-OH | | | 624*a* |
| 6α-Fluoro-$\Delta^{1,4}$-pregnadiene-11α,21-diol-3,20-dione | Hxl: 11α-OH | | | 624*a* |
| 6α-Fluoro-$\Delta^{1,4}$-pregnadiene-11α-ol-3,20-dione | Hxl: 11α-OH<br>D: 1—2 | | | 624*a*<br>759*a* |
| 6α-Fluoro-$\Delta^{1,4}$-pregnadien-11β-ol-3,20-dione | D: 1—2 | | | 759*a* |
| 6α-Fluoro-$\Delta^{1,4}$-pregnadien-21-ol-3,11,20-trione | D: 1—2 | | | 758*a* |
| 9α-Fluoro-$\Delta^{1,4}$-pregnadiene-11β,16α,17α,20β,21-pentaol-3-one | R: 20-CO → 20β-OH | 262—264 | + 15,1 (M) | 420 |
| 9α-Fluoro-$\Delta^{1,4}$-pregnadiene-11β,16α,17α,21-tetraol-3,20-dione, see Triamcinolone | | | | |
| 6α-Fluoro-$\Delta^{1,4}$-pregnadiene-11β,16α,17α-triol-3,20-dione | Hxl: 16α-OH | | | 761 |
| 9α-Fluoro-$\Delta^{1,4}$-pregnadiene-11β,16α,17α-triol-3,20-dione | D: 1—2 | 275—279 | | 632 |
| 6α-Fluoro-$\Delta^{1,4}$-pregnadiene-3,11,20-trione | D: 1—2 | 192—199 | +200 (C) | 759*a* |
| 6α-Fluoro-$\Delta^{4}$-pregnene-11α,21-diol-3,20-dione | Hxl: 11α-OH | | | 624*a* |
| 9α-Fluoro-$\Delta^{4}$-pregnen-14α-ol-3,11,20-trione | Hxl: 14α-OH | | | 738 |
| 9α-Fluoro-$\Delta^{4}$-pregnene-11β,16α,17α,20β,21-pentaol-3-one | Hdg: 1—2<br>(9α-Fluorocortisol) | 227—231 | + 49,5 (M) | 420<br>422 |
| 6α-Fluoro-$\Delta^{4}$-pregnene-11β,16α,17α-triol-3,20-dione | Hxl: 16α-OH | | | 761 |

| *Metabolite* | Type of transformation (*see p. 129*) | m. p. °C | $[\alpha]_D$ | Reference (*see p. 181*) |
|---|---|---|---|---|
| 6α-Fluoro-$\Delta^4$-pregnene-15β,17α,21-triol-3,20-dione | Hxl: 15β-OH | 225—228 | + 79 (E) | 624*a* |
| 9α-Fluoro-$\Delta^4$-pregnene-11β,16α,17,21-tetra-ol-3,20-dione | | | | |
| 16α-17α-acetone ketal 21-acetate | E: 21-OAc | 239—240 | +136 (M) | 192 |
| 9α-Fluoro-$\Delta^4$-pregnene-11β,17α,20β,21-tetra-ol-3-one | R: 20-CO → → 20β-OH | | | 422, 621 |
| Gitoxigenin | Hxl: 16β-OH | 235 | + 34 (M) | 306 |
| 3,16-bisdehydro- | (Gitoxigenin) | 215—220 | | 209*b* |
| 3-dehydro- | O: 3β-OH → 3-CO | | | 209*b*,317 |
| 3-Ketobisnor-$\Delta^4$-cholene-11α,22-diol | Hxl: 11α-OH + R: 22-al → 22-OH | 130—133 | + 78 (C) | 598*d*,945 |
| 3-Ketobisnor-$\Delta^4$-cholene-6β,11α,22-triol | Hxl: 6β-OH | 238—240 | + 22 (M) | 296 |
| | Hxl: 6β,11α-diOH +R: 22-al → 22-OH | | | 298 |
| 3-Ketobisnor-11α-hydroxy-$\Delta^4$-cholenic acid | Hxl: 11α-OH | | | 598*d* |
| 3-Keto-$\Delta^4$-cholenic acid | O: 3β-OH → 3-CO | 185 | | 478 |
| 3-Keto-$\Delta^4$-etienic acid | ($\Delta^4$-Cholestene-3-one) | 245 | | 479 |
| | (3-Keto-$\Delta^4$-cholenic acid) | | | 479 |
| 6α-Methyl-$\Delta^{1,4}$-androstadien-17β-ol-3,11-dione | D: 1—2 | 209—210 | +148 (C) | 598*c* |
| 6α-Methyl-$\Delta^{1,4}$-androstadiene-3,11,17-trione | D: 1—2 | 274—276 | +230 (C) | 598*c* |
| 4-Methyl-$\Delta^4$-androstene-6β,17β-diol-3-one | Hxl: 6β-OH | 218—220 | + 10 (C) | 213 |
| 4-Methyl-$\Delta^4$-androstene-7β,17β-diol-3-one | Hxl: 7β-OH | 190—192 | +105 (C) | 213 |
| 4-Methyl-$\Delta^4$-androstene-11α,17β-diol-3-one | Hxl: 11α-OH | 179,5—180,5 | + 72 (C) | 213 |
| 16-Methylene-$\Delta^4$-androstene-3,17-dione | O: 3β-OH → → 3-CO + I | 161—162 | | 58*a* |
| 17aα-Methyl-$\Delta^4$-D-homo-androstene-11α,17aβ-diol-3,17-dione | (17α-Hydroxyprogesterone) | 261—262 | + 46 (C) | 141, 675 |
| 17β-Methyl-$\Delta^{4,13}$-18-nor-17α-pregnadiene-16α,20-diol-3-one | (16,17-Oxido-progesterone | 172—175 | — 15 (C) | 66, 699 |

| *Metabolite* | Type of transformation (*see p. 129*) | m. p. °C | $[\alpha]_D$ | Reference (*see p. 181*) |
|---|---|---|---|---|
| 17β-Methyl-$\Delta^{5,13}$-18-nor-17α-pregnadiene-3β,16α20,21-tetraol | (16,17-Oxido-$\Delta^{5}$-pregnene-3β,21-diol-20-on 21-acetate) | | | 699 |
| 17β-Methyl-$\Delta^{5,13}$-18-nor-17α-pregnadiene-3β,16α,20-triol | (16,17-Oxido-$\Delta^{5}$-pregnene-3β-ol-20-one) | 202—204 | +189 (E) | 699 |
| 17β-Methyl-$\Delta^{13(18)}$-18-nor-17α-pregnene-16α,20-diol-3-one | (16,17-Oxido-5β-pregnene-3,20-dione) | 223—225 | − 58 (E) | 699 |
| 17β-Methyl-$\Delta^{13}$-18-nor-5α,17α-pregnene-3β,16α,20-triol | (16,17-Oxido-5α-pregnane-3β-ol-20-one) | 200—201 | | 66 |
| 17α-Methyl-$\Delta^{1,4}$-pregnadiene-11α,21-diol-3,20-dione | Hxl: 11α-OH | | | 691 |
| 6α-Methyl-$\Delta^{1,4}$-pregnadien-17α-ol-3,20-dione | D: 1—2 | | | 973 |
| 17α-Methyl-$\Delta^{1,4}$-pregnadien-21-ol-3,20-dione | D: 1—2 | | | 691 |
| 17α-Methyl-$\Delta^{1,4}$-pregnadien-21-ol-3,11,20-trione | D: 1—2 | 183—186 | +120 (M) | 708 |
| 6α-Methyl-$\Delta^{1,4}$-pregnadiene-11β,16α,17α-triol-3,20-dione | Hxl: 16α-OH | | | 763*a* |
| 11α-Methyl-5β-pregnane-11β,17α-diol-3,20-dione | Hxl: 17α-OH | | | 762*a* |
| 16α-Methyl-5β-pregnane-3α,6α,11α-triol-20-one | Hxl: 11α-OH | 237—242 | + 32 (M) | 1003 |
| 17α-Methyl-$\Delta^{4}$-pregnene-11α,21-diol-3,20-dione | Hxl: 11α-OH | | | 691 |
| 16α-Methyl-$\Delta^{1}$-5β-pregnene-17α,21-diol-3,11,20-trione | Hdg: 4—5 | 206—212 | +116 (Di) | 152 |
| 7α-Methylsulfinyl-$\Delta^{4}$-androsten-17β-ol-3-one | (17β-Acetoxy-7α-methylthio-$\Delta^{4}$-androstene-3-one) | 189,5—189,8 | +11 (M) | 193 |
| 17α-Methyltestosterone 1-dehydro- | D: 1—2 | 163—164 | ± 0 (M) | 248, 496, 638*c* |
| 11α-hydroxy- | Hxl: 11α-OH | 250—253 | − 33 (C) | 598*d* |

| *Metabolite* | Type of transformation (*see p. 129*) | m. p. °C | $[\alpha]_D$ | Reference (*see p. 181*) |
|---|---|---|---|---|
| 17α-Methyltestosterone | | | | |
| 1-dehydro- | | | | |
| 11β-hydroxy- | D: 1—2 | | | 638*c* |
| 11-keto- | D: 1—2 | | | 638*c* |
| 16-methylene- | D: 1—2 | 145—148 | —101 (Di) | 58*a* |
| 6β-hydroxy- | Hxl: 6β-OH | 252—253 | + 3 (C) | 121, 578 |
| 11α-hydroxy- | Hxl: 11α-OH | 160—162 | + 62 (V) | 121, 547, 598*d*,656 |
| 16-methylene- | O: 3β-OH → → 3-CO + I | 144—146 | — 5 (Di) | 58*a* |
| 19-Nor-Δ$^{1,4}$-androstadiene-10β,17β-diol-3-one | D: 1—2 | 230—234 | — 28 (M) | 135 |
| 19-Nor-5α-androstan-17β-ol-3,6-dione | (19-Nortestosterone) | | | 554 |
| B-Norandrostane-3β,5α-6β-triol-17-one | (B-Nor-Δ$^5$-androstene-3β-ol-17-one) | 170—173 | + 61 (C) | 338*b* |
| B-Nor-Δ$^5$-androstene-3β,11α-diol-17-one | Hxl: 11α-OH | 110—150 | — 85,5 (C) | 338*b* |
| B-Nor-Δ$^5$-androstene-3β,15α-diol-17-one | Hxl: 15α-OH | 268—270 | — 10 (E) | 338*b* |
| 19-Nor-Δ$^4$-androsten-15α-ol-3,17-dione | Hxl: 15α-OH | 200—201 | +162 (C) | 137 |
| B-Nor-Δ$^4$-androsten-5β-ol-3,17-dione | (B-Nor-Δ$^5$-androstene-3β-ol-17-one) | | | 338*b* |
| 19-Norcortexone | Hxl: 21-OH | | | 522, 591*e*,670 |
| D-Nor-Δ$^{1,4}$-pregnadiene-3,20-dione | D: 1—2 | 193—194 | +110 (C) | 342 |
| D-Nor-Δ$^{1,4}$-pregnadien-11α-ol-3,20-dione | Hxl: 11α-OH | 227—230 | + 77 (C) | 342 |
| 19-Nor-Δ$^{1,3,5(10)}$-pregnatrien-3β-ol-11,20-dione | (19-Nor-11-keto-progesterone) | 229—232 | +324 (Di) | 57 |
| 19-Nor-Δ$^{1,3,5(10)}$-pregnatrien-3β-ol-20-one | (19-Norprogesterone) | 238—240 | +164 (C) | 57, 58 |
| D-Nor-Δ$^4$-pregnene-3,20-dione | (D-Nor-Δ$^5$-pregnene-3β-ol-20-one) | 153—155 | +209 (C) | 342 |
| A-Norprogesterone | | | | |
| 6β-hydroxy- | Hxl: 6β-OH | 217—219 | | 1213 |
| 7α-hydroxy- | Hxl: 7α-OH | 161—161,5 | + 65 (C) | 624*c* |
| 9α-hydroxy- | Hxl: 9α-OH | | | 1214 |
| 11α-hydroxy- | Hxl: 11α-OH | 183—184 | + 93 (C) | 624*d*, 1213 |
| 16α-hydroxy- | Hxl: 16α-OH | 213—214 | + 32 (C) | 624*b* |

| *Metabolite* | Type of transformation (*see p. 129*) | m. p. °C | $[\alpha]_D$ | Reference (*see p. 181*) |
|---|---|---|---|---|
| 19-Norprogesterone | | | | |
| 11α-hydroxy- | Hxl: 11α-OH | 171—173 | + 62 (C) | 57, 58 |
| 11β-hydroxy- | Hxl: 11β-OH | 215—217 | +158 (C) | 57 |
| 14α-hydroxy- | Hxl: 14α-OH | 202—204 | +166 (C) | 57 |
| 19-Nortestosterone | | | | |
| 1-dehydro- | D: 1—2 | | | 591*d* |
| 10β,11β-dihydroxy- | Hxl: 10β,11β-diOH | | | 135 |
| 6β-hydroxy- | Hxl: 6β-OH | 217—219 | — 63 (M) | 138, 327 |
| 10β-hydroxy- | Hxl: 10β-OH | 205—210 | + 70 (C) | 135, 327 |
| 11α-hydroxy- | Hxl: 11α-OH | 167—168 | — 46 (C) | 135, 327 |
| 11β-hydroxy- | Hxl: 11β-OH | 214—219 | + 92 (C) | 135 |
| 12β-hydroxy- | Hxl: 12β-OH | 163—165 | + 92 (C) | 136 |
| 14α-hydroxy- | Hxl: 14α-OH | 192—194 | + 84 (C) | 138 |
| 15α-hydroxy- | Hxl: 15α-OH | 136—139 | + 95 (C) | 137 |
| 16α-hydroxy- | Hxl: 16α-OH | 185—186 | + 30 (C) | 134, 623 |
| 16β-hydroxy- | Hxl: 16β-OH | 154—156 | + 51 (C) | 134 |
| 17α-methyl-11α-hydroxy- | Hxl: 11α-OH | | | 537 |
| Oleandrigenin | | | | |
| 3-dehydro- | O: 3β-OH → 3-CO | 242—249 | | 317 |
| 12β-hydroxy- | Hxl: 12β-OH | 233—238 | + 6 (M) | 326 |
| 9α,11α-Oxido-Δ⁴-androstene-3,17-dione | Ep: 9(11) | 273—275 | +180 (C) | 411 |
| 5,6-Oxido-B-norandrostane-3β,11α-diol-17-one | Ep: 5,6 + Hxl: 11α-OH | 235—240 | — 17,7 (E) | 338*b* |
| 5,6-Oxido-B-norandrostan-3β-ol-17-one | Ep: 5,6 | 192—200 | + 11,5 (C) | 338*b* |
| 14α,15α-Oxido-Δ$^{1,4}$-pregnadiene-17α,21-diol-3,20-dione | D: 1—2 | | | 591*d* |
| 16α,17α-oxido-Δ$^{1,4}$-pregnadiene-17α,21-diol-3,20-dione | D: 1—2 + O: 3β-OH → → 3-CO + I | 206—208 | + 88 (Di) | 988 |
| 16α,17α-Oxido-Δ$^{1,4}$-pregnadien-11α-ol-3,20-dione | D: 1—2 | | | 634 |
| 16α,17α-Oxido-Δ$^{1,4}$-pregnadien-21-ol-3,20-dione | D: 1—2 | | | 634' |
| 14α,15α-Oxido-Δ$^{1,4}$-pregnadiene-11β,17α,21-triol-3,20-dione | D: 1—2 | | | 591*d* |
| 16α,17α-Oxido-5α-pregnan-11α-ol-3,20-dione | Hxl: 11α-OH | 205—207 | + 49 (C) | 212 |

| *Metabolite* | Type of transformation (*see p. 129*) | m. p. °C | $[\alpha]_D$ | Reference (*see p. 181*) |
|---|---|---|---|---|
| 9α,11α-Oxido-$\Delta^4$-pregnene-17α,21-diol-3,20-dione | Ep: 9(11) | 213—215 | + 85 (Di) | 240, 411 |
| 9β,11β-Oxido-$\Delta^4$-pregnene-17α,21-diol-3,20-dione | Ep: 9(11) | 210—211 | + 13 (Di) | 55, 240, 596*a*, 941 |
| 14α,15α-Oxido-$\Delta^4$-pregnene-17α,21-diol-3,20-dione | Ep: 14—15 | 229,6—232,2 | +135 (Di) | 55, 596*a*, 941 |
| 16α,17α-Oxido-$\Delta^4$-pregnene-11α,21-diol-3,20-dione | Hxl: 11α-OH | 218—220 | +127 (M) | 971 |
| 16α,17α-Oxido-$\Delta^4$-pregnene-11β,21-diol-3,20-dione | Hxl: 11β-OH | | | 598*a* |
| 16α,17α-Oxido-$\Delta^4$-pregnene-14α,21-diol-3,20-dione | Hxl: 14α-OH | | | 598*a* |
| 11β,12β-Oxido-$\Delta^4$-pregnene-3,20-dione | Ep: 11—12 | | | 240 |
| 16α,17α-Oxido-$\Delta^4$-pregnen-7β-ol-3,20-dione | Hxl: 7β-OH | 237—239 | +156 (C) | 481 |
| 16α,17α-Oxido-$\Delta^4$-pregnen-11α-ol-3,20-dione | Hxl: 11α-OH | 235—237 | −146 (C) | 50, 335, 681, 807, 907 |
| 14α,15α-Oxido-$\Delta^4$-pregnene-11β,17α,21-triol-3,20-dione | Ep: 14—15 + Hxl: 11β-OH | 225—226,5 | +159 (Di) | 596*a*, 941 |
| 16α,17α-Oxido-$\Delta^4$-pregnene-11β,14α,21-triol-3,20-dione | Hxl: 11β,14α-diOH | | | 598*a* |
| Periplogenin | Hxl: 5β-OH | 245—248 | + 25,3 (M) | 205, 319, 320, 322 |
| 17α-Periplogenin | Hxl: 5β-OH | 232—241 | + 35,1 (M) | 322 |
| Prednisolone | D: 1—2 | 238—241 | +105 (Di) | 226, 579, 588, 591*a*, 633, 717, 810*b*, 944*a*, 952 |
| | D: 1—2, 4—5 | | | 434, 714, 960 |
| | Hxl: 11β-OH | | | 615, 721 |
| | Hxl: 11β-OH + D: 1—2 | | | 157, 447 600 |

| *Metabolite* | Type of transformation (*see p. 129*) | m. p. °C | $[\alpha]_D$ | Reference (*see p. 181*) |
|---|---|---|---|---|
| Prednisolone | | | | |
| 2,6-dimethyl- | D: 1—2 | | | 750 |
| 16β-ethyl- | Hxl: 11β-OH | | | 638 |
| 2-fluoro- | D: 1—2 | | | 637*a* |
| 9α-fluoro- | D: 1—2 | 265—269 | + 109 (C) | 258, 313, 433, 496, 591*d*, 633, 634, 712, 951 |
| 6α-fluoro-16α-hydroxy- | Hxl: 16α-OH | | | 976 |
| 9α-fluoro-14α-hydroxy- | Hxl: 14α-OH | | | 738 |
| 9α-fluoro-16-keto- | D: 1—2 | | | 724 |
| 9α-fluoro-2-methyl- | D: 1—2 | | | 722 |
| 6β-hydroxy- | Hxl: 6β-OH | | | 968 |
| 7α-hydroxy- | Hxl: 7α-OH | | | 612 |
| 14α-hydroxy- | D: 1—2 | 227—229 | +103,5 (Di) | 591*d*, 603, 700 |
| 16α-hydroxy- | D: 1—2 | 231—232 | + 77 (M) | 584, 591*c*, 634, 708*a* |
| | Hxl: 16α-OH | | | 623 |
| 16-keto- | D: 1—2 | | | 724 |
| 6α-methyl- | D: 1—2 | 245—247 | + 83 (Di) | 724*a* |
| 16α-methyl: | D: 1—2 | 199—203 | | 975 |
| | Hxl: 11β-OH | | | 638 |
| 16β-methyl- | Hxl: 11β-OH | 225—228 (21-acetate) | +100 (Di) | 638 |
| 16-methylene- | D: 1—2 | 233—235 | + 22 (Di) | 985*a* |
| Prednisone | D: 1—2 | 233—235 | +172 (Di) | 81, 84, 250, 251*a*, 313, 360, 398*b*, 404, 433, 438, 440, 495, 595, 496, 579, 594, 688, 810*b*, 944*a*, 951, 952, |

| *Metabolite* | Type of transformation (*see p. 129*) | m. p. °C | $[\alpha]_D$ | Reference (*see p. 181*) |
|---|---|---|---|---|
| Prednisone | D: 1—2,4—5 | | | 434, 708, 714 |
| | Hxl: 21-OH | | | 591*b*, 921 |
| 6α-9α-difluoro-16α-hydroxy- | Hxl: 16α-OH | | | 976 |
| 2 -fluoro- | D: 1—2 | 232—236 | | 637*a* |
| 9 α-fluoro- | D: 1—2 | 244—246 | +108 (Di) | |
| 2-methyl- | D: 1—2 | | | 722 |
| 16α-methyl- | Hxl: 21-OH | | | 976*a* |
| 16β-methyl- | D: 1—2 | 200—205 | +198,4 (Di) | 215 |
| Prednisone | | | | |
| 6β-hydroxy- | Hxl: 6β-OH | 222—224 | | 914, 968 |
| 14α-hydroxy- | D: 1—2 | | | 591*d*,603 |
| 2-methyl- | D: 1—2 | | | 722 |
| 4-methyl- | D: 1—2 | | | 741*a* |
| 6α-methyl- | D: 1—2 | | | 964 |
| 16α-methyl- | D: 1—2 | 210—212 | | 975 |
| 16β-methyl- | D: 1—2,4—5 | 200—205 | +198,4 (Di) | 215 |
| $\Delta^{4,17(20)}$-Pregnadiene-6β,21-diol-3,11-dione | Hxl: 6β-OH | 252—254 | +107 (Di) | 169 |
| $\Delta^{4,17(20)}$-Pregnadiene-9α,21-diol-3,11-dione | Hxl: 9α-OH + O: 11β-OH → 11-CO | 219,5—221 | +173 (Di) | 169, 751 |
| $\Delta^{1,4}$-Pregnadiene-11α,20β-diol-3-one | D: 1—2 + R: 20-CO → 20β-OH | 227—229 | + 41 (C) | 594, 701, 931 |
| $\Delta^{1,4}$-Pregnadiene-16α,20β-diol-3-one | D: 1—2 + R: 20-CO → 20β-OH | 226—227 | — 15 (C) | 8, 575 |
| $\Delta^{1,4}$-Pregnadiene-11β,17α-diol-21-thiol-3,20-dione | S: 21—SAc | | | 593*a* |
| $\Delta^{1,4}$-Pregnadiene-17α,21-diol-3,11,20-trione, see Prednisone | | | | |
| $\Delta^{1,4}$-Pregnadiene-3,20-dione | D: 1—2 | 152—153 | +122 (C) | 311, 377, 433, 495, 634 |
| | D: 1—2 + O: 3β-OH → 3-CO | | | 93, 810*b* |
| 6β,17α-dihydroxy- | Hxl: 6β-OH | 245—246 | + 4 (C) | 968 |
| 11β,17α-dihydroxy- | D: 1—2 | | | 591*d* |
| 11α-hydroxy- | D: 1—2 | 228—230 | + 93 (C) | 594, 634, 701 |
| 11β-hydroxy- | D: 1—2 | 232—234 | | 433 |

| *Metabolite* | Type of transformation (*see p. 129*) | m. p. °C | $[\alpha]_D$ | Reference (*see p. 181*) |
|---|---|---|---|---|
| $\Delta^{1,4}$-Pregnadiene-3,20-dione | | | | |
| 17α-hydroxy- | D: 1—2 | 232—234 | + 38,5 () | 591*d* |
| 11-keto- | D: 1—2 | | | 591*d* |
| $\Delta^{4,6}$-Pregnadien-11α-ol-3,20-dione | Hxl: 11α-OH | 160—162 | +111 (C) | 333, 650, 892 |
| $\Delta^{1,4}$-Pregnadien-17α-ol-21-thiol-3,11,20-trione | S: 21-SAc | | | 593*a* |
| $\Delta^{1,4}$-Pregnadien-21-ol-3,11,20-trione | D: 1—2 | 227,5—230,5 | +173 (M) | 313, 433, 720 |
| 9α-fluoro- | D: 1—2 | | | 810*b* |
| 12α-fluoro- | D: 1—2 | | | 730 |
| $\Delta^{1,4}$-Pregnadiene-11β, 17α,20β,21-tetraol-3-one | R: 20-CO → 20β-OH | 244—247,5 (20,21-diacetate) | +110 (Di) | 85, 218, 251 |
| $\Delta^{1,4}$-Pregnadiene-12β, 17α,20β,21-tetraol-3-one | D: 1—2 | 245—247 | | 341 |
| $\Delta^{1,4}$-Pregnadiene-11α, 17α,21-triol-3,20-dione, see Epicortisol, 1-dehydro | | | | |
| $\Delta^{1,4}$-Pregnadiene-11β, 17α,21-triol-3,20-dione, see Prednisolone | | | | |
| $\Delta^{4,9(11)}$-Pregnadiene-14α,17α,21-triol-3,20-dione | Hxl: 14α-OH | 190,8—194,2 | | 535, 702 |
| $\Delta^{1,4}$-Pregnadiene-17α, 20α,21-triol-3,11-dione | R: 20-CO → → 20α-OH | 238—240 | +117 (Di) | 85 |
| $\Delta^{1,4}$-Pregnadiene-17α, 20β,21-triol-3,11-dione | R: 20-CO → 20β-OH | 184—185 | +118 (Di) | 85, 218, 251, 360 |
| | D: 1—2 + R: 20-CO → 20β-OH | | | 81, 82, 439, 440 |
| | D: 1—2 | | | 187*a* |
| $\Delta^{4,6}$-Pregnadiene-17α, 20β,21-triol-3,11-dione | R: 20-CO → 20β-OH | 208—209 | +126 (Di) | 150 |
| $\Delta^{1,4}$-Pregnadiene-17α, 20α,21-triol-3-one | R: 20-CO → 20α-OH | 212—213 | + 13 (C) | 444 |
| $\Delta^{1,4}$-Pregnadiene-17α, 20β,21-triol-3-one | D: 1—2 + R: 20-CO → 20β-OH | 195—196 | + 33 (M) | 307, 436, 444, 484, 810*b* |
| | ($\Delta^5$-Pregnene-3β, 17α,21-triol-20-on 3,21-diacetate) | | | 810*b* |

| *Metabolite* | Type of transformation (*see p. 129*) | m. p. °C | $[\alpha]_D$ | Reference (*see p. 181*) |
|---|---|---|---|---|
| 5α-Pregnane-3β,16α-diol-11,20-dione | Hxl: 16α-OH | 246–248 | | 971*a* |
| 5α-Pregnane-11α,15α-diol-3,20-dione | Hxl: 15α-OH | 208–210 | +105 (C) | 368*a* |
| 5β-Pregnane-11α,15α-diol-3,20-dione | Hxl: 15α-OH | 175–177 | +108,5 (C) | 368*a* |
| 5β-Pregnane-11α,17α-diol-3,20-dione | Hxl: 11α-OH | | | 598*d* |
| 5β-Pregnane-11α,21-diol-3,20-dione | Hxl: 21-OH | | | 677 |
| 5β-Pregnane-11β,21-diol-3,20-dione | Hdg: 4–5 | | | 392 |
| 5α-Pregnane-12β,15α-diol-3,20-dione | Hxl: 12β,15α-diOH | 253–257 | + 70 (M) | 368*a* |
| 5β-Pregnane-12β,15α-diol-3,20-dione | Hxl: 12β,15α-diOH | 255–261 | + 56 (C) | 368*a* |
| 5β-Pregnane-17α,21-diol-3,20-dione | Hdg: 1–2, 4–5 | 189–190 | + 40 (?) | 392 |
| 5α-Pregnane-3β,7β-diol-20-one | Hxl: 7β-OH | 194–195 | + 77,6 (C) | 531, 803*b* |
| 5α-Pregnane-3β,11α-diol-20-one | R: 3-CO → 3β-OH | 179–181 | + 62 (C) | 63 |
| 5α-Pregnane-3β,16α-diol-20-one | Hxl: 16α-OH | 258–260 | | 971*a* |
| 5β-Pregnane-3β,16α-diol-20-one | Hxl: 16α-OH | 198–199 | | 971*a* |
| 5β-Pregnane-3α,21-diol-20-one | Hdg 4–5 | 149–151 | + 99 (C) | 382, 392 |
| 5β-Pregnane-3β,21-diol-20-one | Hdg: 4–5 | | | 392 |
| 5β-Pregnane-17α,21-diol-3,11,20-trione | Hdg: 4–5<br>Hxl: 21-OH<br>S: 21-OAc | | | 392<br>677<br>437 |
| 5α-Pregnane-3,20-dione | Hdg: 4–5 | 199–200 | +127 (E) | 377 |
| 5β-Pregnane-3,20-dione | Hdg: 4–5 | 123 | +105 (C) | 392 |
| 5α-Pregnan-3α-ol-11,20-dione | R: 3-CO → 3α-OH | 160–165 | + 93 (C) | 63 |
| 5β-Pregnan-3α-ol-11,20-dione | R: 3-CO → 3α-OH | 174,5–175 | +112 (E) | 63, 676 |
| 5α-Pregnan-9α-ol-3,20-dione | Hxl: 9α-OH | | | 1001 |
| 5β-Pregnan-9α-ol-3,20-dione | Hxl: 9α-OH | | | 1001 |

| *Metabolite* | Type of transformation (*see p. 129*) | m. p. °C | $[\alpha]_D$ | Reference (*see p. 181*) |
|---|---|---|---|---|
| 5α-Pregnan-11α-ol-3,20-dione | Hxl: 11α-OH<br>Hxl: 11α-OH + Hdg: 4—5 | 198—200 | + 82 (C) | 120<br>331 |
| 5β-Pregnan-11α-ol-3,20-dione | Hxl: 11α-OH | 102—106 | + 83 (C) | 120 |
| 5β-Pregnan-16α-ol-3,20-dione | Hxl: 16α-OH + Hdg: 4—5 | 199—200 | + 90,5 (C) | 328 |
| 5β-Pregnan-21-ol-3,20-dione | Hxl: 21-OH<br>Hdg: 4—5 | 152—154 | | 921<br>392 |
| 5α-Pregnan-3α-ol-20-one | Hdg: 4—5 + R: 3-CO → 3α-OH | 175 | + 96 (C) | 377 |
| 5α-Pregnan-3β-ol-20-one | Hdg: 4—5 + R: 3-CO → 3β-OH | 194—195 | + 95 (C) | 68 |
| 5β-Pregnan-3α-ol-20-one | Hdg: 4—5 + R: 3-CO → 3α-OH | 151—154 | +112 (M) | 382, 392 |
| 5β-Pregnan-3β-ol-20-one | Hdg: 4—5 + R: 3-CO → 3β-OH | 188—192 | + 90 (C) | 392 |
| 5β-Pregnan-15α-ol-3,6,20-trione | Hxl: 15α-OH | 218—220 | + 84 (C) | 25 |
| 5β-Pregnan-16α-ol-3,11,20-trione | Hxl: 16α-OH | | | 624*e* |
| 5α-Pregnane-3β,11β,17α,21-tetraol-20-one | (Cortisol) | 184—186 | | 310 |
| 5β-Pregnane-3α,11β,17α,21-tetraol-20-one | (Cortisol) | | | 392 |
| 5β-Pregnane-3β,11α,20-triol | Hxl: 11α-OH | | | 598*d* |
| 5α-Pregnane-3β,17α,21-triol-11,20-dione | (Cortisone) | 195—200 | | 310 |
| 5β-Pregnane-3α,17α,21-triol-11,20-dione | (Cortisone) | 187—189 | + 88 (C) | 49, 382, 392 |
| 5β-Pregnane-3β,17α,21-triol-11,20-dione | (Cortisone) | 215—225 | + 48 (C) | 392 |
| 5β-Pregnane-11α,17α,21-triol-3,20-dione | Hxl: 11α-OH<br>(Cortexolone) | 220—222 | + 57 (C) | 332<br>332 |
| 5β-Pregnane-11β,17α,21-triol-3,20-dione | (Cortisol) | 192—198 | + 94 (C) | 392 |
| 5β-Pregnane-3,4,5-triol-20-one | (4,5-Oxido-5β-pregnane-3,20-dione) | 233—239 | | 699 |
| 5β-Pregnane-3α,6α,11α-triol-20-one | Hxl: 11α-OH | 223—225 | + 61 (M) | 374, 1003 |
| 5α-Pregnane3β,7β,21-triol-20-one | Hxl: 7β-OH | 205—213 | +106 (E) | 512 |

| *Metabolite* | Type of transformation (*see p. 129*) | m. p. °C | $[\alpha]_D$ | Reference (*see p. 181*) |
|---|---|---|---|---|
| 5α-Pregnane-3β,11β,16α-triol-20-one | Hxl: 16α-OH | 257—260 | | 971*a* |
| 5β-Pregnane-3α,11α,17α-triol-20-one | Hxl: 11α-OH | 184—186 | + 52 (An) | 598*d* |
| 5β-Pregnane-3β,11α,17α-triol-20-one | Hxl: 11α-OH | | | 598*d* |
| 5β-Pregnane-3α,11β,21-triol-20-one | (Corticosterone) | 196—197 | + 88 (C) | 392 |
| 5α-Pregnane-3β,17α,21-triol-20-one | (Cortexolone) | 226—230 | + 44,5 (Di) | 228 |
| 5β-Pregnane-3α,17α,21-triol-20-one | (Cortexolone) | 203—205 | + 62 (C) | 392 |
| $\Delta^{1,4,6}$-Pregnatrien-21-ol-3,20-dione | D: 1—2 | 166—168 | +112 (An) | 496, 708 |
| $\Delta^{1,4,9(11)}$-Pregnatriene-17α,21-diol-3,20-dione | D: 1—2 | | | 591*d* |
| $\Delta^{1,4,14}$-Pregnatriene-17α,21-diol-3,20-dione | D: 1—2 | | | 591*d* |
| $\Delta^{1,4,17(20)}$-Pregnatriene-11β,21-diol-3,20-dione | D: 1—2 | | | 972 |
| $\Delta^{1,4,16}$-Pregnatriene-3,20-dione | D: 1—2 | | | 591*d* |
| $\Delta^{1,4,6}$-Pregnatriene-11α,17α,21-triol-3,20-dione | Hxl: 11α-OH | | | 605 |
| $\Delta^{1,4,9(11)}$-Pregnatriene-16α,17α,21-triol-3-one | ($\Delta^{9(11)}$-5α-Pregnane-3β,16α,17α, 21-tetraol-20-one 3,21-diacetate) | 200—201 (16,21-diacetate) | + 6 (M) | 635*c* |
| $\Delta^4$-Pregnene-11β,21-diol-3,20-dione, see Corticosterone | | | | |
| $\Delta^4$-Pregnene-17α,21-diol-3,20-dione, see Cortexolone | | | | |
| $\Delta^5$-Pregnene-3β,11α-diol-7,20-dione | ($\Delta^5$-Pregnane-3β-ol-20-one) | 228—230 | | 655 |
| $\Delta^4$-Pregnene-11α20β-diol-3-one | R: 20-CO → 20β-OH | 178—179 | + 73 (C) | 388 |
| $\Delta^4$-Pregnene-11α,21-diol-3,20-dione | Hxl: 11α-OH | 158—159 | +171 (E) | 119, 141, 401, 402, 512, 675 |
| | Hxl: 21-OH | | | 299*a*, 591*e*, 670, 677 |

| *Metabolite* | Type of transformation (*see p. 129*) | m. p. °C | $[\alpha]_D$ | Reference (*see p. 181*) |
|---|---|---|---|---|
| Δ4-Pregnene-11α,21-diol-3,20-dione | Hxl: 11α,21-diOH | | | 509, 513 |
| Δ4-Pregnene-16α,20β-diol-3-one | R: 20-CO → 20β-OH | 212—213 | + 44 (C) | 575 |
| Δ4-Pregnene-20α,21-diol-3-one | R: 20-CO → 20α-OH + S: 21-OAc | | | 576 |
| | R: 20-CO → 20α-OH + Hxl: 21-OH | | | 576 |
| (20 R)-Δ4-Pregnene-7ξ,20-diol-3-one-18-oic (18 → 20)-lactone | Hxl: 7ξ-OH | | | 241 |
| (20 R)-Δ4-Pregnene-11α,20-diol-3-one-18-oic (18 → 20)-lactone | Hxl: 11α-OH | 191,5—192 | + 89 (C) | 241 |
| Δ1-5β-Pregnene-17α,21-diol-3,11,20-trione | Hdg: 4—5 | 185—189 | | 152 |
| Δ4-Pregnene-17α,21-diol-3,11,20-trione, see Cortisone | | | | |
| Δ4-Pregnene-21-ol-3,20-dione, see Cortexone | | | | |
| Δ4-17α-Pregnen-11α-ol-3,20-dione | (16-Dehydroprogesterone) | 209—211,5 | — 12 (C) | 295, 803*b* |
| Δ4-Pregnen-20α-ol-3-one | R: 20-CO → 20α-OH | 151—152 | + 90 (E) | 576 |
| Δ4-Pregnen-20β-ol-3-one | R: 20-CO → 20β-OH | 174—175 | + 86 (C) | 142, 575 |
| Δ4-Pregnen-21-ol-3,11,20-trione, see 11-Dehydrocorticosterone | | | | |
| Δ4-Pregnene-11β,16α,17α,21-tetraol-3,20-dione | | | | |
| 16α,17α-acetone ketal 21-acetate | E: 21-OAc | | | 192 |
| Δ4-Pregnene-11α,17α,20β,21-tetraol-3-one | R: 20-CO → 20β-OH | 205—206 | + 58 (E) | 251 |
| Δ4-Pregnene-11β,17α,20β,21-tetraol-3-one | R: 20-CO → 20β-OH | 133—135 | + 85 (Di) | 85, 195, 218, 229, 251, 360 |
| Δ4-Pregnene-17α,19,20α,21-tetraol-3-one | R: 20-CO → 20α-OH | | | 444 |
| Δ4-Pregnene-6β,11α,21-triol-3,20-dione | Hxl: 6β,11α-diOH | 220—226 | +105 (M) | 402 |

| *Metabolite* | Type of transformation (*see p. 129*) | m. p. °C | $[\alpha]_D$ | Reference (*see p. 181*) |
|---|---|---|---|---|
| $\Delta^4$-Pregnene-11α,17α21-triol-3,20-dione, see 11-Epicortisol | | | | |
| $\Delta^4$-Pregnene-11β,17α,21-triol-3,20-dione, see Cortisol | | | | |
| $\Delta^4$-Pregnene-17α,20α,21-triol-3,11-dione | R: 20-CO → 20α-OH | 240—242 | +158 (Di) | 85 |
| $\Delta^4$-Pregnene-17α,20β,21-triol-3,11-dione | R: 20-CO → 20β-OH | 208,5—209,5 | +140 (An) | 85, 218, 251*a*, 439 |
| $\Delta^4$-Pregnene-16α,17α,21-triol-3,20-dione | | | | |
| 16α,17α-acetone ketal 21-acetate | E: 21-OAc | | | 192 |
| $\Delta^5$-Pregnene-3β,7,11α-triol-20-one | Hxl: 7,11α-diOH | 250—255 | − 41 (M) | 648 |
| $\Delta^5$-Pregnene-3β,7α,11α-triol-20-one | Hxl: 7α,11α-diOH | 235—237 | + 77 (M) | 25 |
| $\Delta^4$-Pregnene-11β,17α,20α-triol-2-one | R: 20-CO → 20α-OH | | | 444 |
| $\Delta^4$-Pregnene-17α,20α,21-triol-3-one | R: 20-CO → 20α-OH | 225—227,5 | + 76,2 (C) | 85, 444, 576 |
| $\Delta^4$-Pregnene-17α,20β,21-triol-3-one | R: 20-CO → 20β-OH | 193—194 | + 74 (E) | 85, 217, 229, 251*a*, 307, 483, 810*a* |
| Progesterone | O: 3β-OH → → 3-CO + I | 129—130 | +198 (C) | 42, 125, 203, 271, 329, 589*a*, 601 |
| 2β,15β-dihydroxy- | Hxl: 2β,15β-diOH | 217—219 | − 51,5 (C) | 457, 481 |
| 2β,17α-dihydroxy- | Hxl: 2β-OH | 219—221 | −125 (C) | 457 |
| 6β,11α-dihydroxy- | Hxl: 6β,11α-diOH | 250—253 | +100 (C) | 65, 115, 141, 331, 395, 401, 402, 502, 591, 661, 675 |
| 6β,12α-dihydroxy- | ($\Delta^{4,9(11)}$-Pregnadiene-3,20-dione) | 231—232 | +107 (C) | 1073 |

| *Metabolite* | Type of transformation (*see p. 129*) | m. p. °C | $[\alpha]_D$ | Reference (*see p. 181*) |
|---|---|---|---|---|
| Progesterone | | | | |
| 6β,14α-dihydroxy- | Hxl: 6β,14α-diOH | 246—249 | +124 (C) | 202, 371, 458, 524 |
| 6β,15α-dihydroxy- | Hxl: 6β,15α-diOH | 192—203 | +130 (C) | 154, 456 |
| 6β,16α-dihydroxy- | Hxl: 6β-OH | | | 8 |
| 6β,17α-dihydroxy- | Hxl: 6β-OH | 256—258 | + 4 (C) | 296, 400, 402, 531, 537, 542, 803*b* |
| 7α,11α-dihydroxy- | Hxl: 7α-OH | 233—236 | +124 (C) | 25, 999 |
| 7β,11α-dihydroxy- | Hxl: 7β,11α-diOH | 228—232 | | 999 |
| 7α,14α-dihydroxy- | Hxl: 7α,14α-diOH | 252—255 | +175 (C) | 368, 458, 523, 524, 718, 838 |
| 7α,15β-dihydroxy- | Hxl: 7α,15β-diOH | | +121 (M) | 468, 472 |
| 7β,15β-dihydroxy- | Hxl: 7β,15β-diOH | 230—232 | +130 (C) | 46*a*, 470, 472, 481 |
| | Hxl: 15β-OH | | | 46*a*, 470 |
| 9α,11α-dihydroxy- | Hxl: 9α,11α-diOH | 207—209 | +161 (C) | 25 |
| | Hxl: 11α-OH | | | 25 |
| 9α,14α-dihydroxy- | Hxl: 9α,14α-diOH | 272—273 | +179 (C) | 369, 458 |
| 9α,15α-dihydroxy- | Hxl: 9α,15α-diOH | 240—242 | +212 (C) | 373 |
| | Hxl: 15α-OH | | | 25 |
| | Hxl: 9α-OH | | | 1001 |
| 9α,17α-dihydroxy- | Hxl: 9α-OH | | | 747 |
| 11β,14α-dihydroxy- | Hxl: 11β,14α-diOH | 224—229 | +211 (M) | 523, 524, 718 |
| 11α,15α-dihydroxy- | Hxl: 15α-OH | 182—183 | +180 (M) | 365 |
| 11α,15β-dihydroxy- | Hxl: 11α,15β-diOH | 202—203 | +134 (C) | 595, 611 |
| 11β,15α-dihydroxy- | Hxl: 15α-OH | 173—175 | +230 (C) | 370 |
| 11α,16α-dihydroxy- | Hxl: 11α-OH | | | 8 |
| 11α,17α-dihydroxy- | Hxl: 11α-OH | 220—222 | + 76 (C) | 141, 296, 331, 339, 402, 457, 558, 651, 675, 803*b* |
| | Hxl: 11α,17α-diOH | | | 116, 463*a* |
| 11β,17α-dihydroxy- | Hxl: 11β-OH | 226—228 | +135,5 (An) | 141, 296, 406 |
| | Hxl: 17α-OH | | | 593 |
| 11α,18-dihydroxy- | Hxl: 11α-OH | 195—197 | | 638*a*, 841, 942 |
| 12β,14α-dihydroxy- | Hxl: 12β,14α-diOH | 240—242 | +129 (C) | 373 |
| | Hxl: 12β-OH | | | 25 |

| *Metabolite* | Type of transformation (*see p. 129*) | m. p. °C | $[\alpha]_D$ | Reference (*see p. 181*) |
|---|---|---|---|---|
| Progesterone | | | | |
| 12β,15α-dihydroxy- | Hxl: 12β-OH | 218 | +139 (C) | 365 |
| | Hxl: 12β,15α-diOH | | | 183, 365 |
| | ($\Delta^5$-Pregnene-3β-ol-20-one) | | | 25 |
| 14α,15β-dihydroxy- | Hxl: 14α,15β-diOH | 259—263 | +136 (M) | 472 |
| 15β,17α-dihydroxy- | Hxl: 15β-OH | 258—259 | + 54,3 (C) | 457, 595 |
| 11α,17α-dihydroxy-6α-fluoro- | Hxl: 11α-OH | | | 624*a* |
| 2β-hydroxy- | Hxl: 2β-OH | 191—198 | — 51 (C) | 481 |
| 6β-hydroxy- | Hxl: 6β-OH | 179—180 | +105 (C) | 395, 502, 532 |
| 7α-hydroxy- | Hxl: 7α-OH | 229—230 | +167 (C) | 259, 368, 574, 612, 854, 970 |
| 7β-hydroxy- | Hxl: 7β-OH | 188—191 | +141 (C) | 46*a*, 259, 466 |
| 9α-hydroxy- | Hxl: 9α-OH | 193—194,5 | +188 (C) | 111, 369, 747, 1001 |
| 11α-hydroxy- | Hxl: 11α-OH | 166—167 | +175,9 (C) | 141, 159, 331, 463, 504, 531, 540, 591, 666, 675 |
| 6α-fluoro- | Hxl: 11α-OH | | | 624*a* |
| 11β-hydroxy- | Hxl: 11β-OH | 185—188 | +217 (An) | 341, 406, 524 |
| 12β-hydroxy- | Hxl: 12β-OH | 195—198 | +205 (An) | 341 |
| 14α-hydroxy- | Hxl: 14α-OH | 198—200 | +190 (C) | 369, 458, 524, 543, 570, 664*a*, 905 |
| 15α-hydroxy- | Hxl: 15αOH | 231—232 | +218 (C) | 8, 53, 68, 154, 214*a*, 299, 405, 456, 472, 574, 595, 610, 682, 997 |
| 15β-hydroxy- | Hxl: 15β-OH | 203—205 | +155 (C) | 8, 257, 472, 481, 574, 610 |

| *Metabolite* | Type of transformation (*see p. 129*) | m. p. °C | $[\alpha]_D$ | Reference (*see p. 181*) |
|---|---|---|---|---|
| Progesterone | | | | |
| 16α-hydroxy- | Hxl: 16α-OH | 225—226 | +158 (C) | 328, 330, 501, 623 |
| 17α-hydroxy- | Hxl: 17α-OH | 213—215 | + 97 (C) | 256, 593, 731, 747 |
| 19-hydroxy- | Hxl: 19-OH | 170—171 | +175 (E) | 615 |
| 6β-hydroxy-15-keto- | (Progesterone) | 280—286 | +122 (E) | 481 |
| 7α-hydroxy-11-keto- | (Progesterone) | 229—232 | +215 (M) | 25 |
| 15α-hydroxy-11-keto- | Hxl: 15α-OH | 227—229 | | 373 |
| 15β-hydroxy-11-keto- | Hxl: 15β-OH | 224—228 | +289 (C) | 25 |
| 15α-hydroxy-12-keto- | Hxl: 15α-OH | 235—239 | +261 (C) | 373 |
| | Hxl: 15α-OH + O: 12α-OH → 12-CO | | | 25 |
| 16α-methyl-11α,17α-dihydroxy- | Hxl: 11α,17α-diOH | | | 607 |
| 6α-methyl-11α-hydroxy- | Hxl: 11α-OH | 151—153 | +120 (E) | 723*a* |
| 16α-methyl-11β-hydroxy- | Hxl: 11β-OH | | | 773 |
| 7α,14α,15β-trihydroxy- | Hxl: 7α,14α15β-triOH | 263—265 | +103 (M) | 458, 468 |
| 11β,12β,15α-trihydroxy- | Hxl: 12β,15α-diOH | 229—231 | +148 (C) | 370 |
| 11β,14α,17α-trihydroxy- | Hxl: 11β,14α-diOH | | | 561 |
| 11β,14α,17α-trihydroxy-12α-fluoro- | Hxl: 14α-OH | | | 734, 899 |
| Pyrazolo(4',3':16,17)-$\Delta^4$-androsten-3-one | O: 3β-OH → 3-CO + I | 235—236 | +105 (C) | 58*a* |
| Sarmentogenin | Hxl: 11α-OH | 270—275 | + 21 (M) | 464 |
| *Seco*-derivatives: | | | | |
| 7α-Methyl-1-glycolyl-perhydroindanone-(5)-[β-propionic acid-(4)] | (Cortexone acetate) | 145—147 (acetate) | + 59,5 (C) | 382*a* |
| 3aα-H-4α-[3'-propionic acid]-5α-hydroxy-7-aβ-methyl-hexahydro-1-indanone-δ-lactone | (Testosterone) | 124—127 | + 23,6 (C) | 502*a* |
| 3aα-H-4α-[3'-propionic acid]-5β-hydroxy-7aβ-methyl-hexahydro-1-indanone-δ-lactone | (Testosterone) | 128,5—130,5 | + 52,6 (C) | 502*a* |
| 3aα-H-4α-[3'-propionic acid]-7aβ-methyl-hexahydro-1,5-indanedione | (Testosterone) ($\Delta^4$-Androstene-3,17-dione) | 110—111,5 | +121 (C) | 502*a* 417 |

| *Metabolite* | Type of transformation (*see p. 129*) | m. p. °C | $[\alpha]_D$ | Reference (*see p. 181*) |
|---|---|---|---|---|
| *Seco*-derivatives | | | | |
| 9,10-Seco-$\Delta^{1,3,5(10)}$-androstatriene-3,9α-diol-17-one | (Progesterone) | 195—197 | + 48 (E) | 377, 378 |
| | (Testosterone) | | | 502*a* |
| 9,10-Seco-$\Delta^{1,3,5(10)}$-androstatriene-3,9β-diol-17-one | (Progesterone) | 155—157 | + 90,3 (E) | 377, 378, |
| | (Testosterone) | | | 502*a* |
| 9,10-Seco-$\Delta^{1,3,5(10)}$-androstatriene-3β,17β-diol-9-one | ($\Delta^4$-Androstene-3,17-dione) | 133—134 | + 15,9 (C) | 110 |
| 9,10-Seco-$\Delta^{1,3,5(10)}$-androstatrien-3-ol-9,17-dione | ($\Delta^4$-Androstene-3,17-dione) | 123,5—125 | +100,5 (C) | 106, 110, 417 |
| | (Testosterone) | | | 502*a* |
| | (Progesterone) | | | 378 |
| 9,10-Seco-$\Delta^{1,3,5(10)}$-androstatriene-3,9β,17β-triol | (Testosterone) | 171,5—172,5 | + 19 (E) | 502*a* |
| 9,10-Seco-$\Delta^{1,3,5(10)}$-pregnatriene-3,9-diol-20-one | (Progesterone) | | | 377, 378 |
| 9,10-Seco-$\Delta^{1,3,5(10)}$-pregnatrien-3-ol-9,20-dione | (Progesterone) | | | 377, 378 |
| Testolic acid | ($\Delta^4$-Androstene-3,17-dione) | 141—142 | + 65 (M) | 193*a* |
| | (Progesterone) | | | 193*a* |
| | (Testosterone) | | | 338*a* |
| | (Testololactone) | | | 193*a* |
| Testololactone | (Progesterone) | 207—209 (211—212) | + 43 (C) | 56, 72, 142, 193*a*, 334, 388 |
| Testololactone | (17α-Hydroxyprogesterone) | | | 334 |
| | (Cortexone) | | | 334, 689*a* |
| | (Cortexolone) | | | 334 |
| | ($\Delta^4$-Androstene-3,17-dione) | | | 193*a*, 334 |
| | (Testosterone) | | | 142, 334, 338*a* |
| 1-dehydro- | (Progesterone) | 219—220 | − 49 (C) | 142, 311, 336, 596, 598*e* |

| *Metabolite* | Type of transformation (*see p. 129*) | m. p. °C | $[\alpha]_D$ | Reference (*see p. 181*) |
|---|---|---|---|---|
| Testololactone | | | | |
| 1-dehydro- | (Cortexolone) | | | 142 |
| | ($\Delta^4$-Androstenedione) | | | 83 |
| | (Testosterone) | | | 142, 596 |
| | (Cortexone) | | | 689*a* |
| 1-dehydro-11α-hydroxy- | (Progesterone) | 251—253 | − 58 (C) | 596 |
| 2β-hydroxy- | Hxl: 2β-OH | 180—182 | −182 (C) | 481 |
| 6β-hydroxy- | Hxl: 6β-OH | 235—236 | − 13 (C) | 638*b* |
| 7α-hydroxy- | Hxl: 7α-OH | 275—280 | + 27 (C) | 590 |
| Windaus keto acid | ($\Delta^4$-Cholestene-3-one) | 155 | + 37 (C) | 479 |
| "β"-Sitosten-3-one | ("β"-Sitosterol) | 83 | | 478 |
| Solasodine | | | | |
| 7ξ-hydroxy- | Hxl: 7ξ-OH | 234—238 | − 82,4 (C) | 355 |
| 9α-hydroxy- | Hxl: 9α-OH | 213—215 | −138 (C) | 355 |
| 11α-hydroxy- | Hxl: 11α-OH | 200—203 | −110 (C) | 355 |
| $\Delta^{4,22}$-Stigmastadien-3-one | (Stigmasterol) | 125 | | 478 |
| Strophanthidone | | | | |
| 4-anhydro- | D: 1,2 + O: 3β-OH → 3-CO | 222—224 | | 236, 415 |
| 4-anhydro-19-dihydro | R: 19-al → 19-OH | 247—251 | | 418 |
| Testosterone | R: 17-CO → 17β-OH | 155—156 | +109 (E) | 161, 75, 264, 266, 265, 385, 448 |
| | ($\Delta^5$-Androstene-3β,17β-diol) | | | 125, 126, 302, 478, 528 |
| | ($\Delta^5$-Androstene-3α-ol-17-one) | | | 266, 268 |
| | ($\Delta^5$-Androstene-3,17-dione) | | | 265 |
| | (Progesterone) | | | 160 |
| acetate | R: 17-CO → 17β-OH + E: 17β-OAc | 146—148 | + 91 (C) | 260 |
| | (Progesterone) | | | 140 |
| 1-dehydro- | (Progesterone) | 172—173 | + 23 (C) | 142, 311, 336, 575, 598*e* |
| | D: 1—2 + R: 17-CO → 17β-OH | | | 83, 385, |
| | O: 3β-OH → → 3-CO + I | | | 93 |

| *Metabolite* | Type of transformation (*see p. 129*) | m. p. °C | $[\alpha_D]$ | Reference (*see p. 181*) |
|---|---|---|---|---|
| Testosterone | | | | |
| 1-dehydro- | | | | |
| 11α-hydroxy- | Hxl: 11α-OH | 174—175,5 | − 7 (C) | 945 |
| | (Progesterone) | | | 596 |
| 1α,6β-dihydroxy- | Hxl: 1α,6β-diOH | 256—260 | + 19 (M) | 481 |
| 1α-hydroxy- | Hxl: 1α-OH | 250—254 | +103 (C) | 481 |
| 6β-hydroxy- | Hxl: 6β-OH | 216—222 | + 32 (C) | 121, 187*b*, 481 |
| 9α-hydroxy- | (Progesterone) | 210—211 | +104 (C) | 111, 747 |
| 11α-hydroxy- | Hxl: 11α-OH | 181—181,5 | + 93 (C) | 121, 531, 598*d*, 945 |
| | R: 17-CO → 17β-OH | | | 385 |
| | (11α-hydroxyprogesterone) | | | 72, 160 |
| 11β-hydroxy- | R: 17-CO → 17β-OH | 235—236,5 | +136 (C) | 385 |
| 12β-hydroxy- | Hxl: 12β-OH | 121—125 | +120 (C) | 481 |
| | R: 17-CO → 17β-OH | | | 341 |
| 14α-hydroxy- | Hxl: 14α-OH | 185—187 | +124 (C) | 122, 538 |
| 15α-hydroxy- | Hxl: 15α-OH | 204—206 | +135 (M) | 154, 456 |
| 15β-hydroxy- | Hxl: 15β-OH | 220—222 | + 57 (E) | 481 |
| | R: 17-CO → 17β-OH | | | 187 |
| 16α-hydroxy- | Hxl: 16α-OH | 191—192 | + 80 (C) | 36, 187*b*, 330, 623 |
| | (16α-Hydroxyprogesterone) | | | 575 |
| 16β-hydroxy- | Hxl: 16β-OH | 183,5—185,5 | +103 (C) | 187*b* |
| | Hxl: 16β-OH + R: 17-CO → 17β-OH | | | 112 |
| 11-keto- | R: 17-CO → 17β-OH | 187—188 | + 199 (An) | 385, 1087 |
| | (11-Ketoprogesterone) | | | 72, 160 |
| 16-keto- | ($\Delta^4$-Androstene-3,17-dione) | 161—162,5 | − 66 (C) | 112 |
| Tomatidine | | | | |
| 7α,11α-dihydroxy- | Hxl: 7α,11α-diOH | 266—270 | + 23,3 (E) | 355, 355*b* |
| 7α-hydroxy- | Hxl: 7α-OH | 243—247 | − 3,5 (C) | 355, 355*b* |
| 9α-hydroxy- | Hxl: 9α-OH | 192—195 | + 2,3 (C) | 355, 355*b* |

| *Metabolite* | Type of transformation (*see p. 129*) | m. p. °C | $[\alpha_D]$ | Reference (*see p. 181*) |
|---|---|---|---|---|
| Triamcinolone | Hxl: 16α-OH | 269—271 | + 75 (An) | 461, 623 |
| | D: 1—2 | | | 461, 627*a* |
| | | | | 634 |
| 16α,17α-acetone ketal 21-acetate | E: 21-OAc | | | 192 |
| 6α-chloro- | Hxl: 16α-OH | | | 976 |
| 6-dehydro- | Hxl: 16α-OH | | | 623 |
| 6α-fluoro- | Hxl: 16α-OH | 266—268 | + 43 (?) | 976 |
| 6α-methyl- | D: 1—2 | 242—247 | | 964*a* |

# 12
# LITERATURE

## *a*) Reviews


1. ARNAUDI C.: Oxidations microbiennes des stéroides (33 ref.), Experientia 7: 81—89, *b*951.
2. BERNHAUER K.: Bakterielle Dehydrierungen steroider Hormone (18 ref.), Erg. Enzymforsch. 11: 219—221, 1950.

2a. BUETOW D. E., LEVEDAHL B. H.: Responses of microorganisms to sterols and steroids (293 ref.), Ann. Rev. Microbiol. 18: 167—194, 1964.

3. BOWERS A.: Microbial transformations of steroids, 3rd Nat. Congress of Microbiol. Mexico City 1960.

3*a*) CASAS-CAMPILLO C.: Transformations of steroids by *Actinomycetes* (50 ref.), Rev. Latin Am. Microbiol. 3, Suppl. 6: 1—17 (1960).

4. DEINDOERFERF. H. *et al.*: Fermentation process review, Steroid transformations (56 ref.), Appl. Microbiol. 11: 287—289, 1963.
5. EPPSTEIN S. H. *et al.*: Microbiological transformations of steroids and their applications to the synthesis of hormones (170 ref.), Vitamins Hormones 14: 359—437, 1956.
6. FISCHER F. G.: Die Bedeutung biochemischer Oxydationen und Reduktionen für präparative Zwecke, E. Hydrierungen und Dehydrierungen bei Steroiden, Angew. Chem. 53; 467—470, 1940.
7. FLOREY K.: Hydroxylierung von Steroiden durch biologische Methoden (46 ref.), Chimia 8: 81—87, 1954.
8. FRIED J. *et al.*: The use of microorganisms in the synthesis of steroid hormones and hormone analogues (87 ref.), Rec. Progr. Hormone Res. 11: 149—177, 1955.
9. GHISLANDI V.: I metodi fermentativi per la preparazione di nuovi derivati steroidici (33 ref.), Il Farmaco, Ed. prat. 12: 83—90, 1957.
10. HANČ O.: Microbial transformations of steroid (99 ref.), in Czech, Chemie (Prague) 9: 586—599, 1957.
11. HANČ O., TŮMOVÁ-RIEDL E.: Die mikrobielle Umwandlung von Steroiden (98 ref.), Pharmazie 9: 877—890, 1954.
12. HANČ O.: Transformation of steroids by microorganisms and biochemical reaction mechanisms (42 ref.), Global impacts of applied microbiology, p. 420, Stockholm 1963.

12*a*) HASEGAWA T.: Oxidation of steroids by microorganisms (72 ref.), in Japanese, Nippon Nogei-Kagaku Kaishi 32: 63—70, 1958.

13. KOGAN L. M.: Microbiological transformation of steroids (172 ref.), in Russian, Uspechi Chimii 31: 581—608, 1962.
14. MAMOLI L.: Neue Ergebnisse in der enzymatischen Biochemie der Keimdrüsen Hormone (19 ref.), Öster. chem. Ztg 42: 190—194, 1939; Chimica Industria 21: 70, 1939.

15. Masuo E.: Microbiological oxidation of steroids and its application to industry (26 ref.), in Japanese, Yuki Gosei Kagaku Kyokaishi 18: 718—724, 1960.
16. Meystre Ch. *et. al.*: Nouveaux produits d'oxidation biologique dans la domaine des stéroides, Chimia 9: 119, 1955.
17. Moniz de Aragao R.: Microbial transformation of steroids (138 ref.), Arquiv. Ferment 3: 21, 1961.
18. Neher R.: Chemical and microbiological aspects of steroid syntheses, Antibiotica et Chemotherapia 7: 1, 1960.
19. Peterson D. H.: Microbiological approaches to the preparation of cortisone and hydrocortisone (70 ref.), Research 6: 309 (1953).
20. Peterson D. H.: Microorganisms and steroid transformations, in the book: Perspectives and horizons in microbiology, Chap. 9., Rutgers Univ. Press, New Brunswick 1955.
21. Peterson D. H.: Microbiological conversions of steroids (84 ref.), Record chem. Progr. 17: 211—240, 1956.
22. Peterson D. H.: Microbiological transformation of steroids and their application to the synthesis of hormones (125 ref.), Int. Congr. Biochem., Symp. No 4, Repr. No. 3, P. 82—119, Pergamon Press, London 1959.
23. Peterson D. H.: Introduction of nuclear double bonds into the steroid nucleus by microbial enzymes, Global impacts of applied microbiology, p. 439, Stockholm 1963.
23*a*) Peterson D. H.: Microbial transformations of steroids and their application to the preparation of hormones and derivatives (300 ref.), a chapter in the book: Rainbow C., Rose A. H. (editors) — Biochemistry of industrial microorganisms, Acad. Press, New York 1963, pp. 537—606.
24. Saito H.: Microbial application for steroid syntheses (25 ref.) (in Japanese), J. Jap. Chem. 9: 391—398, 1955.
25. Schubert A. *et al.*: Zur Umwandlung von Steroiden mit Mikroorganismem (49 ref.), Z. Chemie 2: 289, 1962.
25*a*) Shirasaka M. *et al.*: Microbial transformation of steroids, in Japanese, Takamine Kenkyusho Nempo 12: 81—90, 1960.
25*b*) Shirasaka M.: Microbiological conversions of steroid compounds I. II. (61 ref.), in Japanese, Hakko Kyokaishi 18: 81—88, 173—181, 1960.
26. Shull G. M.: Transformation of steroids by molds, Trans. N. Y. Acad. Sci. 19: 147—172, 1956.
27. Stodola F. H.: Chemical transformations by microorganisms, 134 P., Wiley (New York) 1958.
28. Stoudt T. H.: The microbiological transformation of steroids (141 ref.), Adv. appl. Microbiol. 2: 183—222, 1960.
29. Tamm C.: Microbiological transformations of steroids and other natural substances, Planta med. 8: 331—354, 1960.
30. Tamm C.: Umwandlung von Naturstoffen durch mikrobielle Enzyme (131 ref.), Angew. Chemie 74: 225—260, 1962.
31. Titus E.: The metabolism of cardiac lactones by microorganisms (65 ref.), Adv. appl. Microbiol. 3: 279—292, 1961.
32. Vischer E., Wettstein A.: Mikrobiologische Umwandlungen von Steroiden für technische Zwecke, Angew. Chemie 69: 456—463, 1957.
33. Vischer E., Wettstein A.: Enzymic transformations of steroids by microorganisms (183 ref.), Adv. Enzymol. 20: 237—282, 1958.
34. Wettstein A.: Conversion of steroids by microorganisms, Experientia 11: 465—479, 1955.
35. Woodford F. P.: Living reagents (16 ref.), Research 8: 412—418, 1955.

## *b)* Individual papers dealing with microbial transformations of steroids

36. ADAMS W. J. *et al.*: 16-oxo- and 16alpha-hydroxytestosterone, J. chem. Soc. 1956, 297.
37. ADELBERG E. A. *et al.*: Attempted induction of microbial mutants requiring steroidal growth factors, J. Bacteriol. 69: 733, 1955.
38. AGNELLO E. J. *et al.*: A new class of biologically active corticoids, J. am. chem. Soc. 77: 4684, 1955.
39. ARNAUDI C.: Researches on the microbiological degradation of sterols, Appl. Microbiol. 2: 274, 1954.
40. ARNAUDI C.: Sopra le attivitá ed i caratteri di alcuni batteri agenti di bioossidazione, Nuovi Ann. Igiene Microbiol. 2: 85, 1951.
40*a*) ARNAUDI C.: Influenze of the nutritive medium on the microbic degradation of cholesterol, Rend. Ist. Lombardo Sci. 83: 151, 1950.
41. ARNAUDI C.: Oxidation of some sexual hormones by *Micrococcus dehydrogenans,* Boll. sez. ital. Soc. int. Microbiol. 11: 208, 1939.
42. ARNAUDI C.: Microbial oxidation of sterols and of substances belonging to the group of sex hormones. New observations on *Flavobacterium dehydrogenans,* Boll. Ist. sieroterap. milan. 21: 1, 1942.
43. ARNAUDI C.: Intorno alle attivitá biochimiche dei Flavobacterium, Experientia 2: 138, 1946.
44. ARNAUDI C., COLLA C.: Oxydation du cholestérol, Experientia 5: 120, 1949.
45. ASAI T. *et al.*: Hydroxylation of steroids by *Syncephalastrum racemosum,* J. gen. appl. Microbiol (Japan) 5: 127, 1959.
46. ASAI T. *et al.*: Studies on the microbiological hydroxylation of steroids, Nippon Nogeikagaku Kaishi 35: 122, 1961.
46*a*) ASAI T. *et al.*: Hydroxylation of steroids by *Diplodia tubericola,* Agr. biol. Chem. 25, A15. 1961.
47. BABA S., BRODIE H. J., HAYANO M., PETERSON D. H., ŠEBEK K.: The mechanism of enzymatic hydroxylation at an enolizable position, Steroids 1: 151, 1963.
48. BARBER G. W. *et al.*: Preparation of 19-hydroxycorticosterone, J. org. Chem. 25: 1168, 1960.
49. BARKEMEYER H. R. *et al.*: A microbiological reduction of the cortisone A-ring by Streptomyces sp., App. Microbiol. 8: 237, 1960.
50. BARMENKOV A. S. *et al.*: Biological transformation of 16alpha, 17alpha-oxidoprogesterone by *Rhizopus nigricans* (in Russian), Med. Prom. 16, No. 7, p. 45, 1962.
51. BARMENKOV A. S. *et al.*: The improved method for preparation of 11alpha-hydroxyprogesterone (in Russian), Med. Prom. 15, No. 3, p. 39, 1961.
51*a*) BAUMGÄRTEL T., ZAHN D.: Verfahren der Reduktion mit Bakterien im menschlichen Darm, Dtsch Z. Verdaungs. 11: 257, 1951.
51*b*) BELLET P., GÉRARD D.: Sur quelques transformations microbiologiques du digitoxoside, Ann. pharm. franç. 21: 593, 1963.
52. BERNSTEIN S. *et al.*: 15alpha- and 15beta-hydroxy-Reichstein's S substance and transformation products, J. am. chem. Soc. 82: 3685, 1960.
53. BERNSTEIN S. *et al.*: 15alpha- and 15beta-hydroxy-Reichstein's substance S and transformation products, Chemistry Industry 1956, 111.
54. BIESTEK A. *et al.*: Microbiological synthesis of testosterone from dehydroepiandrosterone (in Polish), Acta physiol. polon. 12: 881, 1961.

55. Bloom B. M., Shull G. M.: Epoxidation of unsatured steroids by microorganisms, J. am. chem. Soc. 77: 5767, 1955.
56. Bodánszky A., Kollonitsch J., Wix G.: Mikrobiologische Oxydation von Steroiden durch *Cephalosporia,* Experientia 11: 384, 1955.
57. Bowers A. *et al.*: New Approaches to C-11 oxygenated 19-norpregnanes, J. org. Chem. 27: 361, 1962.
58. Bowers A. *et al.*: Microbiological oxidations of 19-norprogesterone, Tetrahedron 2: 165, 1958.
58*a*) Brueckner K. *et al.*: Reactions with 16-methylene- and 16-hydroxymethylene-17-oxo steroids, Chem. Ber. 94: 2897, 1961.
59. Budzier H. H.: Stimulation von Mikroorganismen durch Sexualhormone, Naturwiss. 43: 68, 1956.
60. Buetow D. E., Levedahl B. H.: The relationships between steroid molecular structure and cell growth, Arch. Biochem. 86: 34, 1960.
61. Buetow D. E., Levedahl B. H.: The action of testosterone at the cell level. Testosterone stimulation of the respiration of *Euglena gracilis,* Arch. Biochem. Biophys. 94: 358, 1961.
61*a*) Butenandt A. *et al.*: Über das Verhalten von delta$^{1}$-ungesättigten Steroidketonen bei der Reduktion mit gärender Hefe, Ber. 73 B, 818, 1940.
62. Butler M., Knight B. C. J. G.: Steroid growth requirement and steroid growth inhibitors of *Mycoplasma,* J. gen. Microbiol. 22: 483, 1960.
63. Camerino B. *et al.*: Selektive Reduktion der Carbonylgruppe von Steroiden durch Hefe, Chimia 7, 234, 1953; Helv. chim. Acta 36: 1945, 1953.
64. Camerino B. *et al.*: Microbiological dihydroxylation of progesterone, Gazz. chim. ital. 83: 684, 1953.
65. Camerino B. *et al.*: Diidrossilazione microbiologica del progesterone, Gazz. chim. ital. 84: 301, 1954.
66. Camerino B., Vercellone A.: Azione del lievito fermentante su 16,17-epossi-20-cheto-steroidi, Gazz. chim. ital. 86: 260, 1956.
67. Camerino B., Modelli R.: Configurazione in C-20 di 17beta-metil-18-nor-delta$^{13}$-17alpha-pregneni ottenuti per riduzione con lievito fermentante, Gazz. chim. ital. 86: 1219, 1956.
68. Camerino B. *et al.*: Ossigenazione di steroidi con *Penicillium notatum Westling,* Gazz. chim. ital. 86: 1226, 1956.
69. Canonica L. *et al.*: Un nuovo prodotto di ossidazione biologica del desossicorticosterone, Gazz. chim. ital. 93: 301, 1963.
70. Canonica L. *et al.*: Nuovi prodotti di ossidazione microbica della sostanze 16alpha-metil-S-Reichstein ad opera di *Curvularia lunata,* Gazz. chim. ital. 93: 368, 1963.
71. Canonica L. *et al.*: Idrossilazione del 3beta-acetossi-14-iso-carda-5,20(22)-dioenolide ad opera del *Mucor griseo-cyanus,* Gazz. chim. ital. 93: 787, 1963.
72. Čapek A., Hanč O. *et al.*: Das Studium der Überführung von Progesteron in Testololacton, Naturwiss. 43: 471, 1956.
73. Čapek A., Hanč O., Pavlů H.: Studies of transformation of steroids by species and strains of the genus *Aspergillus* (in Czech), Čs. Mikrobiol. 2: 168, 1957.
74. Čapek A., Hanč O., Malíková E.: Influence of progesterone on the course of basic fermetation curves in the strain *Rhizopus nigricans* (in Czech), Čs. Mikrobiol. 2: 282, 1957.
75. Čapek A. *et al.*: Studies of the optimum conditions of reduction of delta$^{4}$-androstene-3,17-dione to testosterone by the yeast *Saccharomyces cerevisiae,* Čs. Mikrobiol. 3: 92, 1958 (in Czech).

76. ČAPEK A., PAVLŮ H., HANČ O.: Respirationsmessung als Nachweis der Umwandlung von Steroiden bei Mikroben, Naturwiss. 45: 89, 1958.
77. ČAPEK A., PAVLŮ H., HANČ O.: Einfluss der Steroide auf die Atmung der Mikroorganismem, Folia biol. 4: 337, 1958.
78. ČAPEK A., HANČ O.: Transformation of steroids by different species and strains of *Fusaria*, Folia Microbiol. 5: 251, 1960.
79. ČAPEK A., HANČ O.: Microbial 11beta-hydroxylation of steroids, Folia microbiol. 6: 237, 1961.
80. ČAPEK A., HANČ O.: Transformation of progesterone by various species and strains of *Penicillium*, Folia microbiol. 7: 121, 1962.
81. ČAPEK A., HANČ O., KAKÁČ B., TADRA M.: Dehydrogenation of cortisone in position 1—2, Folia Microbiol. 7: 175, 1962.
82. ČAPEK A., HANČ O.: Transformation of steroids by *Mycobacterium flavum*, Folia Microbiol. 7: 181, 1962.
83. ČAPEK A., HANČ O., TADRA M.: Microbial preparation of 1,4-androstadiene derivatives, Folia microbiol. 8: 120, 1963.
84. ČAPEK A. *et al.*: Microbial preparation of prednisone from cortisone acetate (in Czech), Čs. Farm. 12: 408, 1963.
85. CARVAJAL F. *et al.*: Stereospecific reductions of the 20-carbonyl group, J. org. Chem. 24; 695, 1959.
86. CASAS-CAMPILLO C. *et al.*: Transformation of 19-nor-delta$^{4,7}$-androstadien-3,17-dione to equilin, $7^{th}$ Latin Am. Congr. Chem., Mexico 1960.
87. CASAS-CAMPILLO C., RUIZ J.: 11alpha-Hydroxylation of 19-norprogesterone by *Rhizopus nigricans*, $3^{rd}$ Nat. Congr. Microbiol., Mexico City 1960.
88. CASAS-CAMPILLO C., BEJALIL L. F.: The in vitro effect of 21,21-dimethoxy progesterone on acid-fast microorganisms, Am. Rev. resp. Dis. 83: 272, 1961.
89. CASAS-CAMPILLO C., ESPARZA F.: Steroid transforming activities of hallucinogenic fungi in submerged growth, $8^{th}$ Congr. Microbiol., Montreal 1962, Abstr. p. 71.
90. CASAS-CAMPILLO C. *et al.*: Antimicrobial properties of 21, 21-dimethoxy progesterone and other progesterone analogues, J. Bacteriol. 81: 366, 1961.
91. CHARNEY W.: The microbial 1-dehydrogenation of steroids, Abstr. Com. Intern. Congr. Microbiol., Stockholm 1958, p. 401.
92. CHARNEY W. *et al.*: The microbiological conversion of 4-pregnene-11beta, 17alpha, 21-triol-3,20-dione-11,21-diacetate to compound F, Arch. Biochem. Biophys. 79: 402, 1959.
93. CHARNEY W. *et al.*: The action of *Corynebacterium simplex* on non-corticoid steroid substrates, Tetrahedron 18: 591, 1962.
94. CHATTAWAY F. W. *et al.*: The effect of certain steroids upon the growth of *Trichophyton rubrum*, J. gen. Microbiol. 28: 437, 1962.
95. CHATTAWAY F. W. *et al.*: On the mode of action of 11-deoxycorticosterone on the metabolism of *Trichophyton rubrum*, J. gen. Microbiol. 30: 261, 1963.
96. CHEN J. W., KOEPSELL H. J., MAXON W. D.: Kinetics of 1-dehydrogenation of steroids by *Septomyxa affinis*, Biotechnol. Bioeng. 4, No. 1., p. 65, 1962.
97. COLINGSWORTH D. R. *et al.*: A partial microbiological synthesis of adrenal cortex hormones, J. am. chem. Soc. 74: 2381, 1952.
98. COLINGSWORTH D. R. *et al.*: A partial microbiological synthesis of hydrocortisone, J. biol. Chem. 203: 807, 1953.
99. CONNER R. L., NAKATANI M.: Stigmasterol antagonism of certain growth inhibitors of

100. CONNER R. L.: Inhibiton of growth of *Tetrahymena pyriformis* by certain steroids, J. gen. Microbiol. 21: 180, 1959.
101. CONNER R. L. *et al.*: Influence of certain sterols and 2,4-dinitrophenol on phosphate accumulation and distribution in *Tetrahymena pyriformis*, J. gen. Microbiol. 26: 437, 1961.
102. COREY E. J., GREGORIOU G. A., PETERSON D. H.: The stereochemistry of 11alpha-hydroxylation of steroids, J. am. chem. Soc. 80: 2338, 1958.
103. DELIN S., PORATH J.: Purification of alpha- and betahydroxysteroid dehydrogenases from *Pseudomonas testosteroni* by gel filtration, Biochim. biophys. Acta 67: 197, 1963.
104. DIRSCHERL W., GEISSLER G.: Über die Einwirkung von Diäthylstilböstrol und Steroidhormonen auf die Atmung von Bäckerhefe, Z. Vitamin Hormon Fermentforsch. 11: 10, 1960.
105. DODSON R. M. *et al.*: Microbiological hydroxylation of $C_{19}$ steroids at position $C_1$ and $C_2$, J. am. chem. Soc. 79: 3921, 1957.
106. DODSON R. M., MUIR R. D.: The microbiological aromatization of steroids, J. am. chem. Soc. 80: 5004, 1958.
107. DODSON R. M., MUIR R. D.: Microbiological transformations. The hydroxylation of steroid at C-9, J. am. chem. Soc. 80: 6148, 1958.
108. DODSON R. M. *et al.*: The oxidation of dehydroisoandrosterone at C-7, J. am. chem. Soc. 81: 6295, 1959.
109. DODSON R. M. *et al.*: 1alpha- and 2beta-hydroxylations of $C_{19}$-steroids, J. am. chem. Soc. 82: 4026, 1960.
110. DODSON R. M., MUIR R. D.: The microbiological aromatization of steroids, J. am. chem. Soc. 83: 4627, 1961.
111. DODSON R. M., MUIR R. D.: The hydroxylation of steroids at C-9, J. am. chem. Soc. 83: 4631, 1961.
112. DODSON R. M., MIZURA S.: The oxidation of androstenedione at C-16, J. org. Chem. 27: 698, 1962.
113. DODSON R. M. *et al.*: The 1beta-hydroxylation of androstenedione, J. org. Chem. 27: 3159, 1962.
114. DULANEY E. L.: Hydroxylation of steroids, principally progesterone, by a strain of *Aspergillus ochraceus*, Mycologia 47: 464, 1955.
115. DULANEY E. L. *et al.*: Hydroxylation of progesterone and 11-deoxy-17-hydroxycorticosterone by *Aspergillus* and *Penicillium*, Appl. Microbiol. 3: 336, 1955.
116. DULANEY E. L. *et al.*: Steroid transformation by *Dactylum dendroides* (Bulliard, Fries), Appl. Microbiol. 3: 372, 1955.
117. DULANEY E. L., STAPLEY E. O.: Studies on the transformation of 11-deoxy-17alpha-hydroxycorticosterone to hydrocortisone with a strain of *Curvularia lunata*, Appl. Microbiol. 7: 276, 1959.
118. EDWARD D. G. F., FITZGERALD W. A.: Cholesterol in the growth of organisms of the *Pleuropneumonia* group, J. gen. Microbiol. 5: 576, 1951.
119. EPPSTEIN S. H. *et al.*: Microbiological transformations of steroids. Preparation of 11-epicorticosterone and 6beta-hydroxy-11-deoxycorticosterone, J. am. chem. Soc. 75: 408, 1953.
120. EPPSTEIN S. H. *et al.*: Preparation of 11alpha-hydroxypregnane-3,20-dione and 11alpha-hydroxy-allopregnane-3,20-dione, J. am. chem. Soc. 75: 421, 1953.
121. EPPSTEIN S. H. *et al.*: The oxygenation of androgens by Rhizopus, J. am. chem. Soc. 76: 3174, 1954.
122. EPPSTEIN S. H. *et al.*: Tertiary hydroxylation of steroids by fungi of the order *Mucorales*, J. am. chem. Soc. 80: 3382, 1958.

123. ERCOLI A.: Biochemische Oxydationen in der Reihe der Sterine und der Keimdrüsenhormonen mittels des *Micrococcus dehydrogenans*, Boll. Sci. Fac. chim. Ind. Bologna 1940, 279.
124. ERCOLI A.: Dehydrierung von Äthinylandrostendiol zu Anhydrooxyprogesteron, Biochim. Terap. sper. 28: 125, 1941.
125. ERCOLI A.: Umwandlung des delta$^5$-Androstendiols-3,17 in Testosteron, Z. physiol. Chem. 270: 266, 1941.
126. ERCOLI A., MOLINA L.: The bacteria which oxidize sterols having hormone activity. A new microorganism *Flavobacterium androstenedionicum*, Boll. Ist. Sieroterap. milanese 23: 1944. 158,
127. ERCOLI A., MOLINA L.: Further contributions to the knowledge of *Flavobacterium androstenedionicum*, Boll. Ist. Sieroterap. milanese 23: 175, 1944.
128. ERCOLI A., DE RUGGIERI P.: Phytochemical reduction of 3,6-diketocholanic acid to 3alpha-hydroxy-6-ketocholanic acid, Boll. Soc. ital. Biol. Sper. 28: 611, 1952.
129. EROSHIN V. K. *et al.*: The ability of fungi *Mucorales* to oxidize steroid S (in Russian), Mikrobiologiya 31: 608, 1962.
130. EROSHIN V. K.: The transformation of steroid S by fungi *Absidiae* (in Russian), Med. Prom. 16, No. 11, p. 23, 1962.
131. ESPARZA F.: 11alpha- and 11beta-hydroxylation of steroid S by *Cunninghamella bainieri*, 3$^{rd}$ Nat. Congr. Microbiol. Mexico City 1960.
132. FERRARI R. A., ARNOLD A.: Inhibition of beta-hydroxysteroid dehydrogenase I. Structural characteristics of some steroidal inhibitors, Biochim. biophys. Acta 77: 349, 1963.
133. FERRARI R. A., ARNOLD A.: Inhibition of beta-hydroxysteroid dehydrogenase II. Kinetics and pH-effect, Biochim. biophys. Acta 77: 357, 1963.
134. FLINES J. DE *et al.*: Microbiological conversion of 19-nortestosterone I. 16-hydroxylation, Rec. Trav. chim. 82: 121, 1963.
135. FLINES J. DE *et al.*: 10- and 11-hydroxylations, Rec. Trav. chim. 82: 129, 1963.
136. FLINES J. DE *et al.*: 12-hydroxylation, Rec. Trav. chim. 82: 139, 1963.
137. FLINES J. DE *et al.*: 15alpha-Hydroxylation, Rec. Trav. chim. 82: 143, 1963.
138. FLINES J. DE *et al.*: 6beta- and 14alpha-Hydroxylation, Rec. Trav. chim. 82: 149, 1963.
139. FONKEN G. S., MURRAY H. C.: Dehydrogenation of 5beta-pregnane-3,11,20-trione-20-ethylene ketal by *Septomyxa affinis*, J. org. Chem. 27: 1102, 1962.
140. FONKEN G. S. *et al.*: Pathway of progesterone oxidation by *Cladosporium resinae*, J. am. chem. Soc. 82: 5508, 1960.
141. FRIED J. *et al.*: Hydroxylation in position 11 and synthesis of cortisone from Reichstein's compound S, J. am. chem. Soc. 74: 3962, 1952.
142. FRIED J. *et al.*: Side-chain degradation, ring. D cleavage and dehydrogenation in ring A, J. am. chem. Soc. 75: 5764, 1953.
143. FUKUI T.: Bildung der 7-oxy-3,12-diketocholansäure aus Dehydrocholsäure durch *Bacillus coli communis*, J. Biochem. (Japan) 25: 61, 1937.
144. GAINES D. S., TOTTER J. R.: Pteroylglutamic acid-like effect of dehydroepiandrosterone on growth of certain microorganisms, Proc. Soc. exp. Biol. Med. 74: 558, 1950.
145. GALE P. R. *et al.*: Identification of vitamin $K_{2(35)}$, an apparent cofactor of a steroidal 1-dehydrogenase of *Bacillus sphaericus MB 431*, Biochemistry 1: 788, 1962.
146. GHISLANDI V.: I metodi fermentativi per la preparazione di nuovi derivati steroidici, Il Farmaco, Ed. prat. 12: 83, 1957.
147. GOODMAN J. J. *et al.*: A study of the microbiological reduction of triamcinolone and related delta$^{1,4}$-3,20-diketosteroids, J. biol. Chem. 235: 965, 1965.
148. GOODMAN J. J., SMITH L. L.: Effect of medium composition on isomerization of 9alpha-

fluoro-16alpha-hydroxy hydrocortisone and 9alpha-fluoro-16alpha hydroxy prednisolone (Triamcinolone) during microbiological fermentations, Appl. Microbiol 8: 363, 1960.
149. GOODMAN J. J.; SMITH L. L.: 2beta- and 16alpha-hydroxylation of 9alpha-fluorohydrocortisone by strain of *Streptomyces roseochromogenes,* Appl. Microbiol. 9: 372, 1961.
150. GOULD D. *et al.*: Reduction with hydroxylating organism *Curvularia lunata,* J. org. Chem. 22: 829, 1957.
151. GREENSPAN G. *et al.*: Introduction of 1 and 2beta-hydroxyl groups into Reichstein's compound S, J. am. chem. Soc. 79: 3922, 1957.
152. GREENSPAN G. *et al.*: Stereospecific microbiological reduction of 3-keto-1,4-pregnadienes to 3-keto-1-pregnenes, J. org. Chem. 26: 1676, 1961.
153. GUAL C. *et al.*: Microbiological synthesis of 2-hydroxyandrosta-1,4-diene-3,17-dione, J. org. Chem. 24: 418, 1959.
154. GUBLER A., TAMM Ch.: Hydroxylierung von Androsten-4-dion-3,17, Testosteron und Cortexon durch *Fusarium lini,* Helv. chim. Acta 41: 301, 1958.
154*a*) GUBLER A. *et al.*: Umwandlung von Cardenoliden durch Mikroorganismen. 12beta Hydroxylierung von Digitoxigenin, Helv. chim. Acta 41: 297, 1958.
154*b*) GVOZDIAK P. I.: Microbiological transformation of digitoxigenin by *Fusarium* and *Aspergillus* (in Russian), Med. Prom. 17, No. 11, p. 26, 1963.
155. HAGIWARA H.: Microbiological oxidation of Reichstein's substance S by *Corticium sasakii.* Isolation of hydrocortisone and 11-epihydrocortisone, J. Pharm. Soc. Japan 80: 962, 1960 (Yakugaku Zasshi).
156. HAGIWARA H.: Microbiological oxidation of Reichstein's substance S by *Corticium sasakii.* 17alpha, 19,21-trihydroxy-4-pregnene-3,20-dione, J. Pharm. Soc. Japan 80: 965, 1960.
157. HAGIWARA H.: Microbiological hydroxylation of Reichstein's substance S by *Corticium sasakii.* Isolation of cortisone, 6beta, 17alpha, 21-trihydroxypregnen-3,20-dione and prednisolone, J. Pharm. Soc. Japan 80: 1667, 1960.
158. HALPERIN A. H. *et al.*: Bacterial oxidation of cholic acid, Arch. Biochem. Biophys. 52, 5, 1954.
159. HANČ O., ČAPEK A. *et al.*: Condition studies of 11alpha-hydroxylation of progesterone in unsterile medium (in Czech), Čs. Farm. 6: 119, 1957.
160. HANČ O., ČAPEK A. *et al.*: Direkte Überführung von Progesteron in Testosteron, Arzneimittelforsch. 7: 175, 1957.
161. HANČ O., ČAPEK A. *et al.*: Preparation of testosterone from dehydroepiandrosterone acetate (in Czech), Čs. Farm. 6: 373, 1957.
162. HANČ O. *et al.*: Über die Darstellung von Testosteron und seinem 17alpha-Epimeren, Coll. Czech. chem. Commun. 23: 982, 1958.
163. HANČ O., ČAPEK A.: Spaltung der Seitenkette von Steroiden durch Mikroorganismen, Abstr. Com. Intern. Congr. Microbiol. Stockholm 1958, p. 403.
164. HANČ O., ČAPEK A.: Our experiences in microbial transformations of steroids (in Russian), Pharmacotherapeutica 1950—1959, Coll. of Papers, 1961 Prague, p. 157.
165. HANČ O., ČAPEK A., KAKÁČ B.: Transformation of steroid S (Reichstein) by *Absidia orchidis,* Folia microbiol. 6: 392, 1961.
166. HANČ O.: Microbial biosyntheses of androgens (in Czech), Habilitation thesis, Tech. Univ. Prague 1962.
167. HANČ O., ČAPEK A., TADRA M.: Preparation of cortisol and cortisone from 17alpha-hydroxycortexone by *Absidia orchidis* (in Czech), Čs. Farm. 11: 181, 1962.
168. HANSON F. R. *et al.*: Preparations of 17alpha hydroxycorticosterone, J. am. chem. Soc. 75: 5369, 1953.
169. HANZE A. R., ŠEBEK O. K., MURRAY H. C.: Multiple oxydation of the steroid nucleus, J. org. Chem. 25: 1968, 1960.

170. HARTMAN R. E., HOLMLUND C. E.: Binding of steroids by microorganisms, J. Bacteriol. 84: 1254, 1962.
171. HASEGAWA T. *et al.*: 11beta-Hydroxylation of steroid S Reichstein to cortisol, Bull. agr. Chem. Soc. Japan 21: 390, 1957.
172. HASEGAWA T. *et al.*: Stereochemie des 15alpha- sowie 15beta-Hydroxyprogesterone, Chem. pharm. Bull. Japan 9: 740, 1961.
173. HASEGAWA T., TAKAHASHI T.: Oxidation of steroids by *Polyporus orientalis*, Takeda Kenkyusho Nempo 19: 164, 1960 (in Japanese).
174. HAYAISHI O. *et al.*: 3-hydroxy bile acid dehydrogenase from *Escherichia freundii*, J. Bacteriol. 79: 145, 1960.
175. HAYAISHI O. *et al.*: Reversible enzymatic oxidation of bile acids, Arch. Biochem. Biophys. 56: 554, 1955.
176. HAYAKAWA S.: On beta-oxidation and unsaturation of cholic acid by Actinomyces, Proc. Japan Acad. 30: 128, 1954.
177. HAYAKAWA S.: Formation of 7-hydroxy-3,12-dioxobisnor-4,9(11)-choladienic acid from cholic acid by *Actinomyces*, Proc. Japan Acad. 30: 133, 1954.
178. HAYAKAWA S.: Partial synthesis of methyl-7alpha-acetoxy-3,12-dioxobisnorcholanate from cholic acid, Proc. Japan. Acad. 30: 139, 1954.
179. HAYAKAWA S., SATO Y.: Microbiological transformation of diosgenin, J. org. Chem. 27: 704, 1962.
180. HAYAKAWA S., KUROKAWA K.: Microbiological dehydroxylation of cholic acid, Nature 199: 490, 1963.
180*a*) HAYAKAWA S., SATO Y.: Microbiological transformation of diosgenin, J. org. Chem. 28: 2742, 1963.
181. HAYANO M. *et al.*: Hydroxylation of steroids by microorganism in the presence of $^{18}O_2$, Biochim. biophys. Acta 21: 380, 1956.
182. HAYANO M. *et al.*: Steric considerations in the enzymatic course of the hydroxylation of steroids, J. am. chem. Soc. 80: 2336, 1958.
183. HAYANO M. *et al.*: Enzymic mechanism of the 12beta-hydroxylation of steroids, Biochim. biophys. Acta 32: 269, 1959.
184. HAYANO M. *et al.*: The stereochemical course of enzymatic steroid 1,2-dehydrogenation, Biochem. biophys. Res. Com. 4: 454, 1961.
185. HENNING H. D., ZANDER J.: Verwendung von Hübeners 20beta-Hydroxysteroid-Dehydrogenase bei mikrochemischer Identifizierung und Trennung von Steroiden, Z. physiol. Chem. 330: 31, 1963.
186. HERZOG H. L. *et al.*: 2beta-hydroxylation. Microbial transformation of steroids, J. am. chem. Soc. 79: 3921, 1957.
187. HERZOG H. L. *et al.*: 15beta-hydroxylation, J. org. Chem. 24: 691, 1959.
187*a*) HERZOG H. L. *et al.*: Microbial transformation of steroids. 1-Dehydro analogs of cortical steroids, Tetrahedron 18: 581, 1962.
187*b*) HERZOG H. L. *et al.*: 6beta-hydroxylation and other transformations of testosterone by *Wojnowicia graminis*, J. org. Chem. 25: 2177, 1960.
188. HEUSGHEM C., WELSCH M.: Transformation de l'oestradiol en oestron par un *Streptomyces*, Bull. Soc. chim. biol. 31: 282, 1949.
189. HOEHN W. M. *et al.*: Separation and identification of ketocholanic acids formed during oxidation of cholic acid by *Alcaligens faecalis*, J. biol. Chem. 152: 59, 1944.
190. HOKAMA Y., SALLE A. J.: Studies on the metabolism of bile acids by the members of the *Enterobacteriaceae* and *Streptococcus faecalis*, J. Bacteriol. 75: 130, 1958.

191. Holmlund C. E. *et al.*: 1,2-Dehydrogenation of steroids by *Bacillus sphaericus, Corynebacterium simplex, Nocardia corallina* and *Septomyxa affinis*, I[st] Intern. Fermentation Symposium Roma 1960.
192. Holmlund C. E. *et al.*: Microbiological esterification of steroids, J. am. chem. Soc. 83: 2586, 1961.
192*a*) Holmlund C. E. *et al.*: 2-Fluoroprednisone, J. org. Chem. 27: 2122, 1962.
193. Holmlund C. E. *et al.*: Stereospecific oxidation of a methylthio steroid to sulfoxide by *Calonectria decora*, J. org. Chem. 27: 1468, 1962.
193*a*) Holmlund C. E. *et al.*: Microbial formation and hydrolysis of testololactone, Arch. Biochem. biophys. 103: 105, 1963.
194. Horvath J., Krámli A.: Microbiological oxidation of cholesterol with *Azotobacter*, Nature 160: 639, 1947.
195. Hübener J. H., Lehmann C. O.: Die enzymatische 20-Keto-Reduktion durch Extrakte aus *Streptomyces hydrogenans*, Z. physiol. Chem. 313: 124, 1958.
196. Hübener H. J. *et al.*: 20beta-Hydroxysteroid Dehydrogenase, ein neues kristallines Enzym, Biochim. biophys. Acta 35: 270, 1959.
197. Hübener H. J. *et al.*: 20beta-Hydroxysteroid-Dehydrogenase. Darstellung und Kristallisation, Biochem. Z. 333: 95, 1960.
198. Hughes H. B., Schmidt L. H.: Oxidation of steroids by *Alcaligenes faecalis*, Proc. Soc. exp. Biol. Med. 51: 162, 1942.
199. Hurlock B., Talalay P.: 3alpha-Hydroxysteroids as coenzymes of hydrogen transfer between di- and tri-phosphopyridine nucleotides, J. biol. Chem. 233: 886, 1958.
200. Iizuka H. *et al.*: Microbiological hydroxylation of progesterone by *Aspergillus*, J. gen. appl. Microbiol. (Japan) 4: 67, 1958.
201. Iizuka H., Naito A.: Oxidation of progesterone and Reichstein's compound S by genus *Pseudomonas* and others, J. gen. appl. Microbiol. (Japan) 6: 169, 1960.
202. Iizuka H. *et al.*: Production of 6beta, 14alpha-dihydroxyprogesterone and 6beta-hydroxycompound S by bacteria, Nippon Nogei Kagaku Kaishi 34: 472, 1960 (Bull. agr. chem. Soc. (Japan) 24, A 48, 1960).
203. Iizuka H. *et al.*: Oxidation of steroids by *Bacillus pulvifaciens*, J. gen. appl. Microbiol. (Japan) 7: 118, 1961.
204. Ishii H. *et al.*: Transformation of cardiac aglycones by microorganisms. Formation of 7beta-hydroxydigitoxigenin, J. Pharm. Soc. Japan 80: 1150, 1960.
205. Ishii H. *et al.*: Transformation of cardenolides by microorganisms. Formation of periplogenin from digitoxigenin, Ann. Rep. Shionogi Res. Lab. 11: 59, 1961; J. Pharm. Soc. Japan 81: 153, 1961.
206. Ishii H. *et al.*: Formation of 7beta-hydroxydigitoxigenin by *Rhizopus arrhizus*, J. Pharm. Soc. Japan 81: 805, 1961.
207. Ishii H.: *et al.*: Dihydroxylation of digitoxigenin by *Absidia orchidis*, J. Pharm. Soc. Japan 81: 1051, 1961.
208. Isono M., Abe M.: 1-dehydrogenation of hydrocortisone by bacteria. Nippon Nogeikagaku Kaishi 35: 672, 1961 (Agr. biol. Chem. Japan 25: A 56, 1961).
209. Isono M., Abe M.: Oxidation of hydrocortisone by the mixed culture of different bacteria, Nippon Nogeikagaku Kaishi 35: 677, 1961 (Agr. biol. Chem. Japan 25: A 56, 1961).
209*a*) Juhasz G., Tamm Ch.: 7beta-Hydroxylierung von Digitoxigenin durch *Aspergillus oryzae, Rhizopus arrhizus* und *Trichothecium roseum* (Helv. chim. Acta 44: 1063, 1961).
209*b*) Kamiya T., Yamano T.: Oxidation of hydroxyl group of gitoxigenin, Chem. Pharm. Bull. (Japan) 9: 579, 1961.
210. Kapow E. O., Petsiavas B. N.: Effect of physical variables on microbiological transformation of steroids, Ind. Eng. Chem. 48: 2213, 1956.

211. KAWAHARA F. S., SHU-FANG WANG, TALALAY P.: Cristalline delta$^5$-3-ketosteroid isomerase, J. Biol. Chem. 235, PC1, 1960; 237: 1500, 1962.
212. KENNEY H. E. *et al.*: 11alpha-Hydroxylation of tigogenin derivatives, J. am. chem. Soc. 82: 3689, 1960.
213. KIRK D. N. *et al.*: The microbiological hydroxylation of 4-methyltestosterone with *Rhizopus nigricans*, J. chem. Soc. 1960, 3872.
214. KITA D. A. *et al.*: Introduction of the 16alpha-hydroxyl group into oestrogens by *Streptomyces*, Nature 190: 627, 1961.
214*a*) KLUGER B., SIEBERT R., SCHUBERT A.: Mikrobielle Oxydation von Steroiden in 15alpha-Stellung durch *Fusarien*, Naturwiss. 44: 40, 1957.
215. KLUEPFEL D., CORONELLI C.: The microbiological dehydrogenation of 16beta-methyl-5alpha-dehydrocortisone, Experientia 18: 441, 1962.
216. KOEPSELL H. J.: 1-Dehydrogenation of steroids by *Septomyxa affinis*, Biotechnol. Bioeng. 4: 57, 1962.
217. KOGAN L. M. *et al.*: Preparation of delta$^4$-pregnane-17alpha, 20beta, 21-triol-3-one by *Actinomyces* (in Russian) Izv. Akad. Nauk SSSR (chim.) 1962, 302.
218. KOGAN L. M.: *et al.*: Reduction of 17alpha, 21-dihydroxy-20-keto steroids by *Actinomyces albus 3006* (in Russian). Izv. Akad. Nauk SSSR (chim.) 1963, No 2, p. 328.
219. KONDO E., MASUO E.: On the oxidation of Reichstein's substance S by *Gloeosporium* and *Glomerella species*, Nippon Nogei-Kagaku Kaishi 34: 759, 1960 (Bull. agr. chem. Soc. Japan 24, A 68, 1960).
220. KONDO E.: On the oxidation of Reichstein's substance S by *Helmintosporium species*, Nippon Nogei-Kagaku Kaishi 34: 762, 1960 (Bull. agr. chem. Soc. Japan 24, A 68, 1960).
221. KONDO E. *et al.*: On the oxidation of Reichstein's substance S by *Cercospora species*, Nippon Nogei-Kagaku Kaishi 34: 844, 1960 (Bull. agr. chem. Soc. Japan 24, A 75, 1960).
222. KONDO E., MASUO E.: Introduction of double bond into ring A, Nippon Nogei-Kagaku Kaiski 34: 847, 1960 (Bull. agr. chem. Soc. Japan 24: A 75, 1960).
223. KONDO E.: On the production of testololactone by *Aspergillus oryzae* (in Japanese), Ann. Rep. Shionogi Res. Lab. 10: 91, 1960.
224. KONDO E.: On the oxidation of steroids by *Fusarium* species (in Japanese), Ann. Rep. Shionogi Res. Lab. 10: 95, 1960.
225. KONDO E., MASUO E.: On the production of 11-epiprednisolone (in Japanese), Ann. Rep. Shionogi Res. Lab. 10: 103, 1960.
226. KONDO E., MASUO E.: "Pseudo-crystallofermentation" of steroid. A new process for preparing prednisolone by a microorganism, J. gen. appl. Microbiol. (Japan) 7: 113, 1961.
227. KONDO E., MITSUGI T.: On the oxidation of Reichstein's substance S by *Curvularia* species, Nippon Nogei-Kagaku Kaishi 35: 521, 1961 (Agr. biol. Chem. Japan 25, A 44, 1961).
228. KONDO E. *et al.*: 1-Dehydrogenation and 20-carbonyl reduction of steroids by *Streptomyces* species, Agr. biol. Chem. 26: 16, 1962 (Japan).
229. KONDO E.: *et al.*: Reduction of ring A of steroids by *Streptomyces aureus*, Agr. biol. Chem. (Japan) 26: 22, 1962.
230. KONDO E.: Steroid 1-dehydrogenation by a crude enzyme preparation from *Arthrobacter simplex*, Agr. biol. Chem. (Japan) 27: 69, 1963.
231. KRÁMLI A., HORVÁTH J.: Microbiological oxidations of steroids, Nature 162, 1948.
232. KRÁMLI A. *et al.*: Microbiological oxidation of sterols, Nature 163: 219, 1949.

233. KRASILNIKOV N. A. *et al.*: Transformation of cortisone to prednisone by *Mycobacterium sp. No. 193* (in Russian), Dokl. Akad. Nauk SSSR 128; 1063, 1959.
234. KRASILNIKOV N. A. *et al.*: 1,2-Dehydrogenation of hydrocortisone by *Mycobacterium sp. No. 193* (in Russian), Dokl. Akad. Nauk SSSR 128: 836, 1959.
235. KRIDER M. M. *et al.*: Hydrolysis of 5alpha-22a-spirostane glycosides by fungal saponases, J. am. chem. Soc. 76: 3515, 1954.
235*a*) KUCAJEVA A. G., ČAPEK A.: Streoid transformation activity of *Actinomycetes* as characteristics for their differentiation (in Russian), Mikrobiologya 32: 995, 1963.
236. KUPCHAN S. M. *et al.*: Microbiological transformations. The aromatization of ring A of Strophantidin, J. am. chem. Soc. 84: 1752, 1962.
237. KUROSAWA Y.: Influence of steroid on the amylase and protaese formation by *Aspergillus* species, Nippon Nogei-Kagaku Kaishi 31: 474, 1957 (in Japanese).
238. KUROSAWA Y.: Metabolic products of testosterone by *Aspergillus oryzae*, Nippon Nogei-Kagaku Kaishi 31: 478, 1957 (in Japanese).
238*a*) KUROSAWA Y.: 11alpha-Hydroxylation of steroids by *Metarrhizium anisopliae*, Nippon Nogei Kagaku Kaishi 32: 515, 1958.
239. KUROSAWA Y. *et al.*: The incorporation of $O^{18}$ into progesterone by *Calonectria decora*, Agr. Biol. Chem. (Japan) 25: 424, 1961.
240. KUROSAWA Y. *et al.*: The epoxidation of unsaturated steroids, Agr. Biol. Chem. (Japan) 25: 838, 1961.
241. LÁBLER L., ŠORM F.: Microbiological hydroxylation of 18-substituted steroids, Coll. czech. chem. Com. 27: 276, 1962.
242. LANGBEIN G.: Beitrag zur Konstitutionsermittlung des 12beta, 15beta-Dioxyprogesterons, Angew. Chem. 69: 725, 1957.
243. LASKIN A. I., THOMA R. W., TBEJO W. H.: 1,2-Dehydrogenation of steroids by *Nocardia*, Bacteriol. Proc. 1960, 34.
244. LESTER G., STONE D., HECHTER O.: The effects of deoxycorticosterone and other steroids on *Neurospora crassa*, Arch. Biochem. Biophys. 75: 196, 1958.
245. LESTER G., HECHTER O.: Effect of deoxycorticosterone on the growth of microorganisms, J. Bacteriol. 76: 365, 1958.
246. LESTER G., HECHTER O.: Growth regulation in *Penicillium puberulum* by estradiol-17beta and deoxycorticosterone, J. Bacteriol. 81: 803, 1961.
247. LEVY H. R., TALALAY P.: Enzymatic introduction of double bonds into steroid ring A, J. am. chem. Soc. 79: 2658, 1957.
248. LEVY H. R., TALALAY P.: Bacterial oxidation of steroids. Ring A dehydrogenations by intact cells, J. biol. Chem. 234: 2009, 1959.
249. LEVY H. R., TALALAY P.: Studies on the enzymatic mechanism of ring A dehydrogenation, J. biol. Chem. 234: 2014, 1959.
250. LINDNER F. *et al.*: Dehydrierung von delta$^4$-3-Ketosteroiden zu 1,4-Dien-3-Ketonen durch Bakterien der *Subtilis-Gruppe*, Naturwiss. 43: 39, 1956.
251. LINDNER F. *et al.*: Gewinnung von 20beta-Hydroxysteroiden aus 17alpha, 21-Dihydroxy-20-ketosteroiden durch mikrobiologische Hydrierung mit *Streptomyces hydrogenans*, Z. physiol. Chem. 313: 117, 1958.
251*a*) LINDNER F. *et al.*: Chemische und mikrobiologische Arbeiten auf dem Gebiet der Nebennierenrindenhormone, Medizin und Chemie 6: 193, 1958.
252. LORCK H., BREMER O.: Microbial 1-dehydrogenation of steroid hormones, 7$^{th}$ Intern. Congress Microbiol. Montreal 1962, Abstr. p. 72.
253. LUEDEMANN G. *et al.*: The transformation of Reichstein's compound S by *Scenedesmus sp.*, J. org. Chem. 26: 4128, 1961.

254. LUEDEMANN G. *et al.*: 6beta-Hydroxylation of steroid S Reichstein by *Chlorella sp.*, Bacteriol. Proc. 1960, 34.
255. MCALEER W. J., DULANEY E. L.: Transformation of progesterone to 11-deoxycorticosterone by *Wojnowicia graminis*, Arch. Biochem. Biophys. 62: 109, 1956.
256. MCALEER W. J., DULANEY E. L.: Transformation of progesterone to 17alpha-hydroxyprogesterone by *Trichoderma viride*, Arch. Biochem. Biophys. 62: 111, 1956.
257. MCALEER W. J. *et al.*: Hydroxylation of progesterone by *Bacillus cereus* and *Bacillus megaterium*, Arch. Biochem. Biophys. 73: 127, 1958.
258. MCALEER W. J. *et al.*: 1-hydroxylation of 9alpha-fluorocortisone, J. org. Chem. 23: 508, 1958.
259. MCALEER W. J. *et al.*: Microbially produced 7alpha- and 7beta-hydroxy-delta$^4$-3-keto steroids, J. org. Chem. 23: 958, 1958.
260. MCGUIRE J. S. *et al.*: Formation of testosterone acetate by *Saccharomyces fragilis*, Biochim. biophys. Acta 45: 392, 1960.
261. MACHIDA M.: Metabolism of cholic acid by microorganism, J. Biochem. (Japan) 40: 435, 1953.
262. MAGANA I.: Transformation of 19-nortestosterone to estrone and estradiol by *Corynebacterium simplex*, 3$^{rd}$ Nat. Congress Microbiol. Mexico City 1960.
263. MAMOLI L., VERCELLONE A.: Über die biochemische Hydrierung des Dehydro-androsteron, Z. physiol. Chem. 245: 93, 1937.
264. MAMOLI L., VERCELLONE A.: Biochemische Umwandlung von delta$^4$-Androstendion in delta$^4$-Testosteron, Ber. 70: 470, 1937.
265. MAMOLI L., VERCELLONE A.: Biochemische Umwandlung von delta$^5$-Androstendion in Isoandrostandiol und delta$^4$-Testosteron, Ber. 70: 2079, 1937.
266. MAMOLI L., VERCELLONE A.: Biochemische Umwandlung von Dehydroandrosteron in das Androstendion Ber. 71: 154, 1938.
267. MAMOLI L., VERCELLONE A.: Bakterielle Oxydation von Dehydroandrosteron zu Androstendion, Ber. 71: 1686, 1938.
268. MAMOLI L.: Biochemische Umwandlung von Dehydroandrosteron in Testosteron, Ber. 71: 2278, 1938.
269. MAMOLI L.: Biochemische Umwandlung von Östron-estern in Östradiol, Ber. 71: 2696, 1938.
270. MAMOLI L., SCHRAMM G.: Über bakterielle Hydrierung des Testosterons und Androstandions, Ber. 71: 2698, 1938.
271. MAMOLI L.: Bakterielle Dehydrierung von Pregnenolon zu Progesteron, Ber. 71: 2701, 1938.
272. MAMOLI L. *et al.*: Über die biochemische Dehydrierung von Keimdrüsenhormonen mit einem reinen Bakterienstamm, Naturwiss. 27: 319, 1939.
273. MAMOLI L. *et al.*: Hydrierungen in der Gruppe der Steroide mit Hilfe eines *Bacillus* der Art *putrificus*, Z physiol. Chem. 261: 287, 1939.
274. MAMOLI L.: Bacterial oxidation of methylandrostenediol to methyltestosterone (in Italian), Gazz. chim. ital. 69: 237, 1939.
275. MAMOLI L.: Über biochemische Dehydrierungen in der Cortingruppe, Ber. 72: 1863, 1939.
276. MANN K. M. *et al.*: Effect of enzyme inhibitors on the microbiological oxidation of steroids, Feder. Proc. 14: 251, 1955.
277. MANN K. M. *et al.*: Studies on microbiological oxidation of steroids by *Cunninghamella blakesleeana*. Effect of alcohols and phenol, Appl. Microbiol. 3: 14, 1955.
278. MARCUS O. I., TALALAY P.: Induction and purification of alpha- and beta-hydroxysteroid dehydrogenases, J. biol. Chem. 218: 161, 1956.

279. Marx W., Wainfan E.: Stimulation of bacterial oxidation by thyroxine and some related compounds, 3rd intern. Congress Biochem. Bruxelles 1955, Abstr. p. 64.
280. Masao I., Matadzo A.: Dehydrogenation of hydrocortisone by bacteria (in Japanese) Nippon Nogei-Kagaku Kaishi 35: 672, 1961.
281. Masuo E.: Microbiological oxidation of steroids, Yuki Gosei Kagaku Kyokaishi 18: 718, 1960 (in Japanese).
282. Mateles R. I., Fuld G. J.: Studies on continuous microbiological transformation of progesterone, 136th Meeting ACS 1959 Atlantic City.
283. Matkovics B.: Über die Wirkung von Sterinen auf Mikroorganismen, Naturwiss. 44: 400, 1957.
284. Matkovics B., Pulay G.: Die Wirkung von Sterinen auf Mikroorganismen, Z. Bakteriol. Parasitenkunde 2, 3. Abt.: 440, 1958.
285. Matkovics B., Konschynsky A.: Die Wirkung des Cholesterins auf das Redoxpotential und die Penicillinproduktion des *Penicillium chrysogenum Q 176* Stammes, Z. Bakteriol. Parasitenkunde 2, 3. Abt.: 540, 1958.
286. Matkovics B.: Die Wirkung des Ergosterins auf das Redoxpotential und Trockensubstanzgehalt von *Penicillium chrysogenum Q 176*-Kulturen, Z. Bakteriol. Parasitenkunde 2, 3. Abt.: 543, 1958.
287. Matkovics B., Sipos G.: Über die Wirkung von Sterinen auf das Wachstum der Mikroorganismen, Acta Univ. Szeged. Acta biol. 5: 25, 1959.
288. Matkovics B.: Über die Einwirkung von Sterinen auf das Wachstum von Mikroorganismen, Pharmazie 15; 122, 1960.
289. Maxwell E. S. *et al.*: Antibiotic effect of steroids on *Saccharomyces fragilis* and the isolation of a resistant mutant, J. Bacteriol. 80: 1, 1960.
290. Mazur R. H., Muir R. D.: Microbiological dehydrogenation of C-ring azasteroids, J. org. Chem. 28; 2442, 1963.
291. Meeks R. C. *et al.*: A convenient method for preparing of 5alpha- and 5beta-1-dehydrosteroids, Chemistry Industry 1958, 391.
292. Meier K. E., Ivens K.: Einfluss von Hormonen auf Bakterienwachstum, Naturwiss. 41: 284, 1954.
293. Meier K. E., Lenk R.: Einfluss von Sexualhormonen auf Bakterien, Naturwiss. 42: 129, 1955.
294. Meier R., Schuler W.: Vergleich der Wirkung von Steroiden auf den Hefestoffwechsel, Helv. physiol. pharmacol. acta 14: 279, 1956.
295. Meister P. D. *et al.*: The preparation of 11alpha-hydroxy-17alpha-progesterone, J. am. chem. Soc. 75: 55, 1953.
296. Meister P. D. *et al.*: The oxygenation of 17alpha-hydroxyprogesterone to 6beta, 17alpha-dihydroxyprogesterone and 11alpha, 17alpha-dihydroxyprogesterone, J. am. chem. Soc. 75: 416, 1953.
297. Meister P. D. *et al.*: 17alpha-hydroxylation, J. am. chem. Soc. 76: 4050, 1954.
298. Meister P. D. *et al.*: The transformation of 3-ketobisnor-4-cholen-22-al to 11alpha, 22-dihydroxybisnor-4-cholen-3-one by *Rhizopus*, J. am. chem. Soc. 76: 5679, 1954.
299. Meystre Ch., Vischer E., Wettstein A.: Mikrobiologische Hydroxylierung von Cortexon in der 7alpha-, 15alpha- oder 15beta-Stellung, Helv. chim. Acta 38: 381, 1955.
299*a*) Meystre Ch., Vischer E., Wettstein A.: Microbiological hydroxylation of steroids in the 17alpha- and 21-positions, Helv. chim. Acta 37: 1548, 1954.
300. Meystre Ch. *et al.*: Nouveaux produits d' oxydation biologique dans la domaine des steroides, Chimia 9: 119, 1955.
301. Mininger R. J., Wall. M.: Steroidal sapogenins. Attempted microbiological oxidation, Arch. Biochem. Biophys. 60: 427, 1956.

302. MOLINA L., ERCOLI A.: The bacteria which oxidize sterols having hormone activity (in Italian), Boll. Ist. sieroterap. milan. 23: 164, 1944.
303. MOLINA L.: Bacteriological characters of a strain of *Corynebacterium mediolanum* (in Ixalian), Boll. Ist. steroterap. milan. 23: 168, 1944.
304. MOLINA L.: Further studies on *Flavobacterium androstenedionicum* (in Italian), Boll. Ist. sieroterap. milan. 23: 172, 1944.
305. MURRAY H. C., ŠEBEK O. K.: The dehydrogenation of steroids by *Septomyxa affinis*, Bacteriol. Proc. 1960, A 22.
306. NAWA H. *et al.*: Microbiological transformations of cardiac aglycones, Nature 184: 469, 1959.
307. NAWA H. *et al.*: Transformation of steroids by *Pseudomonas*, Tetrahedon 4: 201, 1958.
308. NESEMANN G. *et al.*: 20beta-Hydroxysteroid-dehydrogenase. Züchtung von *Streptomyces hydrogenans* und Induktion des Enzyms, Biochem. Z. 333: 88, 1960.
309. NESEMANN G., HÜBENER H. J., SCHMIDT-THOMÉ J.: 20beta-Hydroxysteroid-dehydrogenase. Einfluss der Steroidkonzentration auf die Induktion des Enzyms, Biochem. Z. 336: 329, 1962.
310. NGUYEN-DANG T.: Réduction de la cortisone et du cortisol par une bactérie du genre *Catenabacterium*, Ann. pharm. franç. 20: 556, 1962.
311. NISHIKAWA M. *et al.*: Microbiological oxidation of progesterone by *Fusarium solani*, Pharm. Bull. (Japan) 3: 322, 1955.
312. NISHIKAWA M., HAGIWARA H.: 19-Hydroxylation of Reichsten's compound S by *Corticium sasakii*, Chem. Pharm. Bull. (Japan) 6: 226, 1958.
313. NOBILE A. *et al.*: Microbiological transformation of steroids. Delta$^{1,4}$-dien-3-ketosteroids, J. am. chem. Soc. 77: 4184, 1955.
313*a*) NOGUCHI S.: Hydrolysis of steroidal esters by malt enzyme. Selective hydrolysis of steroidal acetates, J. pharm. Soc. (Japan) 81: 369, 1961.
314. NORMAN A. *et al.*: Microbial transformation products of cholic acid in the rat, Biochim. biophys. Acta 29: 467, 1958.
315. NORMAN A. *et al.*: In vitro formation of deoxycholic and lithocholic acid by human intestinal microorganisms, Proc. Soc. exp. Biol. Med. 110: 552, 1962.
316. NOZAKI Y. *et al.*: Effect of structural changes in the substrate molecule on the bioconversions of cardiac aglycones, Ann. Rep. Shionogi Res. Lab. 11; 9, 1961.
317. NOZAKI Y.: Microbial dehydrogenation of 3-hydroxyl group of *Digitalis* cardiac aglycone, Agr. Biol. Chem. (Japan) 25: 461, 1961.
318. NOZAKI Y., OKUMURA T.: Microbial 1beta-hydroxylation of digitoxigenin, Agr. Biol. Chem. (Japan) 25: 515, 1961.
319. NOZAKI Y.: Influence of pH on the bioconversions of cardiac aglycones by *Rhizopus arrhizus* and *Mucor parasiticus*, Agr. Biol. Chem. (Japan) 25: 559, 1961.
320. NOZAKI Y.: Bioconversion of digitoxigenin by *Absidia orchidis*: 1beta-hydroxylation, Agr. Biol. Chem. (Japan) 25: 879, 1961.
321. NOZAKI Y.: Bioconversion of digitoxigenin by *Absidia orchidis*. Selective inhibiton of hydroxylation reaction by the substrate itself, Agr. Biol. Chem. (Japan) 25: 884, 1961.
322. NOZAKI Y.: Transformations of cardio-active steroids by microorganisms. Bioconversion of 3-epidigitoxigenin and 17alpha-digitoxigenin, Agr. Biol. Chem. (Japan) 25: 890, 1961.
323. NOZAKI Y., MASUO E., SATOH D.: Acceleration of microbial hydroxylations of digitoxigenin by the addition of $C_{21}$ delta$^4$-3-keto steroids, Agr. Biol. Chem. (Japan) 26: 399, 1962.
324. NUSSBAUM A. L. *et al.*: Microbiological stereo-selectivity towards steroidal mixture epimeric at C(20), Chemistry Industry 1960, 836.

325. O'CONNELL P. W. *et al.*: Studies on microbiological oxidation of steroids by *Cunninghamella blakesleeana H-334*. Composition of medium, Appl. Microbiol. 3: 16, 1955.
326. OKADA M. *et al.*: Bioconversion of cardiac aglycones by *Gibberella saubinettii (Mont.) Sacc.*, Chem. Pharm. Bull. (Japan) 8: 530, 1960.
327. PEDERSON R. L. *et al.*: Microbiological transformations of steroids. The preparations of a tertiary hydroxysteroid 10-hydroxy-19-nortestosterone, J. am. chem. Soc. 78: 1512, 1956.
328. PERIMAN E. *et al.*: Microbiological hydroxylation of progesterone, J. am. chem. Soc. 74: 2126, 1952.
329. PERLMAN D.: Microbiological conversion of pregnenolone to progesterone, Science 115: 529, 1952.
330. PERLMAN E. *et al.*: Antibiotic and vitamin B 12 production by a steroid oxidizing *Actinomycete*, J. Bacteriol. 69: 347, 1955.
331. PETERSON D. H., MURRAY H. C.: Microbiological oxygenation of steroids at carbon 11, J. am. chem. Soc. 74: 1871, 5933, 1952.
332. PETERSON D. H. *et al.*: Microbiological transformation of steroids. The 11-epimer of compound F and other new oxygenated derivatives of Reichstein's compound S. A new route to cortisone, J. am. chem. Soc. 75: 412, 1953.
333. PETERSON D. H. *et al.*: Preparation of 11alpha-hydroxy-6-dehydroprogesterone, J. am. chem. Soc. 75: 419, 1953.
334. PETERSON D. H. *et al.*: Degradation of $C_{21}$-steroids to $C_{19}$-ketones and to testololactone, J. am. chem. Soc. 75: 5768, 1953.
335. PETERSON D. H. *et al.*: Oxygenation of 16alpha, 17alpha-oxidoprogesterone to 11alpha-hydroxy-16alpha, 17alpha-oxidoprogesterone by *Rhizopus nigricans Ehrb.*, J. am. chem. Soc. 77: 4428, 1955.
336. PETERSON D. H. *et al.*: Metabolism of progesterone by *Cylindrocarpon radicicola* and *Streptomyces lavendulae*, J. Bacteriol. 74: 684, 1957.
337. PETERSON D. H. *et al.*: Preparation of ring-substituted methyl derivatives of estrone and estradiol by *Septomyxa affinis*, Chemistry Industry 1960, 1301.
338. PORTMAN O. W. *et al.*: Alteration of bile salts by bacteria, Proc. Soc. exp. Biol. Med. 109: 959, 1962.
338*a*) PRAIRIE R. L., TALALAY P.: Enzymatic formation of testololactone, Biochemistry 2: 203, 1963.
338*b*) PROCHÁZKA Ž. *et al.*: Microbial oxygenation of B-Norsteroids, Coll. czech. chem. Commun. 26: 2068, 1961.
339. RAAB W.: Antibacterial action of phenanthrene-related substances, Science 103: 670, 1946.
340. RAO P. G.: Microbiolagical transformations of steroids with *Fusarium roseum*, Indian J. Pharm. 25: 131, 1963.
341. RASPÉ G. KIESSLICH.: 12beta-Hydroxy-Steroide, Naturwiss. 48: 479, 1961.
342. REIMANN H. *et al.*: Chemical and microbiological transformation of some D-norsteroids Chemistry Industry 1963, 334.
343. REISS F.: The effect of hormones on the growth of *Trychophyton purpureum* and *Tr. gypseum*, J. Invest. Dermatol. 8: 245, 1947.
344. REPKE K.: Die Kinetik der Umsetzung von Östron und Östradiol-17beta durch Hefe-Arch. exp. Pathol. Pharmakol. 230: 178, 1957.
345. REUSSER F. *et al.*: Continuous microbiological transformation of steroids, Appl. Microbiol. 9: 346, 1961.
346. RINGOLD H. J., HAYANO M., STEFANOVIC V.: Concerning the stereochemistry and mechanism of the bacterial C-1,2-dehydrogenation of steroids, J. biol. Chem. 238: 1960, 1963.

347. ROBERTS H. R., ROBISON R. S.: Rapid procedure for monitoring steroid transformations, Nature 194: 88, 1962.
348. RODWELL A. W.: The steroid growth-requirement of *Mycoplasma mycoides*, J. gen. Microbiol. 32: 91, 1963.
349. ROLAND J. F., WEINER B. A.: Bio-oxygenation of progesterone by mushrooms, Science 121: 803, 1955.
350. ROTHROCK J. W. *et al.*: Isolation of diosgenin by microbiological hydrolysis of saponin, Arch. Biochem. Biophys. 57: 151, 1955.
351. RUBIN B. A. *et al.*: 11beta-Hydroxylation of steroids by *Curvularia lunata*, Bacteriol. Proc. 1956, 33.
352. SALMONY D.: The effect of oestrogens and chemically relatedcompounds on the respiration of yeast and on oxidative phosphorylation, Biochem. J. 62: 411, 1956.
353. SANTER M. *et al.*: Steroid metabolism by a species of *Pseudomonas*, J. biol. Chem. 198: 397, 1952.
354. SATO Y. *et al.*: Untersuchungen auf dem Gebiet der mikrobiologischen Umsetzung. $C_1$-Dehydrierung von Reichsteins Substanz S, Hydrocortison, Pregnenolon und Dehydroepiandrosteron durch *Bacillus pulvifaciens*, Chem. pharm. Bull. (Japan) 9: 932, 1961.
355. SATO Y., HAYAKAWA S.: Microbiological hydroxylation of steroidal alcaloids, J. org. Chem. 26: 4181, 1961.
355*a*) SATO Y., HAYAKAWA S.: The microbiological hydroxylation of solasodine, J. org. Chem. 28: 2739, 1963.
355*b*) SATO Y., HAYAKAWA S.: The microbiological hydroxylation of tomatidine, J. org. Chem. 29: 198, 1964.
356. SCHAFFNER C. P., GREENSPAN G.: Oxygenation of steroids by *Rhizoctonia sp.*, Com. Intern. Congress Microbiol. Stockholm 1958, Abstr. p. 404.
357. SCHATZ A. *et al.*: The ability of soil microorganism to decompose steroids, J. Bacteriol. 58: 117, 1949.
358. SCHIESSER A.: Activity of deoxycorticosterone for microbes. Activity of *Nocardia*, Ann. Microbiol. 7: 1, 1956.
358*a*) SCHIESSER A.: Microbial cleavage of digitonin, Ricerca Sc. 22: 449, 1952.
358*b*) SCHIESSER A.: Hydrolytic and oxidative activity of *Flavobacteria*, Ricerca Sci. 23: 638, 1953.
358*c*) SCHIESSER A.: Microbial activities on digitonin, Ann. Microbiol. 5: 79, 1953.
359. SCHMIDT L. H. *et al.*: Action of *Alcaligenes faecalis* on cholic acid, J. biol. Chem. 145: 229, 1942.
360. SCHMIDT-THOMÉ J.: Spezifische Reaktionen mit Hilfe von Mikroorganismen, Angew. Chem. 69: 238, 1957.
361. SCHMIDT-THOMÉ J. *et al.*: 20beta-Hydroxysteroid-dehydrogenase. Substratspezifität der Oxydoreduktion und Induktion bei *Streptomyces hydrogenans*, Biochem. Z. 336: 322, 1962.
362. SCHNEIDER W. P., MURRAY H. C.: 6alpha, 16alpha-dimethyl analogues of cortical hormones (11beta-Hydroxylation by *Curvularia lunata*), Chemistry Industry 1960, 1163.
363. SCHÖNHEIMER R. *et al.*: Über die Bedeutung gesättigter Sterine im Organismus, Z. physiol. Chem. 192: 73, 1930.
364. SCHRAMM G., MAMOLI L.: Bemerkung zur biologischen Bildung des epi-Ätiocholandiols, Ber. 71: 1322, 1938.
365. SCHUBERT A. *et al.*: Mikrobielle Hydroxylierung von Steroiden in 12- und 15-Stellung, Chem. Ber. 90: 2576, 1957.
366. SCHUBERT A.: Mikrobielle Hydroxylierung von Progesteron in 9-Stellung, 4. Intern. Congress Biochem. Wien 1958 Sec. 10, p. 123.

367. Schubert A., Siebert R.: Die Umwandlung von hydroaromatischen Verbindungen insbesondere von Steroiden mit *Calonectria decora,* 4. Intern. Congress Biochem., Wien 1958, Sec. 10, p. 123.
368. Schubert A. *et al.*: Mikrobiologische Hydroxylierung von Progesteron in 7alpha-Stellung, Naturwiss. 45: 264, 1958.
368*a*) Schubert A., Siebert R.: Fermentative oxidation of steroids in the 12- and 15-position Chem. Ber. 91: 1856, 1958.
369. Schubert A. *et al.*: Mikrobielle Hydroxylierung von Steroiden in 9- und 14-Stellung, Chem. Ber. 91: 2549, 1958.
370. Schubert A., Siebert R.: Bildung eines vicinalen cis-Diols der Steroid-Reihe auf mikrobiellem Wege, Angew. Chem. 70: 742, 1958.
371. Schubert A. *et al.*: Mikrobielle Hydroxylierung von Progesteron in 6- und 14-Stellung, Z. Naturforsch. 15b: 269, 1960.
372. Schubert A.: Zur mikrobiellen Hydroxylierung von Pregnanderivaten mit *Calonectria decora,* 5$^{th}$ Intern. Congress Biochem. Moscou 1961, Sec. 14, p. 306.
373. Schubert A. *et al.*: Mikrobiologische Hydroxylierungen von Steroiden, Z. Naturforsch. 17b: 436, 1962.
374. Schubert A. *et al.*: Mikrobielle Hydroxylierung von 3alpha, 6alpha-Dihydroxy-5beta-pregnan-20-on mit *Calonectria decora* in 11alpha-Stellung, Tetrahedron 18: 993, 1962.
375. Schubert K. *et al.*: Über die Aromatisierung von C-19-Steroiden durch *Mycobacterium smegmatis,* Z. Naturforsch. 15b: 584, 1960.
376. Schubert K., Böhme K. H., Hörhold C.: Über die Bildung phenolischer Steroide durch Mycobakterien, Ber. dtsch. Akad. Wissenschaft. 2: 308, 1960.
377. Schubert K. *et al.*: Bildung einer Ketosäure durch mikrobiellen Abbau von Progesteron, Z. Physiol. Chem. 325: 260, 1961.
378. Schubert K. *et al.*: Bildung von Aromatisierungs- und Hydrierungsprodukten aus Progesteron durch *Mycobacterium smegmatis,* Z. Naturforsch. 16b: 595, 1961.
379. Schubert K.: Mikrobieller Abbau von Steroiden, 5$^{th}$ Intern. Congress Biochem. Moscou 1961, Sec. 14, p. 306.
380. Schubert K. *et al.*: Der Hydrierungsverlauf bei 17-Ketosteroiden mit Darmbakterien unter anaeroben Bedingungen, Z. Naturforsch. 17b: 84, 1962.
381. Schubert K. *et al.*: Stereospezifische Hydrierung von delta$^{1,4}$-Androstadien-3,17-dion zu delta$^{1}$-5beta-Androsten-3,17-dion und 5beta-Androstan-5alpha-ol-17-on mit *Clostridium paraputrificum* unter anaeroben Bedingungen, Z. physiol. Chem. 332: 310, 1963.
382. Schubert K. *et al.*: Selektive Hydrierung von Steroidhormonen durch Clostridien unter anaeroben Bedingungen, Z. Naturforsch. 18b: 284, 1963.
382*a*) Schubert K. *et al.*: Mikrobieller Abbau von Desoxycorticosteron zu 7a-Methyl-1-glykolyl-perhydroindanon-(5)-(beta-propionsäure-(4)), Z. Naturforsch. 18b: 988, 1963.
383. Schuytema E. C. *et al.*: Transformations of progesterone by *Basidiomycetes,* Appl. Microbiol. 11: 256, 1963.
384. Šebek O. K. *et al.*: Specificity in the conversion of steroids by the *Trichomonads,* Bacteriol. Proc. 1957, 80.
385. Šebek O. K., Michaelis R. M.: Interconversions of 17-keto- and 17beta-hydroxyl groups in steroids by *Protozoa,* Nature 179: 210, 1957.
386. Šebek O. K.: Microbial preparation of intermediates for the syntheses of steroid hormones, Intern. Congress Microbiol. Stockholm 1958, Abstr. Com. p. 405.
387. Šebek O. K. *et al.*: Alternative routes for the preparation of 11alpha-hydroxy-4-androstene-3,17-dione by fungi, Bacteriol. Proc. 59$^{th}$ Gen. Meeting, 1959.
388. Šebek O. K., Reineke L. M., Peterson D. H.: Intermediates in the metabolism of steroids by *Penicillium lilacinum,* J. Bacteriol. 83: 1327, 1962.

389. SEHGAL S. N. *et al.*: Transformation of steroids with spores of microorganisms, 8[th] Intern. Congress Microbiol. Montreal 1962, Abstr. p. 71.
390. SHIRASAKA M. *et al.*: Microbiological hydroxylation of Reichstein's compound S by *Sclerotinia sp.*: 2beta-hydroxylation, Bull. agr. chem. Soc. Japan 22: 273, 1958.
391. SHIRASAKA M. *et al.*: Microbiological hydroxylation of progesterone and 17alpha-hydroxyprogesterone by *Sclerotinia libertiana*: 2beta-hydroxylation, Bull. agr. chem. Soc. Japan 23: 245, 1959.
392. SHIRASAKA M., TSURUTA M.: The microbial reduction of delta$^4$-3-ketone pregnene compounds by a fungus *Alternaria bataticola,* Arch. Biochim. Biophys. 85: 277, 1959.
393. SHIRASAKA M. *et al.*: 11-Hydroxylation of Reichstein's compound S by microbes, Ann. Repts Takamine Lab. 12: 81, 1960.
394. SHIRASAKA M., TSURUTA M.: Microbiological 2beta-hydroxylation of deoxycorticosterone and corticosterone, Arch. Biochem. Biophys. 87: 338, 1960.
395. SHIRASAKA M., TSURUTA M.: 11alpha-hydroxylation of steroids by *Streptomyces* species, Nature 185: 845, 1960.
396. SHIRASAKA M. *et al.*: Microbiological 11beta-hydroxy dehydrogenation of corticosterone, Nature 186: 390, 1960.
397. SHIRASAKA M., OZAKI M.: 16alpha-Hydroxylation of steroids by *Streptomyces sp.*, J. Ferment. Assoc. Japan. 19: 389, 1961.
398. SHIRASAKA M.: Hydroxylation of steroids by *Sclerotinia libertiana*, Chem. Pharm. Bull. (Japan) 9: 54, 1961.
399. SHIRASAKA M.: Hydroxylation of steroids by *Absidia regnieri*, Chem. Pharm. Bull (Japan) 9: 59, 1961.
400. SHIRASAKA M.: Hydroxylation of steroids by *Botrytis cinerea*, Chem. Pharm. Bull. (Japan) 9: 152, 1961.
401. SHIRASAKA M. *et al.*: 11alpha-Hydroxylation of steroids by *Gloeosporium kaki* and *Glomerella lagenarium,* Chem. Pharm. Bull. (Japan) 9: 159, 1961.
402. SHIRASAKA M. *et al.*: Hydroxylation of steroids by *Sclerotinium hydrophilum,* Chem. Pharm. Bull. (Japan) 9: 196, 1961.
403. SHIRASAKA M.: Hydroxylation of steroids by *Stachylidium bicolor,* Chem. Pharm. Bull. (Japan) 9: 203, 1961.
404. SHIRASAKA M. *et al.*: 1-Dehydrogenation of steroids by fungi, Chem. Pharm. Bull. (Japan) 9: 207, 1961.
405. SHIRASAKA M. *et al.*: Hydroxylation of steroids by *Gibberella saubinetii*: 6beta- and 15alpha-Hydroxylation, Chem. PHarm. Bull. (Japan) 9: 238, 1961.
406. SHULL G. M., KITA D. A.: Introduction of the 11beta-hydroxyl group into $C_{21}$ steroids, J. am. chem. Soc. 77: 763, 1958.
407. SHULL G. M., BLOOM B. M.: Microbiological epoxidation of unsatured steroids, 3[rd] Congress Biochem., Bruxelles 1955, Abstr. p. 48.
408. SHULL G. M.: Transformation of steroids by molds, Trans. N. Y. Acad. Sci. 19: 147, 1956.
409. SIEBERT R.: Mikrobielle Umsetzung von Reichstein Substanz S zu Epi-F, 5[th] Intern. Congress Biochem., Moscou 1961, Sec. 14, p. 280.
409*a*) SIH Ch. J., WEISENBORN F. L.: The use of metabolic inhibitors in the study of microbial transformations of organic compounds, J. am. chem. Soc. 82: 2653, 1960.
410. SIH Ch. J.: Enzymatic preparation of 1-dehydro-steroids, J. pharm. Sci. 50: 712, 1961.
411. SIH Ch. J.: Microbial epoxidation of steroids, J. Bacteriol. 84: 382, 1962.
412. SIH Ch. J., BENNETT R. E.: Steroid 1-dehydrogenase of *Nocardia restrictus*, Biochim. biophys. Acta 56: 584, 1962.
413. SIH Ch. J.: Mechanisms of steroid oxidation by microorganisms Biochim. biophys. Acta 62: 541, 1962.

414. SIH Ch. J., LAVAL J.: Enzymic analysis of functional groups on the steroid nucleus, Biochim. biophys. Acta 64: 409, 1962.
415. SIH Ch. J. *et al.*: Microbiological transformation of strophanthidin, J. med. pharm. Chem. 5: 629, 1962.
416. SIH Ch. J. *et al.*: Purification and properties of steroid esterase from *Nocardia restrictus*, J. biol. Chem. 238: 566, 1963.
417. SIH Ch. J., WANG K. C.: Isolation and characterization of 3alpha-H-4alpha-(3'-propionic acid)-7beta-methylhexahydro-1,5-indanedione, Mechanisms of steroid oxidation by microorganisms, J. am. chem. Soc. 85: 2135, 1963.
418. SIH Ch. J. *et al.*: Microbiological transformations. Reduction of the steroid C-19 aldehyde group, J. org. Chem. 28: 854, 1963.
418*a*) SIH Ch. J., RAHIM A. M.: Mechanisms of steroid oxidation by microorganisms, Enzymatic mechanisms of ring A aromatization, J. pharm. Sci. 52: 1075, 1963.
419. SINGH K., SEHGAL S. N., VEZINA C.: Steroid hydroxylation and carbohydrate utilization by conidia of *Aspergillus ochraceus*, $8^{th}$ Intern. Congress Microbiol., Montreal 1962, Abstr. p. 71.
420. SMITH L. L. *et al.*: Microbiological reduction of triamcinolone, J. am. chem. Soc. 82: 1437, 1960.
420*a*) SMITH L. L. *et al.*: 6beta-Hydroxylation of 9alpha-fluorohydrocortisone, J. org. Chem. 26: 974, 1961.
421. SMITH L. L. *et al.*: 2beta-hydroxylation of 9alpha-fluorohydrocortisone by *Streptomyces reseochromogenes*, J. org. Chem. 26: 2859, 1961.
422. SMITH L. L. *et al.*: 16alpha-hydroxysteroids. Carbonyl reduction by *Streptomyces roseochromogenes*, Biochemistry 1; 353, 1962.
423. SMITH P. F.: Fate of ergosterol and cholestanol in *Pleuropneumonia*-like organisms, J. Bacteriol. 84: 534, 1962.
424. SNOG-KJAER S. *et al.*: On the formation of coprosterol in the intestine, Experientia 11: 316, 1955.
425. SOBEL H. *et al.*: The assimilation of cholesterol by *Mycobacterium smegmatis*, J. Bacteriol. 57: 377, 1949.
426. SOKOLOVA L. V. *et al.*: The preparation of epicortisol from compound S (Reichstein) acetate by *Thiegemella hyalospora* (in Russian), Med. Prom. 16, No. 10, p. 26, 1962.
427. SOKOLOVA L. V. *et al.*: The preparation of prednisone from cortisone by *Mycobacterium* $B_5$ (in Russian), Med. Prom. 15, No. 11, p. 29, 1961.
428. SQUIRE E. W., SQUIRE E. M.: Steroid effect on bacterial growth, J. Bacteriol. 55: 766, 1948.
429. ŠROGL M. *et al.*: Oscillopolarographic detection of transformation of steroids by *Actinomycetes*, Folia microbiol. 8: 237, 1963.
430. STADTMAN T. C. *et al.*: Studies on the microbiological degradation of cholesterol, J. biol. Chem. 206: 511, 1954.
430*a*) STEFANOVIC V. *et al.*: Some observations of the delta$^1$-dehydrogenation of steroids by *Bacillus sphaericus*, Biochim. biophys. Acta 71: 429, 1963.
431. STITCH S. R. *et al.*: Microbiological oxidation of testosterone, $4^{th}$ Intern. Congress Biochem. Wien, 1958, Sec. 9, p. 114.
432. STONE D. *et al.*: Hydroxylation of deoxycorticosterone with *Neurospora crassa*, J. am. chem. Soc. 77: 3926, 1955.
433. STOUDT T. H. *et al.*: The microbial 1-dehydrogenation of some delta$^4$-3-ketosteroids, Arch. Biochem. Biophys. 59: 304, 1955.
434. STOUDT T. H. *et al.*: The microbial dehydrogenation of some pregnanes and allopregnanes to 1,4-pregnadienes, Arch. Biochem. Biophys. 74: 280, 1958.

435. SUGAWARA S. *et al.*: Hydroxylation of Reichstein's compound S by *Bacillus cereus,* Arch. Biochem. Biophys. 80: 383, 1959.
436. SUTTER D. *et al.*: Action of several bacterial species on 4-pregnene-17alpha, 20-diol-3,20-dione, J. org. Chem. 22: 578, 1957.
437. SUVOROV N. N. *et al.*: Microbiological deacetylation of corticosteroid 21-acetates (in Russian), Dokl. Akad. Nauk SSSR 132: 1325, 1960.
438. SUVOROV N. N. *et al.*: Microbiological transformation of cortisone by *Mycobacterium* $B_5$ (in Russian), Med. Prom. 14, No. 1, p. 22, 1960.
438*a*) SUVOROV N. N. *et al.*: Microbiological 20alpha-reduction of ketosteroids by *Bacillus megaterium* (in Russian), Dokl. Akad. Nauk SSSR 152: 1130, 1963.
439. SZPILFOGEL S. A. *et al.*: Microbiological preparation of substance U from cortisone, Rec. Trav. chim. 75: 402, 1956.
440. SZPILFOGEL S. A. *et al.*: 1,2-dehydrogenation of Reichstein's substances U and E, Rec. Trav. chim. 75: 1227, 1956.
441. TAKAHASHI T., HASEGAWA T.: The steroid oxidizing property of *Rhizoctonia solani* (in Japanese), Nippon Nogei-Kagaku Kaishi 35: 1394, 1961; Agr. biol. Chem. Japan 26, A 4, 1962.
442. TAKAHASHI T., HASEGAWA T.: The steroid oxidizing property of *Pellicularia filamentosa* group (in Japanese), Nippon Nogei-Kagaku Kaishi 35: 1399, 1961; Agr. biol. Chem. Japan 26, 1962.
443. TAKAHASHI T. *et al.*: 1-Dehydrogenation and 20-hydrogenation by *Stereum fasciatum,* Nippon Nogei-Kagaku Kaishi 36: 67, 1962; Agr. biol. Chem. Japan 26, A 10, 1962.
444. TAKAHASHI T., UCHIBORI Y.: Stereospecific reduction of the 20-carbonyl group by fungi, Agr. biol. Chem. Japan 26: 89, 1962.
445. TAKAHASHI T.: Nuclear magnetic resonance (NMR) spectra of 19-hydroxysteroids and their application to microbiological 19-hydroxylation of steroids, Agr. biol. Chem. Japan. 27: 633, 1963.
446. TAKAHASHI T.: 19-Hydroxylation of 17alpha, 20alpha, 21-trihydroxypregn-4-en-3-one by *Pellicularia filamentosa* f. s. *microsclerotia,* Agr. biol. Chem. Japan 27: 639, 1963.
447. TAKEDA R. *et al.*: Transformation of Reichstein's substance S to prednisolone by *Pseudomonas,* Tetrahedron Letters 1959, No. 18, p. 17.
448. TALALAY P., DOBSON N. M.: Purification and properties of a beta-hydroxysteroid dehydrogenase, J. biol. Chem. 205: 823, 1953.
449. TALALAY P., MARCUS P. I.: Alpha- and beta-hydroxysteroid dehydrogenases, Feder. Proc. 13: 308, 1954.
450. TALALAY P. *et al.*: The reaction catalyzed by a beta-hydroxysteroid dehydrogenase, J. biol. Chem. 212: 801, 1955.
451. TALALAY P., MARCUS P. I.: Specifity, kinetics and inhibition of alpha- and beta-hydroxysteroid dehydrogenases, J. biol. Chem. 218: 675, 1956.
452. TALALAY P., LEVY H. R.: The steric and molecular specificity of steroid dehydrogenases, Ciba Found. Study group 2., 1959, p. 53.
453. TAMM Ch., GUBLER A.: Mikrobiologische Umwandlung von herzaktiven Glykosiden und Aglykonen, Chimia 13: 116, 1959.
453*a*) TAMM Ch., GUBLER A.: Umwandlung von Cardenoliden durch Mikroorganismen. Diginatigenin aus Gitoxigenin, Helv. chim. Acta 41: 1762, 1958.
454. TAMM Ch., GUBLER A.: Umsetzung von Aglykonen und Glykosiden mit *Fusarium lini,* Helv. chim. Acta 42: 239, 1959.
455. TAMM Ch., GUBLER A. Umwandlung von Krötengiften durch Mikroorganismen, 12beta-Hydroxylierung von Bufalin, Helv. chim. Acta 42: 473, 1959.

456. TAMM Ch. *et al.*: Über die Substratspezifität der Hydroxylierung von Androstan- und Pregnanderivaten durch *Fusarium lini*, Helv. chim. Acta 46: 889, 1963.

457. TANABE K. *et al.*: Microbial oxidation of steroids by *Sclerotinia libertiana*, Chem. pharm. Bull. (Japan) 7: 804, 1959.

458. TANABE K. *et al.*: Microbial oxidation of progesterone by *Absidia regnieri*, Chem. pharm. Bull. (Japan) 7: 811, 1959.

459. TELLER J. D., BONGIOVANNI A. M.: Alteration of steroid dehydrogenase synthesis in a mutant of *Pseudomonas testosteroni*, Nature 197: 1112, 1963.

460. TESTA E.: Hydroxylation microbiologique de delta$^1$-corticosteroides, 16$^e$ Congr. Intern. chim., Paris 1957, Res. Com. II. p. 270.

460*a*) TESTA E.: Microbiological hydroxylation of delta$^1$-corticosteroids, Ann. Chim. (Rome) 47: 1132, 1957.

461. THOMA R. W. *et al.*: Oxidation of steroids by microorganisms. 16alpha-hydroxylation of 9alpha-fluorohydrocortisone and 9alpha-fluoroprednisolone by *Streptomyces roseochromogenes*, J. am. chem. Soc. 79: 4818, 1957.

462. THOMA R. W. *et al.*: 1,2-Dehydrogenation of steroids by *Mycobacterium* and *Nocardia*, 1$^{st}$ Intern. Ferment. Symp. Roma 1960.

462*a*) THOMA R. W. *et al.*: Steroid-ring dehydrogenases in microbial cells and cell-free systems, Sci. Repts Ist. Super. Sanità, 1: 326, 1961.

463. TIMOFEEVA A. G. *et al.*: Microbiological preparation of 11alpha-hydroxyprogesterone (in Russian), Med. Prom. 11, No. 7, p. 23, 1957.

463*a*) TIMOFEEVA A. G. *et al.*: Comparative study of steroid transformation by some molds of the imperfect group, Izvest. Akad. Nauk USSR 26: 574, 1961 (in Russian).

464. TITUS E. *et al.*: Production of a new cardiac aglycone by microbiological hydroxylation of digitoxigenin, J. biol. Chem. 235: 3399, 1960.

465. TORI K., KONDO E.: 8beta-Hydroxylation of Reichstein's substance S by microorganism, Tetrahedron Letters 1963, 645.

466. TSUDA K. *et al.*: Hydroxylierung von Progesteron und Reichsteins Substanz S durch *Diplodia tubericola*, J. gen. appl. Microbiol. (Japan) 5: 1, 1959.

466*a*) TSUDA K. *et al.*: 9alpha, 15beta-Dihydroxylation of progesterone by *Syncephalastrum sp.*, Chem. pharm. Bull (Japan) 6: 387, 1958.

467. TSUDA K. *et al.*: Hydroxylierung von Progesteron und Reichsteins Substanz S durch *Achromobacter kashiwasekiensis* nov. sp., J. gen. appl. Microbiol. (Japan) 5: 7, 1959.

468. TSUDA K.: Hydroxylierung von Progesteron in der 7alpha, 14alpha und 15beta-Stellung durch *Syncephalastrum racemosum Cohn*, Chem. pharm. Bull. (Japan) 7: 369, 1959.

469. TSUDA K. *et al.*: 15alpha-Hydroxylierte Reichsteins Substanz S, Chem. pharm. Bull. (Japan) 7: 534, 1959.

470. TSUDA K. *et al.*: Über 7beta-16beta-dihydroxypregn-4-en-3-, 20-dion, Chem. pharm. Bull. (Japan) 8: 626, 1960.

471. TSUDA K. *et al.*: Isolation of 22-dehydrocholesterol from *Hypnea japonica*, J. am. chem. Soc. 82; 1442, 1960.

472. TSUDA K. *et al.*: Hydroxylierung von Progesteron in der 7alpha, 7beta, 14alpha, 15alpha oder 15beta-Stellung durch *Helmintosporium sativum*, Chem. pharm. Bull. (Japan) 9: 735, 1961.

473. TSUDA K. *et al.*: $C_1$-Dehydrierung von Reichsteins Substanz S, Hydrocortison, Pregnenolon und Dehydroepiandrosteron durch *Bacillus pulvifaciens*, Chem. pharm. Bull. (Japan) 9: 925, 1961.

474. TUMOVÁ E., SIBLÍKOVÁ O., HANČ O.: Biochemical oxydation of pregnane steroids (in Czech), Čs. Farm. 4: 65, 1955.

475. TURFITT G. E.: The microbiological degradation of steroids. The sterol content of soils, Biochem. J. 37: 115, 1943.
476. TURFITT G. E.: The cholesterol-decomposing organisms of soil, J. Bacteriol. 47: 487, 1944.
477. TURFITT G. E.: Oxidation of cholesterol by *Proactinomyces sp.*, Biochem. J. 38: 492, 1954.
478. TURFITT G. E.: Oxidation of hydroxysteroids to keto-derivatives by *Proactinomyces sp.*, Biochem. J. 40: 79, 1946.
479. TURFITT G. E.: Fission of the steroid molecule, Biochem. J. 42: 376, 1948.
480. TURFITT G. E.: Steroid utilization by the microflora of soils, J. Bacteriol. 54: 557, 1947.
481. TWEIT R. C. *et al.*: The substrate specificity of hydroxylations by a *Penicillium sp. ATCC 12 556*, J. org. Chem. 27: 3654, 1962.
481*a*) TWEIT R. C. *et al.*: Configuration of 7-hydroxy-delta$^4$-3-oxosteroids, J. org. Chem. 26: 2856, 1961.
482. UCHIBAYASHI M.: Transformation of steroids by *Pseudomonas*, Chem. Pharm. Bull. (Japan) 8: 112, 1960.
483. UCHIBAYASHI M.: Transformation of steroids by *Pseudomonas*, Chem. Pharm. Bull. (Japan) 8: 117, 1960.
484. UCHIBAYASHI M.: Formation of prednisolone from Reichstein's substance S by *Pseudomonas*, Chem. Pharm. Bull. (Japan) 8: 235, 1960.
485. URECH J. *et al.*: Substratspezifische Hydroxylierung von Steroiden mittels Pilz-Stämmen der Gattung *Gibberella*, Helv. chim. Acta 43; 1077, 1960.
486. URECH J. *et al.*: Mikrobiologische Unwandlung eines 11,18-dioxygenierten Progesteron-Derivates, Chimia 15: 573, 1961.
487. URECH J. *et al.*: Préparation et activité biologique de quelques dérivés hydroxylés en 14alpha de l'hydrocortisone et de la prédnisolone, Gaz. chim. ital. 93: 133, 1963.
488. VALÇAVI U. *et al.*: Riduzione microbiologica della sostanze S di Reichstein, Gazz. chim. ital. 91: 958, 1961.
489. VERCELLONE A., MAMOLI L.: Über die biochemische Hydrierung des Androstendions, Z. physiol. Chem. 248: 277, 1937.
490. VERCELLONE A., MAMOLI L.: Über biochemische Dehydrierung in der Reihe des Keimdrüsenhormons. Weiterer Beitrag zur Genese der Sexualhormone, Ber. 71: 152, 1938.
491. VÉZINA C., SEHGAL S. N., SINGH K.: Hydroxylation of progesterone by conidia of *Aspergillus ochraceus*, Appl. Microbiol. 11: 50, 1963.
492. VISHNIAC H. S.: Activity of steroids as growth factors for a *Labyrinthula species*, J. gen. Microbiol. 12: 464, 1955.
493. VISCHER E., WETTSTEIN A.: Mikrobiologische Reaktionen. Seitenketten-Abbau und Dehydrierung bei Steroiden, Experientia 9: 371, 1953.
494. VISCHER E. *et al.*: 16-alfa-Oxy-cortexon. Mikrobiologische Reaktionen. Helv. chim. Acta 37: 321, 1954.
495. VISCHER E. *et al.*: Mikrobiologische Herstellung von 1-Dehydro-Steroiden, Helv. chim. Acta 38: 835, 1955.
496. VISCHER E. *et al.*: Herstellung weiterer 1-Dehydro-Steroide auf mikrobiologischem Wege, Helv. chim. Acta 38: 1502, 1955.
497. VISCHER E. *et al.*: Mikrobiologische Spaltung razemischer Steroide. Synthese von d-Aldosteron, Experientia 12: 50, 1956.
498. VISCHER E.: Microbiological conversion of 11-deoxy-corticosterone, Intern. Congress Microbiol., Stockholm 1958, Abstr. Com. p. 407.
499. VISCHER E., WETTSTEIN A.: Enzymatische Hydrierung im Ring A von Steroiden mittels Streptomyzeten, Experientia 16: 355, 1960.

500. VONDROVÁ O., HANČ O.: 16alpha-Hydroxylation of progesterone by a strain of *Actinomyces globosus,* Folia Microbiol. 5: 247, 1960.
501. VONDROVÁ O., ČAPEK A.: Transformation reactions of steroids as a diagnostic feature in the classification of *Actinomycetes,* Folia Microbiol. 8: 117, 1963.
502. VONDROVÁ O. *et al.*: Hydroxylation of the steroid molecule by strain of *Streptomyces fradiae,* Folia microbiol. 8: 176, 1963.
502*a*) WANG K. C., SIH Ch. J.: Mechanism of steroid oxidation by microorganisms, Seco-intermediates, Biochemistry 2: 1238, 1963.
503. WANG SHU-FANG, KAWAHARA F. S., TALALAY P.: The mechanism of the delta$^{5}$-3-ketosteroid isomerase reaction. Absorption and fluorescence spectra of enzyme-steroid complexes, J. biol. Chem. 238: 576, 1963.
504. WEAVER E. A. *et al.*: Effect of concentration on the microbiological hydroxylation of progesterone, Appl. Microbiol. 8: 345, 1960.
505. WECHTER W. J., MURRAY H. C.: Microbiological 11alpha-hydroxylation of 6beta-hydroxy-3,5-cyclopregnan-20-one, Chemistry Industry 1962, 411.
506. WEHRLI H. *et al.*: 11,20-Dihydroxy-18,20-cyclopregnan-Verbindungen, Helv. chim. Acta 44: 1927, 1961.
507. WAINFAN E. *et al.*: Metabolism of cholesterol by intestinal bacteria in vitro, J. biol. Chem. 207: 843, 1954.
508. WEISS-BERG E., TAMM Ch.: Über die Steroidoxydase von *Fusarium lini* (Bolley), Helv. chim. Acta 46: 1166, 1963.
508*a*) WEISS-BERG E., TAMM Ch.: Hydroxylierung und Dehydrierung durch *Psilocybe semperviva* und *Ps. mexicana,* Helv. chim. Acta 46: 2435, 1963.
509. WEISZ E. *et al.*: Die Bildung von 11-Epicorticosteron auf mikrobiologischem Wege, Naturwiss. 43: 39, 1956.
510. WELSCH M., HEUSGHEM C.: Streptomycés transformant l'oestradiol en oestrone, C. r. Soc. biol. 142: 1074, 1948.
511. WETTSTEIN A.: Phytochemische Hydrierung von Oestron zu alpha-Oestradiol, Helv. chim. Acta 22: 250, 1939.
512. WETTSTEIN A. *et al.*: Biologische Hydroxylierungen von Steroiden, Experientia 8: 422, 1952.
513. WIX G. *et al.*: Verwandlung des Progesterons in 11-Epicorticosteron mit Hilfe von *Aspergillus niger,* Acta microbiol. Acad. Sci. Hung. 4/1, 9, 1957.
514. WIX G.: Versuche zur Isolierung von Lithocholsäureabbauendem Mikroorganismus, Acta microbiol. Acad. Sci. Hung. 2: 369, 1955.
515. WIX G., BODÁNSZKY A., KOLLONITSCH J.: Oxydation von Steroiden durch *Trichoderma viride,* Acta microbiol. Acad. Sci. Hung. 3: 333, 1956.
516. WIX G. *et al.*: Die Wirkung der Ultraviolettbestrahlung auf die sterinoxydierenden Eigenschaften des Stammes *Aspergillus niger* No 47, 4$^{th}$ Intern. Congress Biochem. Wien 1958, Sec. 15, p. 189.
517. WIX G., ALBRECHT K.: A new method for the microbiological oxidation of steroids, Nature 183: 1279, 1959.
518. WIX G., ALBRECHT K.: 1,2-Dehydrogenation of steroids by *Fusarium caucasicum,* Acta microbiol. Acad. Sci. Hung. 8: 339, 1961.
519. WIX G.: Mikrobiologische Darstellung von delta$^{1,4}$-Androstadien-3,17-dion aus verschiedenen Pregnan-Derivaten, Dtsch. Akad. Landwirtschaftswiss., Tagungsber. 27, p. 229, 1961.
520. WIX G., ALBRECHT K.: The influence of some derivatives of naphtalene on the microbiological production of delta$^{1,4}$-androstadienedione with *Fusarium caucasicum,* 5$^{th}$ Intern. Congress Biochem., Moscou 1961, Sec. 14, p. 277.

521. WIX G., ALBRECHT K.: Inhibition of microbiological breakdown of delta [1,4]-androstadienedione with cyclic compounds, J. biochem. microbiol. Technol. Eng. 3: 325, 1961.
522. ZAFFARONI A. *et al.*: Microbiological hydroxylation of steroids in position 21, Experientia 11: 219, 1955.
523. ZETSCHE K.: Der Einfluss von Fermentinhibitoren und Spurenelementen auf die Steroidhydroxylierung durch *Curvularia lunata,* Naturwiss. 47: 232, 1960.
524. ZETSCHE K.: Chemical physiological investigations concerning the hydroxylation of steroids by fungi of the genus *Curvularia,* Arch. Mikrobiol. 38; 237, 1961.
525. ZETSCHE K.: Die mikrobiologische Umwandlung von Reichsteins Substanz S-Azetat in Hydrocortison, Naturwiss. 48: 407, 1961.
526. ZIMMERMANN W., MAY G.: Bekterielle Oxidation von Steroiden, Zbl. Bakteriol. I. Abt., 151: 462, 1944.
527. ZUIDWEG M. H. J. *et al.*: Formation of hydrocortisone by hydroxylation of Reichstein's compound S with an enzyme preparation from *Curvularia lunata,* Biochim. biophys. Acta 58: 131, 1962.

## *c)* Patent literature:

(Arranged according to the state and in the increasing order according to the number.)

U.S. Patents (incl. 1962)

528. Schering co. (KOESTER H., MAMOLI L., VERCELLONE A.): Keto steroids, 2 236 574 (1941).
529. Schering co. (MAMOLI L.): Keto steroids, 2 341 110 (1941).
530. E. Gamble Deaconess Home Assoc. (SCHMIDT L. E., HUGHES H. B.): Conversion of hydroxysteroids to ketosteroids, 2 360 447 (1944).
531. Upjohn Co. (MURRAY H. C., PETERSON D. H.): Oxygenation of steroids by Mucorales fungi, 2 602 769 (1952).
532. Upjohn Co. (MURRAY H. C., PETERSON D. H.): Steroids, 2 649 400 (1953).
533. Upjohn Co. (HAINES W. J., COLINGSWORTH D. R.): Microbial hydroxylation of steroids by *Aspergillus ochraceus,* 2 649 401 (1953).
534. Upjohn Co. (MURRAY H. C., PETERSON D. H.): Biological oxidation of steroids, 2 649 402 (1953).
535. Pfizer Co. (SHULL G. M., KITA D. A., HEIGHTS R.): Oxygenation of steroids, 2 658 023 (1953).
536. Upjohn Co. (MURRAY H. C., PETERSON D. H.): 11alpha, 17alpha, 21-Trioxypregnan-3,20-dione, 2 659 743 (1953).
537. Upjohn Co. (MURRAY H. C., PETERSON D. H.): 11alpha-Oxy-17alpha-methyl-testosterone and his ester, 2 660 586 (1953).
538. Upjohn Co. (MURRAY H. C., PETERSON D. H.): 14alpha-Hydroxy androstenediones, 2 662 089 (1953).
539. Searle Co. (HECHTER O. M. *et al.*): Biochemical 11-oxygenation of steroids, 2 666 016 (1954).
540. Upjon Co. (MURRAY H. C., PETERSON D. H.): 11alpha-Hydroxy progesterone and esters, 2 666 070 (1954).
541. Upjohn Co. (MURRAY H. C., PETERSON D. H.): 11alpha-Hydroxy-17alpha-isoprogesterone and esters, 2 666 071 (1954).
542. Upjohn Co. (MURRAY H. C., PETERSON D. H.): 6beta-17alpha-Dihydroxy-4-pregnene-3,20-dione and esters, 2 670 357 (1954).

543. Upjohn Co. (MURRAY H. C., PETERSON D. H.): 14alpha-Hydroxyprogesterone, 2 670 358 (1954).
544. Upjohn Co. (MURRAY H. C., PETERSON D. H.): 6beta, 17alpha, 21-Trihydroxy-4-pregnene-3,20-dione and esters, 2 671 096 (1954).
545. Upjohn Co. (MURRAY H. C. *et al.*): 6,11alpha-Dihydroxyprogesterone, 2 672 466 (1954).
546. Upjohn Co. (MURRAY H. C., PETERSON D. H.): 21-Acetoxy-14alpha, 17alpha-dihydroxy-4-pregnene-3,20-dione, 2 673 866 (1954).
547. Upjohn Co. (MEISTER P. D.): 11-Keto-17alpha-methyltestosterone, 2 678 933 (1954).
548. Upjohn Co. (HAINES W. J., KUIZENGA M. H.): Biooxidation of steroids in 11-position, 2 683 724 (1954).
549. Upjohn Co. (MURRAY H. C., PETERSON D. H.): 6beta-Hydroxytestosterone and 6beta-Hydroxy-10-normethyltestosterone and their esters, 2 683 725 (1954).
550. Upjohn Co. (MURRAY H. C., PETERSON D. H.): 6beta-Hydroxy-17alpha-methyltestosterone, 2 686 791 (1954).
551. Upjohn Co. (MURRAY H. C., PETERSON D. H.): 11alpha-Hydroxytestosterone and 11alpha-hydroxy-10-nortestosterone, 2 686 792 (1954).
552. Upjohn Co. (MURRAY H. C., PETERSON D. H.): 11alpha-Hydroxy-6-dehydroprogesterone and esters, 2 691 030 (1954).
553. Upjohn Co. (MURRAY H. C., PETERSON D. H.): 4-Bisnorcholen-11alpha, 22-diol-3-on, 2 691 031 (1954).
554. Upjohn Co. (MURRAY H. C., PETERSON D. H.): 11beta-Hydroxy-10-normethylandrostan-3,6-dione, 2 692 273 (1954).
555. Upjohn Co. (MURRAY H. C., PETERSON D. H.): Oxygenation of steroids with Neurospora, 2 695 260 (1954).
556. Upjohn Co. (EPPSTEIN S. H., LEIGH H. M.): Selective oxidation of the 17beta-hydroxyl group of 6beta-hydroxy testosterone, 2 697 715 (1954).
557. Upjohn Co. (EPPSTEIN S. H., LEIGHT H. M.): 17alpha-Methyl-17beta-hydroxy androstane-3,6-dione, 2 697 716 (1954).
558. Upjohn Co. (MURRAY H. C., PETERSON D. H.): 11alpha, 17alpha-Dihydroxy progesterone, 2 698 331 (1954).
559. Upjohn Co. (MURRAY H. C., PETERSON D. H.): 3beta, 7,11alpha-Trihydroxy-5-pregnene-20-one and esters thereof, 2 702 809 (1955).
560. Upjohn Co. (MURRAY H. C., PETERSON D. H.): 3-Hydroxy-5-pregnene-7,11,20-triones and esters thereof, 2 702 810 (1955).
561. PFIZER Co. (SHULL G. M. *et al.*): 11beta, 14alpha, 17alpha-Trihydroxyprogesterone, 2 702 812 (1955).
562. Upjohn Co. (MURRAY H. C., PETERSON D. H.): 3beta, 11alpha-Dihydroxy-5-pregnane-7,20-dione and esters, 2 703 326 (1955).
563. Upjohn Co. (MURRAY H. C., PETERSON D. H.): 21-Acyloxy-4-pregnene-3,6,20-triones, 2 703 327 (1955).
564. Upjohn Co. (MURRAY H. C., PETERSON D. H.): 4-Pregnene-8,17alpha, 21-triol-3,20-dione, 2 703 806 (1955).
565. Searle Co. (DODSON R. M., BERGSTROM C. G.): 9beta, 11beta: 16alpha, 17alpha-Bisepoxyprogesterone, 2 705 711 (1955).
566. O. Mathieson chem. Corp. (PERLMAN D. *et al.*.): 16alpha-Hydroxy progesterone and their esters, 2 709 705 (1955).
567. Upjohn Co. (MURRAY H. C., PETERSON D. H.): 17-Ketosteroids, 2 721 828 (1955).
568. Pfizer Co. (SHULL G. M. *et al.*): Oxygenation of steroids in the 11-position, 2 721 163 (1955).

569. Upjohn Co. (MURRAY H. C., PETERSON D. H.): 14alpha-Hydroxy-11-deoxycorticosterone and 21-acylates, 2 727 711 (1955).
570. Upjohn Co. (MURRAY H. C., PETERSON D. H.): Steroids (14alpha-Hydroxyprogesterone by *Helicostylum piriforme*), 2 735 800 (1956).
571. Upjohn Co. (EPPSTEIN S. H., LEIGH H. M.): 17beta-Acyloxyandrostane-3,6-dione, 2 742 486 (1956).
572. O. Mathieson chem. Co. (FRIED J., THOMA R. W.): 1-Dehydrotestololactone, 2 744 120 (1956).
573. Pfizer Co. (SHULL G. M. *et al.*): 11beta, 14alpha, 17alpha, 21-Tetrahydroxyprogesterone, 2 745 784 (1956).
574. O. Mathieson chem. Corp. (FRIED J. *et al.*): 7- and 15-Hydroxyprogesterone, 2 753 290 (1956).
575. O. Mathieson chem. Corp. (FRIED J., THOMA R. W.): Production of delta$^{1}$-bonds in the ring A of the cyclopentanopolyhydrophenanthrene nucleus by Streptomyces lavendulae, 2 756 179 (1956).
576. Upjohn Co. (MURRAY H. C., PETERSON D. H.): 17-Oxosteroids, (Oxidation by *Penicillium*), 2 762 757 (1956).
577. Pfizer Co. (SHULL G. M. *et al.*): 11beta-Oxygenation of steroids, 2 765 258 (1956).
578. Upjohn Co. (MURRAY H. C., PETERSON D. H.): 6-Oxo-17alpha-methyltestosterone, 2 771 474 (1956).
579. Pfizer Co. (SHULL G. M.): Preparation of delta$^{1,4}$-3-oxo steroids from delta$^{4}$-3-oxo steroids by *Protaminobacter,* 2 776 927 (1957).
580. Ciba pharm. Prod. Inc. (WETTSTEIN A. *et al.*): 21-Hydroxysteroids 2 778 776 (1957).
581. Farbwerke Höchst A. G. (SCHMIDT-THOMÉ J.): 17-Oxosteroids (The preparation of epiandrosterone), 2 783 252 (1957).
582. Pfizer Co. (SHULL G. M. *et al.*): 7,14alpha, 17alpha, 21-Tetrahydroxyprogesterone, 2 783 255 (1957).
583. Secy. Agr. USA (KRIDER M. M. *et al.*): Microbial hydrolysis of steroidal saponins, 2 784 144 (1957).
584. Am. Cyanamid Co. (BERNSTEIN S. *et al.*): 16alpha-Hydroxy-1,4-pregnadienes, 2 789 118 (1957).
585. Am. Cyanamid Co. (FELDMAN L. I., ALLEN W. S.): 11beta-Hydroxylation of steroids, 2 789 940 (1957).
586. O. Mathieson chem. Corp. (THOMA R. W. *et al.*): Production of 11beta-hydroxy steroids by *Colletotrichum,* 2 793 162 (1957).
587. O. Mathieson chem. Corp. (THOMA R. W. *et al.*): 11beta-Hydroxylation of pregnadienes, 2 793 163 (1957).
588. O. Mathieson chem. Corp. (FRIED J., THOMA R. W.): Prednisolone, 2 793 164 (1957).
589. Upjohn Co. (MANN K. M. *et al.*:) Recovery of hydrocortisone, 2 794 816 (1957).
589*a*) TALALAY P.: Selective oxidation-reduction of beta-hydroxy and oxosteroids contg 3beta- and 17beta-groups, 2 796 382 (1957).
590. O. Mathieson chem. Corp. (THOMA R. W., FRIED J.): Testololactones, 2 801 251 (1957).
591. Merck Co. (DULANEY E. L., MCALEER W. J.): 11alpha-Hydroxylation of steroids by *Aspergillus ochraceus,* 2 802 775 (1957).
591*a*) Searle Co. (DODSON R. M., MUIR R. D.): 1-Hydroxy-4-androstene-3,17-dione and its esters, 2 805 231 (1957).
591*b*) HAGEMANN G. *et al.*: Hydroxylation of steroids in the 21-position, 2 805 978 (1957).
591*c*) Am. Cyanamid Co. (BERNSTEIN S., ALLEN W. S.): 16alpha-Hydroxy-1,4-pregnadienes, 2 806 043 (1957).

591*d*) Pfizer Co. (SHULL G. M.): Preparation of 3-oxo-delta$^{1,4}$-steroids from 3-oxo-delta$^{4}$-steroids with Micromonospora, 2 809 919 (1957).

591*e*) Syntex (ZAFFARONI A., RUBIN B. A.): 21-Hydroxylation of cyclopentanophenanthrene derivatives by *Aspergillus niger,* 2 812 285 (1957).

592. Syntex S. A. (ZAFFARONI A., RUBIN B. A.): Process for the introduction of an 11alpha-hydroxy group into cyclopentanophenanthrene derivatives by *Cunninghamella echinulata,* 2 812 286 (1957).

593. Merck Co. (DELANEY E .L., MCALEER W. J.): 17alpha-Oxygenation of steroids by *Sporormia,* 2 813 060 (1957).

593*a*) Schering Co. (NUSSBAUM A. L.): Steroids, 2 814 632 (1957).

594. O. Mathieson chem. Corp. (KROLL H. A. *et al.*): Production of 3-keto-delta$^{1,4}$-pregnadienes by *Bacterium cyclooxydans,* 2 822 318 (1958).

595. Searle Co. (MUIR R. D., DODSON R. M.): Process for the oxygenation of steroids with *Nigrospora,* 2 823 170 (1958).

596. O. Mathieson chem. Corp. (FRIED J., THOMA R. W.): Synthesis of steroids of the 1-dehydrotestololactone series, 2 823 171 (1958).

596*a*) Pfizer Co. (SHULL G. M., BLOOM B. M.): Epoxidation of steroids, 2 830 935 (1958).

597. Pfizer Co. (DAVISSON J. W.): 11-Hydroxylation of 17alpha-hydroxy steroids by the genus *Stachylidium,* 2 830 937 (1958).

598. Pfizer Co. (SHULL G. M., KITA O. A.): Reduction of 20-keto-steroids to 20-hydroxy derivatives, 2 831 876 (1958).

598*a*) Pfizer Co. (AGNELLO E. J. *et al.*): 14alpha, 16alpha-Dihydroxycorticosterone and its 21-esters, 2 835 683 (1958).

598*b*) Schering Co. (NOBILE A.): Production of dienes by *Corynebacteria,* 2 837 464 (1958).

598*c*) Upjohn Co. (ROSSELET J. P.): 3-Oxo-6-methyl-delta$^{1,4}$-steroids, 2 842 566 (1958).

598*d*) Upjohn Co. (MEISTER P. D., WEINTRAUB A.): 11alpha-Hydroxylation of steroids by *Sporotrichum,* 2 877 162 (1959).

598*e*) Upjohn Co. (WEINTRAUB A. *et al.*): 1,2-Dehydrogenation of steroids with *Septomyxa,* 2 902 410 (1959).

599. Upjohn Co. (MURRAY H. C., ŠEBEK O. K.): Use of second steroid to accelerate 1-dehydrogenation of substrate steroid, 2 902 411 (1959).

600. Pfizer Co. (SHULL G. M.): Prednisolone, 2 908 616 (1959).

601. O. Mathieson chem. Corp. (PERLMAN D.): Microbiological oxidation of pregnenolone to progesterone, 2 915 439 (1959).

602. Upjohn Co. (MURRAY H. C., MEISTER P. D.): Manufacture of 17alpha-hydroxy steroid by genus *Trichothecium,* 2 925 366 (1960).

603. Pfizer Co. (SHULL G. M., KITA D. A.): delta$^{1,2}$-14alpha-Hydroxy derivatives of cortisone and hydrocortisone, 2 932 606 (1960).

604. Am. Cyanamid Co. (GOODMAN J. J.): Process for the oxidation of pregnenes, 2 938 834 (1960).

605. Merck Co. (MANNHARDT H. J., METZ H.): 11-alpha-Hydroxy steroids, 2 950 226 (1960).

606. Schering Corp. (CHARNEY W.): delta$^{1,4}$-steroids, 2 951 016 (1960).

607. Merck Co.: Steroid production with *Dactylium dendroides,* 2 954 326 (1960).

608. O. Mathieson chem. Corp. (THOMA R. W., FRIED J.): 9alpha-Fluoro-11beta-hydroxy (or 11-oxo) androstenediones (by *Cylindrocarpon radicicola* from progesterone derivatives), 2 955 075 (1960).

609. Schering Corp. (CHARNEY W. *et al.*): Hydroxylated steroids and methods for their manufacture using *B. megaterium,* 2 958 631 (1960).

610. Merck Co. (DULANEY E. L. *et al.*): 15-Hydroxy progesterone and derivatives thereof, 2 960 434 (1960).

611. O. Mathieson chem. Corp. (Donin M. N. *et al.*): Synthesis of steroids by *Aspergillus giganteus*, 2 960 435 (1960).
612. O. Mathieson chem. Corp.: Synthesis of steroid by *Diploic hatalensis*, 2 960 436 (1960).
613. Pfizer Co. (Nickell L. G.): Transformation of steroids with *Sansevieria zeylanica*, 2 962 422 (1960).
614. Am. Cyanamid Co. (Feldman L. I. *et al.*): Process for the 1-hydroxylation of steroids by *Mortierella*, 2 962 423 (1960).
615. Takeda pharm. Ind. Ltd. (Hasegawa T. *et al.*): Enzymic axidative production of 11- or 19-position hydroxylated steroids (by *Corticium*), 2 966 444 (1960).
616. Ciba A. G.: Process for the microbiological reduction of steroids with *Streptomyces*, 2 972 568 (1961).
617. O. Mathieson chem, Corp. (Laskin A. I. *et al.*): Synthesis of steroids with *Kabatiella phoradendri*, 2 977 286 (1961).
618. Upjohn Co. (Fonken G. S., Murray H. C.): Microbiological conversion of pregnane-series steroids, 2 981 659 (1961).
619. Am. Cyanamid Co. (Blank R. H. *et al.*): Monohydroxylation of steroids (by *Basidiomycetes*), 2 982 695 (1961).
620. Am. Cyanamid Co. (Goodman J. J., Matrishin M.): 16alpha, 17alpha-Dihydroxy-20-keto steroid production by fermentation, 2 982 693 (1961).
621. Merck Corp. (Stoudt T. H. *et al.*): Steroid production with *Streptomyces* and *Cladosporium*, 2 982 694 (1961).
622. Schering Corp. (Carvajal F.): 11alpha-Hydroxylation of steroids by *Glomerella*, 2 985 563 (1961).
623. Pfizer Co. (Kita D. A.): Oxygenation of steroids with *Streptomyces halstedii*, 2 991 230 (1961).
624. Takeda pharm. Ind. Ltd. (Terumichi J.): Dehydrogenation of steroids by *Azotobacter* microorganisms, 2 992 973 (1961).
624*a*) O. Mathieson chem. Corp. (Fried J., Principe P. A.): 6alpha-Halo-11alpha-hydroxy pregnane steroids, 3 004 047 (1959).
624*b*) O. Mathieson chem. Corp. (Lerner L. J. *et al.*): Microbiological synthesis of A-Norpregnane steroids, 3 005 017 (1959).
624*c*) O. Mathieson chem. Corp. (Lerner L. J. *et al.*): Microbiological synthesis of A-Norpregnane steroids, 3 005 018 (1959).
624*d*) O. Mathieson chem. Corp. (Thoma R. W., Weisenborn F. S.): 11alpha-Hydroxy-A-Norprogesterone preparation by fermenation, 3 005 028 (1959).
624*e*) Upjohn Co. (Fonken G. S., Murray H. C.): 16alpha-Hydroxy-5beta-pregnane-3,11,20-trione, 3 007 950 (1960).
624*f*) Schering Corp. (Nobile A.): Ethinyl androstadienes, 3 010 957 (1961).
625. Schering Corp. (Ilavsky J. *et al.*): 11alpha-Hydroxylation of steroids by *Beauveria*, 3 013 945 (1961).
625*a*) Schering Corp. (Charney W. *et al.*): 15beta-Hydroxy adrenocorticoids, 3 014 051 (1961).
626. Merck Corp. (Stoudt T. H.): Ring-A dehydrogenation of steroids with *Nocardia* microorganisms, 3 016 335 (1962).
627. Secy. Agr. USA (Weaver E. A.): Method of increasing microbial activities (Microbial hydroxylation of steroids by *Aspergillus ochraceus*), 3 019 170 (1962).
627*a*) O. Mathieson chem. Corp. (Ross J. W.): 1-Dehydrosteroids, 3 022 226 (1962).
628. Searle Co. (Muir R. D., Dodson R. M.): Microbiological aromatization of steroids, 3 023 229 (1962).
629. Soc. farm. Italia: Production of hydrocortisone by multiple fermentation, 3 030 278 (1962).

630. Wisconsin Alumni Res. Found. (KNIGHT S. G.): Conversion of steroids with mold spores, 3 031 379 (1962).
630*a*) Pfizer Co. (SARDINAS J. L.): Hydroxylating steroids, 3 032 475 (1962).
631. Ciba A. G.: Process for the manufacture of 11alpha-hydroxy-6alpha-halogen-pregnenes, 3 033 759 (1962).
632. Am. Cyanamid Co. (FELDMAN L. I. *et al.*): Production of 1,4-pregnadienes by bacterial fermentation, 3 037 912 (1962).
633. Am. Cyanamid Co. (FELDMAN L. I. *et al.*): Microbiological dehydrogenation of delta$^{4}$-steroids of the pregnene series using *Bacterium mycoides*, 3 037 913 (1962).
634. Am. Cyanamid Co. (FELDMAN L. I. *et al.*): Bacterial production of triamcinolone by bacterial fermentation, 3 037 914 (1962).
635. Upjohn Co. (HANZE A. R. *et al.*): 9alpha-Hydroxylation of delta$^{4}$-3-keto pregnane series steroids by fungi of the *Cunninghamella* and *Helicostylum* genera, 3 038 913 (1962).
635*a*) Pfizer Co. (SHULL G. M.): 19-Hydroxypregnenes, 3 039 926 (1962).
635*b*) Pfizer Co. (SHULL G. M.): 19-Hydroxylated 4-pregnene derivatives, 3 040 038 (1962).
635*c*) Am. Cyanamid Co. (ORIGONI V. E. *et al.*): delta$^{1,4}$-3-Oxosteroids of the pregnane series, 3 047 468 (1962).
636. O. Mathieson chem. Corp. (SIH Ch. J., BEBBETT R. E.): Dehydrogenation of steroids by dehydrogenase and hydrogen carrier, 3 047 469 (1962).
637. Am. Cyanamid Co. (PRUESS L. M. *et al.*): Microbiological 11alpha hydroxylation of 16alpha, 17alpha-epoxy pregnenes, 3 047 470 (1962).
637*a*) Am. Cyanamid Co. (HOLMLUND Ch. E. *et al.*): Pregnanes, 3 047 569 (1962).
638. Schering Corp. (ILAVSKY J., HERZOG H.): 11-Hydroxylation of steroids by *Phoma* micro-organisms, 3 054 725 (1926).
638*a*) CIBA A. G. (WETTSTEIN A., VISCHER E.): 11alpha, 18-Dihydroxyprogesterone, 3 055 806 (1962).
638*b*) O. Mathieson chem. Corp. (FRIED J., THOMA R. W.): Microbial process for 6beta-hydroxytestololactone, 3 058 889 (1962).
638*c*) Schering Corp. (NOBILE A.): 17alpha-Alkyl-1-dehydrotestosterone, 3 060 092 (1962).
639. O. Mathieson chem. Corp. (LASKIN A. I.): 2beta-Hydroxy steroids and microbiological method of producing same, 3 063 992 (1962).

## British Patents

640. Upjohn Co.: 21-Acyloxy-8,17alpha-Dihydroxy-4-pregnene-3,20-diones, 716 075 (1954).
641. Upjohn Co.: Steroids (Transformations by spores of *Helicostylum pyriforme*), 720 429 (1954).
642. Upjohn Co.: Steroids (Oxygenation by *Rhizopus arrhizus*), 720 430 (1954).
643. Upjohn Co.: Steroids (Oxygenation by *Rhizopus nigricans*), 720 509 (1954).
644. Upjohn Co.: Steroids (Oxygenation of progesterone by *Rh. arrhizus*), 723 159 (1955).
645. Upjohn Co.: Steroids (6-Hydroxy-11-keto progesterone), 723 748 (1955).
646. Upjohn Co.: Steroids (14alpha-Hydroxy-4-androstene-3,17-dione), 723 749 (1955).
647. Upjohn Co.: Oxidation of steroids, 724 094 (1955).
648. Upjohn Co.: Steroids, 724 141 (1955).
649. Upjohn Co.: Steroids, 724 142 (1955).
650. Upjohn Co.: Steroids, 724 143 (1955).
651. Upjohn Co.: Steroids, 724 144 (1955).
652. Upjohn Co.: Steroids, 724 145 (1955).
653. Upjohn Co.: Steroids, 724 146 (1955).

654. Upjohn Co.: Steroids, 724 147 (1955).
655. Upjohn Co.: Steroids, 724 148 (1955).
656. Upjohn Co.: Steroids, 724 149 (1955).
657. Upjohn Co.: Steroids 724 532 (1955).
658. Upjohn Co.: Steroids 726 938 (1955).
659. Upjohn Co.: Steroids 726 939 (1955).
660. Upjohn Co.: Steroids 727 811 (1955).
661. Upjohn Co.: Steroids 727 812 (1955).
662. Upjohn Co.: Steroids 728 150 (1955).
663. Upjohn Co.: Steroids 728 230 (1955).
664. Squibb Sons: Biological oxidation of steroids, 732 557 (1955).
664*a*) Upjohn Co. (Murray H. C., Peterson D. H.): 14alpha-Hydroxyprogesterone, 734 798 (1955).
665. Upjohn Co.: Steroids 735 877 (1955).
666. Pfizer Co.: Steroids 740 858 (1955).
667. Upjohn Co.: Steroids 741 120 (1955).
668. Upjohn Co.: Steroids 743 777 (1956).
669. O. Matheison chem. Corp.: Biosynthesis of hydrocortisone, 749 414 (1956).
670. Syntex S. A.: Corticosterones, 749 943 (1956).
671. Merck. Co.: Recovery of steroids from fermentation liquors, 752 006 (1956).
672. Upjohn Co.: Oxygenation of steroids by *Mucorales* fungi, 756 434 (1956).
673. Merck Co.: Oxygenated steroid compounds, 759 731 (1956).
674. Pfizer Co.: Steroids, 765 509 (1957).
675. O. Matheison chem. Corp.: Microbiological oxidation of steroids, 765 867 (1957).
676. Farmaceutici Italia S. A.: Selective reduction of oxo steroids with yeast, 765 888 (1957).
677. Merck Co.: Oxygenated steroid compounds, 767 360 (1957).
678. Merck Co.: Oxygenated steroid compounds, 767 361 (1957).
679. Pfizer Co.: 11beta-Hydroxylation of steroids, 769 998 (1957).
680. Pfizer Co.: Microbiological oxygenation of steroids, 770 000 (1957).
681. Upjohn Co.: Steroid compounds, 772 598 (1957).
682. Upjohn Co.: Steroids, 773 628 (1957).
683. Upjohn Co.: Steroids, 773 928 (1957).
684. Upjohn Co.: Steroids, 774 064 (1957).
685. Pfizer Co.: 11beta-Hydroxylation of steroids, 775 943 (1957).
686. Pfizer Co.: Steroid compounds and the preparation thereof, 783 048 (1957).
687. Pfizer Co.: Improvements in or relating to steroid compounds and the preparation thereof, 787 410 (1957).
688. Pfizer Co.: Steroids and the preparation thereof, 789 363 (1958).
689. Am. Cyanamid Co: Oxidation of steroids, 789 862 (1958).
689*a*) Ciba A. G.: Oxidation products of steroids, 792 803 (1958).
690. Pfizer Co.: Preparation of steroid compounds, 794 158 (1958).
691. Schering Corp.: Improvements in or relating to steroid compounds and to processes for their manufacture, 794 385 (1958).
692. Schering Corp.: 9alpha-Halo steroids and process for manufacturing same, 794 468 (1958).
693. Pfizer Co.: Improvements in or relating to steroids and the preparation thereof, 795 106 (1958).
694. Schering Corp.: Manufacture of 4-pregnene-17alpha, 21-diol-20-ones, 796 769 (1958).
695. Pfizer Co.: Steroid compounds and the preparation thereof, 800 644 (1958).
696. Pfizer Co.: Steroid compounds and the preparation thereof, 800 797 (1958).

697. Lab. Franç. Chimiothérapie: Improvements in or relating to the steroid compounds, 807 203 (1959).
698. Schering Corp.: Steroid compounds and processes for the manufacture thereof, 807 227 (1959).
699. Farmaceutica Italia: New steroid compounds and method for the transformation of epoxy-ketones by the action of fermentating yeast, 813 317 (1959).
700. Pfizer Co.: Steroid compounds and the preparation thereof, 814 000 (1959).
701. O. Mathieson chem. Corp.: Synthesis of steroids, 814 279 (1959).
702. Pfizer Co.: Steroid compounds and the preparation thereof, 814 878 (1959).
703. Ciba Ltd.: Manufacture of polyoxygenated steroids, 816 922 (1959).
704. Ciba Ltd.: Process for splitting steroid racemates, 823 235 (1959).
705. Ciba Ltd.: Process for splitting steroid racemates, 823 719 (1959).
706. Merck Co.: Process for preparing steroid compounds, 824 348 (1959).
707. Am. Cyanamid Co.: New steroids of the delta$^{1,4}$-pregnadiene series and the preparation of delta$^{1,4}$D-pregnadienes, 824 351 (1959).
708. Ciba Ltd.: Dehydro-compounds of the steroid series and process for their manufacture, 825 210 (1959).
708*a*) Am. Cyanamid Co.: Preparation of delta$^{1,4}$-steroids, 826 395 (1960).
709. Ciba Ltd.: Process for the manufacture of polyoxygenated dehydro-steroids, 827 182 (1960).
710. Pfizer Co.: Preparation of steroid compounds, 829 260 (1960).
711. Upjohn Co.: Steroids and the preparation thereof, 829 903 (1960).
712. Merck Co.: Delta$^{1}$-steroids and their preparation by a microbiological process, 830 921 (1960).
713. Pfizer Co.: Steroid compounds and the preparation thereof, 831 073 (1960).
714. Pfizer Co.: Process for the preparation of steroid compounds, 831 256 (1960).
715. Pfizer Co.: Steroid process, 832 116 (1960).
716. Scherico.: Hydroxylated steroids and methods for their manufacture, 832 325 (1960).
717. Shionogi Co.: On improvement in the fermentative preparation of prednisolone, 833 595 (1960).
718. Merck Co.: Steroids, 835 577 (1960).
719. Scherico Ltd.: Improvements relating to 1,4-pregnadienes and the preparation thereof, 835 700 (1960).
720. Pfizer Co.: Steroid compounds and the preparation thereof, 836 724 (1960).
721. O. Mathieson chem. Corp.: Improvements relating to the production of steroids, 837 281 (1960).
722. Merck Co.: Steroid compounds, 838 228 (1960).
723. Merck Co.: Production of steroids, 839 257 (1960).
723*a*) Brit. Drug Houses: 6-Methyl steroid compounds, 840 246 (1960).
724. Merck Co.: Steroid compounds, 841 998 (1960).
725. Schering Corp.: Improvements in or relating to steroid compounds, 843 216 (1960).
726. Schering Corp.: Process for the production of pregnadienes, 843 217 (1960).
727. Brit. Drug Houses Ltd.: 6-Methyl steroid compounds, 840 246 (1960).
728. Schering Corp.: Production of pregnadienes, 843 218 (1960).
729. Merck Co.: Steroid compounds, 845 295 (1960).
730. Merck Co.: Steroid compounds, 846 909 (1960).
731. Merck Co.: Production of steroid compounds, 846 914 (1960).
732. Lepetit S. p. A.: Oxygenation of steroids, 847 610 (1960).
733. Merck Co.: Steroids, 847 844 (1960).

734. Ciba Ltd.: Hydroxy-halogen-pregnenes and process for their manufacture, 847 951 (1960).
735. Lepetit S. p. A.: Improvements in or relating to the manufacture of steroids, 848 914 (1960).
736. Ciba Ltd.: Hydroxy-pregnanes and process for their manufacture, 849 558 (1960).
737. Takeda pharm. Ind.: Hydroxylation of steroids, 849 843 (1960).
738. Ciba Ltd.: Hydroxy-halogen-pregnenes and process for their manufacture, 850 368 (1960).
739. Pfizer Co.: Steroid compounds and process of preparing same, 850 951 (1960).
740. Upjohn Co.: Improvements in or relating to steroids and the manufacture thereof, 851 960 (1960).
741. Upjohn Co.: Improvements in or relating to steroids and the manufacture thereof, 851 990 (1960).
741*a*) Syntex S. A.: 4-Alkyl(aralkyl)-1,4-pregnadiene derivatives, 852 288 (1960).
742. Pfizer Co.: Steroid compounds and preparation thereof, 852 647 (1960).
742*a*) Upjohn Co.: Improvements in or relating to steroids and the manufacture thereof, 853 981 (1960).
743. Brit. Drug Houses Ltd.: Improvements in or relating to 16-substituted steroid compounds, 857 114 (1961).
744. Takeda pharm. Ind.: A process for the transformation of steroids, 859 694 (1961).
745. Upjon Co.: Improvements in or relating to the manufacture of steroids, 861 470 (1961).
746. Lepetit S. p. A.: Improvements in and relating to the manufacture of a steroid, 861 799 (1961).
747. Searle Co.: A process for the preparation of 9alpha-hydroxy steroids, 862 701 (1961).
748. Pfizer Co.: Steroid recovery process, 864 094 (1961).
749. Upjohn Co.: Improvements in or relating to steroids and the manufacture thereof, 872 201 (1961).
750. Upjohn Co.: Improvements in or relating to steroids and the manufacture thereof, 872 210 (1961).
751. Upjohn Co.: Improvements in or relating to steroids and the manufacture thereof, 872 806 (1961).
752. Pfizer Co.: Preparation of steroids, 874 188 (1961).
753. Merck Co.: Steroid compounds, 877 086 (1961).
754. Upjohn Co.: Improvements in or relating to the preparation of steroids, 877 636 (1961).
755. Upjohn Co.: Improvements in or relating to steroids and the manufacture thereof, 878 705 (1961).
756. Upjohn Co.: Improvements in or relating to steroids and the manufacture thereof, 878 706 (1961).
757. Upjohn Co.: Improvements in or relating to steroids and the manufacture thereof, 878 707 (1961).
758. Merck A. G.: Steroids, 878 742 (1961).
758*a*) Upjohn Co.: Steroids, 881 261 (1961).
759. Pfizer Co.: Steroid compounds, preparation thereof and compositions thereof, 881 334 (1961).
759*a*) Upjohn Co.: 1-Dehydro-6-fluoro-oxygenated progesterone and derivatives, 882 604 (1961).
760. Upjohn Co.: Improvements in or relating to steroids and the manufacture thereof, 883 212 (1961).
761. Upjohn Co.: Improvements in or relating to steroids and the manufacture thereof, 883 479 (1961).

762. Wisconsin Alumni Res. Found.: Improvements in a microbiological process for the preparation of steroids and ketones, 883 500 (1961).
762*a*) Upjohn Co.: Improvements in or relating to steroids and the manufacture thereof, 883 711 (1961).
763. Upjohn Co.: Improvements in or relating to steroids and the manufacture thereof, 884 276 (1961).
763*a*) Upjohn Co.: Improvements in or relating to steroids and the manufacture thereof, 884 278 (1961).
764. Upjohn Co.: Improvements in or relating to steroids and the manufacture thereof, 888 994 (1962).
765. Soc. farm. Italia: Production of hydrocortisone, 889 378 (1962).
766. Am. Cyanamid Co.: Process for hydroxylation of steroids, 889 853 (1962).
767. Am. Cyanamid Co.: 3-Keto-4-pregnenes and method of preparation, 890 565 (1962).
768. Syntex S. A.: New pregnadiene compounds, 890 834 (1962).
769. Merck Co.: Steroid compounds and their preparation, 891 594 (1962).
770. O. Mathieson chem. Corp.: A-Nor-steroids, 891 825 (1962).
771. Schering A. G.: Manufacture of a 12-hydroxy-steroid, 892 589 (1962).
772. Upjohn Co.: Improvements in or relating to steroids and the manufacture thereof, 893 148 (1962).
773. Merck Co.: Steroid compounds and their preparation, 896 103 (1962).
774. Merck Co.: Steroid compounds and their preparation, 896 104 (1962).
775. Merck A. G.: Process for the production of 11alpha-hydroxy-steroids, 896 913 (1962).
776. Upjohn Co.: Improvements in or relating to steroids and the manufacture thereof, 898 292 (1962).
777. Upjohn Co.: Improvements in or relating to steroids and the manufacture thereof, 898 492 (1962).
778. Scherico Ltd.: Novel steroids and processes for the manufacture thereof, 901 092 (1962).
779. Takeda pharm. Ind.: Compounds of the estratriene series and their production, 901 104 (1962).
780. Takeda pharm. Ind.: A process for the preparation of delta$^{1,4}$-3-keto-19-hydroxysteroids and esters, 901 105 (1962).
781. Merck Co.: Steroid compounds, 901 306 (1962).
782. Syntex S. A.: Process for the manufacture of halogen-pregnane compounds, 902 028 (1962).
783. Syntex S. A.: Process for the manufacture of hydroxy-pregnane compounds, 902 029 (1962).
784. Pfizer Co.: Steroids and preparation thereof, 902 373 (1962).
785. Merck Co.: Steroid compounds, 905 890 (1962).
786. Farbwerke Höchst A. G.: Process for the manufacture of delta$^{1}$-dehydrosteroids, 908 976 (1962).
787. Am. Home Prod. Corp.: Improvements in an relating to the hydroxylation of steroids, 912 274 (1962).
788. Upjohn Co.: Improvements in or relating to the preparation of steroids, 912 965 (1962).
789. Chinoin: Process for the microbiological oxidation of steroids, 918 736 (1963).
790. Chinoin: Improvements in or relating to the preparation of steroid compounds, 918 738 (1963).
791. Chinoin: Process for the preparation of androsta-1,4-diene-3,17-dione, 918 960 (1963).
792. Upjohn Co.: Improvements in or relating to the manufacture of steroids, 919 686 (1963).
793. Ciba A. G.: Process for the manufacture of 11alpha-hydroxy-17alpha-methyl testosterone, 921 424 (1963).

794. Syntex S. A.: Cyclopentanophenanthrene derivatives and process for the production thereof, 921 533 (1963).
795. Lovans kem. Fabrik: Method for the production of 1-dehydrosteroids, 922 525 (1963).
796. Upjohn Co.: Improvements in or relating to steroids and the preparation thereof, 923 336 (1963).
797. Ciba A. G.: Process for the microbiological reduction of steroids, 923 587 (1963).
798. Upjohn Co.: Improvements in or relating to steroids and the manufacture thereof, 923 947 (1963).
799. Ciba A. G.: New pregnan-18-acid derivatives and pharmaceutical preparations containing them, 924 151 (1963).
800. Merck Co.: Steroid compounds, 924 368 (1963).
801. Ciba A. G.: New halogen-pregnenes and a process for their manufacture, 926 472 (1963).

## French Patents

802. Upjohn Co. (Murray H. C., Peterson D. H.): Procédé d'oxydation de stéroides et produits préparables par ce procédé, 1 162 915 (1958) and 8 supplements.
803. Ciba A. G.: Procédé pour introduire de l'oxygène dans des stéroides, 1 161 349 (1958)
804. Skytte Jensen B. B.: Procédé de préparation d'hormones corticales surrénales, 1 167 432 (1958).
805. O. Mathieson chem. Corp.: Procédé de préparation d'hydrocortisone par biosynthèse, 1 167 465 (1958).
806. O. Mathieson chem. Corp.: Procédé de préparation de stéroides par synthèse, 1 167 917 (1958).
807. Soc. civ. rech. tech. (Vismara F., Ercoli A.): Procédé de préparation de 16alpha, 17alpha-epoxyde-5alpha et 5beta-pregnanes-3,11,20-triones, 1 167 919 (1958).
808. O. Mathieson chem. Corp.: Procédé pour la synthèse de stéroides, 1 169 860 (1959).
809. Merck Co.: Procédé d'oxygénation de stéroides par action microbienne, 1 171 909 (1959).
810. Pfizer Co.: Perfectionnements apportés à l'oxydation de composés stéroides, 1 173 122 (1959).
810*a*) Ciba A. G.: Procédé pour dédoubler des racémates en leurs constituants, 1 173 913 (1959).
810*b*) Schering Co.: Préparation de prégnadiénes, 1 176 095 (1959).
811. Upjohn Co.: Procédé d'oxydation des stéroides par fermentation, 1 180 954 (1959).
812. Ciba A. G.: Procédé d'oxydation biochimique de composés de l'androstaneou de prégnane par une culture d'un eumycéte ou les enzymes qu'elle renferme, 1 185 301 (1959).
813. Pfizer Co.: Perfectionnements apportés aux procédés pour obtenir des produits de fermentation du type corticostéroide, 1 209 063 (1960).
814. Am. Cyanamid Co.: Procédé de préparation de 3-céto-4-prégnenes, 1 210 454 (1959).
815. Pfizer Co.: Perfectionnements apportés aux méthodes pour préparation d'époxydes de composés stéroides, 1 211 325 (1960).
816. Syntex S. A.: Procédé pour introduire par voie microbiologique un hydroxyla en position 16alpha de delta$^{4}$-3,20-dicéto-17alpha-hydroxy-halogéno-prégnénes, 1 212 852 (1960).
817. Takeda pharm. Ind.: Procédé d'oxygénation des stéroides en position 11, 1 215 550 (1960).
818. Syntex S. A.: Nouveaux 6alpha-chloro- ou 6alpha-fluoro-delta$^{1,4}$-3,20-dioxo-prégnadiénes et procédé pour leur préparation, 1 215 564 (1960).

819. Schering A. G. Procédé pour introduire un groupe hydroxylique en position 12 dans le composé S du Reichstein, 1 216 040 (1960).
820. Am. Cyanamid Co.: Procédé d'oxydation des 11-désoxy-stéroides, 1 217 613 (1960).
821. Lab. Franc. chimiotérap.: Procédé d'oxydation de stéroides de la série du prégnane en position 21, 1 222 408 (1960).
822. Merck Co.: Procédé de fabrication d'une nouvelle 16alpha-alkyl-11alpha, 17alpha-dihydroxy-4-prégnéne-3,20-dione, 1 225 664 (1960).
823. Chinoin: Procédé d'oxydation microbiologique de stéroides, 1 227 602 (1960).
824. Chinoin: Procédé de préparation de la delta$^{1,4}$-androstadiénedione à partir de mélanges des stéroides, 1 228 633 (1960).
825. Chinoin: Procédé de préparation du delta$^{1,4}$-androstadiénedione- à partir de stéroides ayant une double liaison en position delta$^{16}$, 1 228 727 (1960).
826. Am. Cyanamid Co.: Préparation de nouveaux stéroides de la série du 3,20-dicéto-delta$^{1,4}$-prégnadiéne, 1 233 508 (1960).
827. Abildgaard K.: Procédé de préparation de composés 1-déhydrocorticostéroides, 1 235 907 (1960).
828. Wisconsin Alumni Res. Found.: Procédé de production microbiologique des composés chimiques au moyen de spores, notamment de stéroides et des cétones, 1 236 235 (1960).
829. Pfizer Co.: Perfectionnements apportés aux procédés d'oxydytion de composés stéroides du type 3-céto-delta$^{4}$, 1 252 611 (1960).
830. O. Mathieson chem. Corp.: Procédé de préparation de 6alpha-halogéno-9(11)-déhydro-stéroides, 1 253 483 (1961).
831. Ciba A. G.: Procédé de préparation par voie microbiologique de delta$^{4}$-6alpha-fluoro-ou de delta$^{4}$-6alpha-chloro-16-alcoyl-11alpha, 17alpha, 21-trihydroxy-prégnéne-3,20-diones, 1 255 540 (1961).
832. Merck Co.: Procédé de fabrication de composés 16-méthyl-11-oxo-1,4-prégnadiéne-17alpha, 21-diol-3,20-dione, 1 257 634 (1961).
833. Upjohn Co.: Préparation de 6alpha-fluoro-11-oxygéné-16alpha-hydroxy-4-prégnéne-3,20-diones, 1 257 642 (1961).
834. Merck Co.: Production par fermentation de dérivés 11alpha-hydroxyles de la série de la 16alpha-alkyl-4-prégnéne-3,20-dione, 1 261 181 (1961).
835. O. Mathieson chem. Corp.: Procédé de déhydrogénation sélective du noyau A des stéroides, 1 262 182 (1961).
836. O. Mathieson chem. Corp.: Synthése de 1-déhydrostéroides de la série de prégnadiéne, 1 265 153 (1961).
837. Upjohn Co.: Procédé de déshydrogénation des stéroides, 1 267 107 (1961).
838. Merck Co.: Procédé de préparation des 14alpha-hydroxy-progéstérones, 1 267 758 (1961).
839. Merck Co.: Préparation de 21-hydroxy-stéroides, 1 267 759 (1961).
840. Chinoin: Procédé de préparation de l'androstadiénedione, 1 267 992 (1961).
841. Ciba A. G.: Procédé de préparation de la 11alpha, 18-dihydroxyprogéstérone, 1 268 024 (1961).
842. Ciba A. G.: Procédé de préparation de nouveaux dérivé de prégnane-18-oiques, qui sont oxygénés en position 16, 1 269 581 (1961).
843. Ciba A. G.: Procédé pour la préparation de nouveaux dérivés de prégnan-18-oigues, qui sont oxygénés en position 16, 1 269 582 (1961).
844. Merck Co.: Procédé d'introduction de double liaison delta$^{1}$ par fermentation dans les composés stéroides, 1 270 504 (1961).
845. Merck A. G.: Procédé d'introduction d'une double liaison en delta$^{1}$ dans des composés stéroides, 1 270 505 (1961).

846. Merck Co.: Procédé de préparation de 2-alcoyl-1,4-prégnadiéne-17alpha-ol-3,20-diones, 1 270 506 (1961).
847. Merck Co.: Procédé de préparation de 17-hydroxystéroides par oxydation microbienne. 1 270 507 (1961).
848. Upjohn Co.: Procédé de préparation de la 1-déshydro-6-méthyl-cortisone, de la 1-déshydro-6-méthyl-hydrocortisone et leurs esters, 1 270 508 (1961).
849. Ciba A. G.: Procédé pour réduire sélectivement, par voie microbiologique, le groupement 3-cétonique dans des stéroides, 1 271 047 (1961).
850. O. Mathieson chem. Corp.: Méthode de préparation des 9alpha-halogéno-11beta-hydroxy (ou 11-céto)-A-Norprogestérones, 1 271 105 (1961).
851. Merck Co.: Procédé pour la fabrication de stéroides déshydrogénés par fermentation, 1 271 457 (1961).
852. Ciba A. G.: Procédé de préparation de la 11alpha-hydroxy-17alpha-méthyl-testostérone, 1 271 489 (1961).
853. O. Mathieson chem. Corp.: Procédé de préparation de la A-Nor-11-désoxycortisostérone et de ses 21-esters, 1 272 523 (1961).
854. Am. Cyanamid Co.: Procédé d'oxydation en position 7 des 7-désoxystéroides de la série du prégnane, 1 272 917 (1961).
855. Am. Cyanamid Co.: Procédé de préparation de nouveaux stéroides fluorés de la série du prégnane et de l'androstane, 1 274 415 (1961).
856. Syntex S. A.: Procédé de préparation de composés de la série du 2-céto-17-hydroxyle-6-fluoro-prégnane, 1 276 661 (1961).
857. Pfizer Co.: Perfectionnements apportés aux procédés pour préparer des composés stéroides à partir de 3-céto-stéroides à noyau saturé, 1 279 210 (1961).
858. Takeda pharm. Ind.: Procédé d'oxygénation de stéroides en position 19 par voie microbiologique, 1 281 867 (1961).
859. Takeda pharm. Ind.: Procédé d'oxydation de stéroides de la série du prégnane par des microorganismes, 1 283 159 (1961).
860. Upjohn Co.: Procédé de préparation de la 6-hydrocarbyl-11-oxygéné-9alpha, 21-difluoro-17alpha-hydroxy-4-pregnéne-3,20-dione, 1 284 512 (1962).
861. Am. Cyanamid Co.: Procédé de production biologique des dérivés 1-hydroxydes 9-halogénostéroides de la série du prégnane, 1 288 022 (1962).
862. Syntex S. A.: Procédé de préparation de composés delta$^4$-3,20-dicéto-11,17,21-trihydroxy-6alpha-chloro-prégnéniques, 1 288 727 (1962).
863. Syntex S. A.: Procédé de préparation de composés delta$^4$-6-halogéno-3,20-dioxo-prégnéniques oxygénés en position 11 et des 1-dihydro-composés correspondants, 1 292 911 (1962).
864. Searle Co.: Procédé de fabrication de stéroides, notamment de la série des 3-oxo-1,4-diénes par action déshydrogénante de microorganismes sur un stéroide non saturé en position 4, 1 295 314 (1962).
865. Am. Cyanamid Co.: Procédé d'hydroxylation des stéroides, 1 300 723 (1962).
866. Pfizer Co.: Perfectionnements apportés aux procédés pour la préparation de composés stéroides, 1 302 925 (1962).
867. Organon: Stéroides oxygénés en position 16, 1 305 377 (1962).
868. Schering A. G.: 16alpha-Méthylstéroides hydroxylés en 11beta ou 14alpha et leur préparation, 1 305 401 (1962).
869. Pfizer Co.: Perfectionnement apportés aux procédés d'oxydation par fermentation de certains composés stéroides, du type 3-céto-4 non saturé, 1 305 953 (1962).
870. Upjohn Co.: Procédé de production de 11alpha, 17alpha-dihydroxy-progestérone, 1 306 853 (1962).

871. Upjohn Co.: Procédé de production de la 16alpha-hydroxy-5beta-prégnane-3,11,20-trione, 1 306 854 (1962).
872. Merck Co.: Procédé d'introduction d'une double liaison en delta[1] dans des composés stéroides, 1 320 703 (1963).
873. Upjohn Co.: Procédé de production de delta[1]-stéroides par l'action biologique du champignon Septomyxa, 1 320 708 (1963).
874. Merck Co.: Procédé de préparation de composés stéroides tels que la 11beta, 17alpha, 21-trihydroxy-16alpha-alcoyl-4-prégnéne-3,20-dione, 1 322 105 (1963).
875. Ciba A. G.: Procédé pour l'oxydation microbiologique de stéroides, 1 325 302 (1963).

Swiss Patents

876. Upjohn Co. (Eppstein S. H., Leigh H. M.): Verfahren zur Gewinnung von nicht als Arzneimittel zu verwendenden Steroiden aus wässrigen Medien, 327 117 (1958).
877. Farmaceutici Italia (Camerino B., Vercellone A.): Selektive Reduktion von Ketosteroiden, 327 124 (1958).
878. Upjohn Co. (Murray H. C., Peterson D. H.): Verfahren zur Herstellung von 11-Oxysteroiden, 329 194 (1958).
879. Upjohn Co. (Murray H. C., Peterson D. H.): Verfahren zur Einführung von Hydroxylgruppe in Steroide, 331 690 (1958).
880. Ciba A. G. (Wettstein A., Vischer E.): Verfahren zur Herstellung von ungesättigten 17-Keto-steroiden, 331 974 (1958).
881. Upjohn Co. (Haines W. J., Colingsworth D. R.): Verfahren zur Herstellung von 11-Oxy-steroiden, 332 650 (1958).
882. Upjohn Co. (Murray H. C., Peterson D. H.): Verfahren zur Herstellung von Testosteronderivaten, 333 077 (1958).
883. Upjohn Co. (Murray H. C., Peterson D. H.): Verfahren zur Herstellung von 11alpha, 17alpha-Dihydroxy-progesteron, 333 720 (1958).
884. Pfizer Co.: Procédé de préparation d'un nouveaux stéroide, 333 729 (1955).
885. Upjohn Co. (Murray H. C., Peterson D. H.): Verfahren zur Herstellung eines 11alpha-Oxy-steroides, 334 829 (1959).
886. Upjohn Co. (Murray H. C., Peterson D. H.): Verfahren zur Herstellung eines Androstendions, 334 918 (1959).
887. Upjohn Co. (Murray H. C., Peterson D. H.): Verfahren zur Herstellung eines Dioxyprogesterons, 334 919 (1959).
888. Upjohn Co. (Murray H. C., Peterson D. H.): Verfahren zur Herstellung eines Trioxypregnenderivates, 334 920 (1959).
889. Upjohn Co. (Murray H. C., Peterson D. H.): Verfahren zur Herstellung von Testosteronderivaten, 334 921 (1959).
890. Ciba A. G.: Wettstein A. *et al.*: Verfahren zur Herstellung von neuen Produkten enzymatischer Oxydation, 335 386 (1959).
891. Upjohn Co. (Murray H. C., Peterson D. H.): Verfahren zur Herstellung eines 14-Oxy-steroids, 335 488 (1959).
892. Upjohn Co. (Murray H. C., Peterson D. H.): Verfahren zur Herstellung eines Progesteronderivates, 335 489 (1959).
893. Upjohn Co. (Murray H. C., Peterson D. H.): Verfahren zur Herstellung von Cholenderivaten, 335 490 (1959).
894. Upjohn Co. (Murray H. C., Peterson D. H.): Verfahren zur Herstellung eines Pregnanderivates, 335 491 (1959).

895. Ciba A. G. (WETTSTEIN A. *et al.*): Verfahren zur Herstellung von 17-Oxysteroiden, 335 492 (1959).
896. Ciba A. G. (WETTSTEIN A. *et al.*): Verfahren zur Herstellung von Dehydroverbindungen der Steroidreihe, 338 447 (1959).
897. Merck Co. (MCALEER W., DULANEY E. L.): Verfahren zur Herstellung von Polyhydroxysteroiden mittels hydroxylierender Enzyme verschiedener Arten von Mikroorganismen, 338 827 (1959).
898. Ciba A. G. (WETTSTEIN A. *et al.*): Verfahren zur Herstellung von Oxypregnanen, 339 620 (1959).
900. Scherico Ltd. (CHARNEY W. *et al.*): Procédé de préparation de 3-céto-stéroides du groupe du prégnane, 339 923 (1959).
901. Ciba A. G. (WETTSTEIN A., VISCHER E.): Verfahren zur Razematspaltung, 341 493 (1959).
902. Upjohn Co. (MURRAY H. C., PETERSON D. H.): Verfahren zur Herstellung von Testosteronderivaten, 341 818 (1959).
903. Scherico Ltd. (NOBILE A., CHARNEY W., SUTTER D.): Procédé de préparation de 3-cétostéroides du groupe du prégnane, 341 820 (1959).
904. Upjohn Co. (MURRAY H. C., PETERSON D. H.): Verfahren zur Herstellung eines Corticosteronderivates, 341 821 (1959).
905. Upjohn Co. (MURRAY H. C., PETERSON D. H.): Verfahren zur Herstellung von 14alpha-Oxy-progesteron, 341 822 (1959).
906. Upjohn Co. (MURRAY H. C., PETERSON D. H.): Verfahren zur Herstellung von 11-Oxysteroiden, 341 823 (1959).
907. Upjohn Co. (MURRAY H. C., PETERSON D. H.): Vefahren zur Herstellung eines Progesteronderivates, 342 217 (1959).
908. Upjohn Co. (MURRAY H. C., PETERSON D. H.): Verfahren zur fermentativen Oxydation von Steroiden, 343 390 (1960).
909. Ciba A. G. (WETTSTEIN A., VISCHER E., MEYSTRE Ch.): Verfahren zur Aufspaltung von d, 1-Steroiden, 344 055 (1960).
910. Pfizer Co. (SHULL G. M., KITA D. A.): Procédé de déshydrogénation d'un 3-céto-stéroide non saturé en position 4(5), 344 999 (1960).
911. Farbwerke Höchst A. G. (LINDNER F. *et al.*): Verfahren zur Herstellung von 1,4-Dien-3-ketonen der Steroidreihe, 346 542 (1960).
912. Am. Cyanamid Co. (FELDMAN L. I., ALLEN W. S.): Procédé pour transformer un 11-désoxy-stéroide en 11-oxo-stéroide correspondant, 348 158 (1960).
913. Merck Co. (STOUDT T. H.): Verfahren zur mikrobiologischen delta$^1$-Dehydrierung von Steroiden, 349 978 (1960).
914. Scherico Ltd. (CARVAJAL F): Procédé de préparation de dérivés du prégnane 6beta-hydroxylés, 350 286 (1961).
915. Ciba A. G. (WETTSTEIN A. *et al.*): Verfahren zur Herstellung von 21-Oxy-steroiden, 351 597 (1961).
916. Upjohn Co. (ŠEBEK O. K., SPERO G. B.): Verfahren zur Herstellung von Steroiden, 351 598 (1961).
917. Ciba A. G. (WETTSTEIN A. *et al.*): Verfahren zur Herstellung polyhydroxylierter Steroide, 352 329 (1961).
918. Ciba A. G. (WETTSTEIN A. *et al.*): Verfahren zur Herstellung von 1-Dehydro-oxysteroiden, 352 330 (1961).
919. Lovens kem. Fabrik (LORCK H. O. B., FRANK E. R.): Verfahren zur Herstellung von 1(2)-Dehydro-corticosteroiden, 353 357 (1961).

920. Merck Co. (STOUDT T. H.): Verfahren zur delta[1]-Dehydrierung von Steroiden, 355 778 (1961).
921. Lab. Franc. chimiothérapie (GUY HAGEMANN *et al.*): Procédé de préparation de stéroides 21-hydroxylés, 356 450 (1961).
922. Pfizer Co. (SHULL G. M., KITA D. A.): Procédé de préparation de dérivés de cortisone, 356 765 (Suppl. to 344 999) (1961).
923. Merck Co. (STOUDT T. H.): Verfahren zur Dehydrierung von Steroiden, 357 392 (1961).
924. Am. Cyanamid Co. (BLANK R. H., BERNSTEIN S. *et al.*): Procédé de préparation de delta$^{1,4}$-prégnadiénes, 359 700 (1962).
925. Upjohn Co. (WEINTRAUB A. *et al.*): Verfahren zur Herstellung von delta[1]-Steroiden, 363 341 (1962).
926. Ciba A. G. (WETTSTEIN A. *et al.*): Verfahren zur Aufspaltung von d,l-Steroiden, 364 503 (1962)-Suppl. to 344 055.
927. Ciba A. G. (WETTSTEIN A. *et al.*): Verfahren zur Aufspaltung von d,l-Steroiden, 364 504 (1962) — Suppl. to 344 055.
928. Upjohn Co. (WEINTRAUB A. *et al.*): Verfahren zur Herstellung von delta[1]-Steroiden, 364 775 (1962).
929. Upjohn Co. (MEISTER P. D., WEINTRAUB A.): Verfahren zur Herstellung von Steroiden, 366 534 (1963).
930. Upjohn Co. (MEISTER P. D., WEINTRAUB A.): Verfahren zur Herstellung von Steroiden, 366 536 (1963).
931. O. Mathieson chem. Corp. (KROLL H. A. *et al.*): Verfahren zur Herstellung von Steroiden, 367 495 (1963).
932. Ciba A. G. (WETTSTEIN A., VISCHER E.): Verfahren zur Darstellung von 11alpha-Oxy-17alpha-methyl-testosteron, 368 166 (1963).
933. Takeda pharm. Ind. (HASEGAWA T. *et al.*): Verfahren zur Herstellung von 11- und 19-Hydroxysteroiden, 368 796 (1963).
934. Schering A. G. (RASPÉ G.): Verfahren zur Einführung einer Hydroxylgruppe in die 12-Stellung von Reichsteins Verbindung S, 369 129 (1963).
935. Merck Co.: Verfahren zur Herstellung von 16alpha-Alkyl-11alpha, 17alpha-dioxy-4-pregnen-3,20-dionen, 369 450 (1963).
936. Merck A. G. (HEINZ-JÜRGEN MANNHARDT, METZ H.): Verfahren zur Herstellung von 11alpha-Hydroxysteroiden, 370 778 (1963).

German (West) Patents

937. Upjohn Co. (HAINES W. J., COLINGSWORTH D. R.): Verfahren zur Herstellung von vorzugsweise in 11-Stellung oxydierten Steroiden, 921 695 (1954).
937*a*) LINDNER F., KEHL H., SCHMIDT-THOMÉ J.: Verfahren zur Herstellung von 4-Pregnen-11-beta, 17alpha, 21-triol-3,20-dion und 11alpha-Epimer, 1 009 627 (1957).
938. Ciba A. G. (WETTSTEIN A. *et al.*): Dehydro-Verbindungen der Steroid-Reihe, 1 020 329 (1957)
939. Pfizer Co.: Verfahren zur Herstellung von 11beta-Hydroxysteroiden, 1 029 824 (1958).
940. Pfizer Co. (SHULL G. M.): Biologische Dehydrogenation von Steroiden, 1 030 341 (1958).
941. Pfizer Co. (SHULL G. M., BLOOM B. M.): Verfahren zur Herstellung von Steroidepoxyden, 1 035 134 (1958).
942. Ciba A. G.: Verfahren zur Herstellung von in 11-Stellung durch eine sauerstoffhaltige Gruppe substituierten Steroiden, 1 036 852 (1958).

943. Lab. Franc. chimiothérap. (HAGEMANN G. *et al.*): Verfahren zur Oxydation von gesättigten und ungesättigten 21-Hydroxysteroiden der Pregnanreihe, 1 037 450 (1958).

944. Am. Cyanamid Co. (FELDMAN L. I., ALLEN W. S.): Verfahren zur Herstellung von 4-Pregnen-11beta, 17alpha, 21-triol-3,20-dion und dessen Estern, 1 038 039 (1958).

944*a*) Pfizer Co. (SHULL G. M., KITA D. A.): Mikrobielle 1-Dehydrogenation von Androstan Steroiden und Pregnan Steroiden, 1 045 400 (1958).

945. Upjohn Co. (MEISTER P. D., WEINTRAUB A.): Verfahren zur fermentativen Hydroxylierung der 11alpha-Stellung von Steroiden mit bis zu 22 C-Atomen und/oder zum Abbau der 17-ständigen Seitenkette dieser Steroide, 1 048 579 (1959).

946. Ciba A. G. (WETTSTEIN A. *et al.*): 9alpha-Halo-14-hydroxypregnan-Verbindungen und Derivaten, 1 049 858 (1959).

949. Pfizer Co. (SHULL G. M.): Prednisolon, 1 050 335 (1959).

950. Am. Cyanamid Co. (BERNSTEIN S., ALLEN W. S.): Verfahren zur Herstellung von 1- und 4-Stellung ungesättigten Steroidverbindungen, 1 053 501 (1959).

951. Scherico Ltd. (CHARNEY W. *et al.*): Verfahren zur Herstellung von 1,4-Pregnadienen, 1 059 906 (1959).

952. Lovens kem. Fabrik (HENNING O., LORCK B., FRANK E. R.): Verfahren zur Herstellung von 1(2)-Dehydro-corticosteroidverbindungen, 1 060 391 (1959).

953. Merck A. G. (BRUCKNER K., METZ H.): Verfahren zur Herstellung von Reichsteins-Substanz-S, 1 060 860 (1959).

954. Pfizer Co. (DAVISSON J. W.): Verfahren zur 11-Hydroxylierung von Steroidverbindungen, 1 062 699 (1959).

955. Ciba A. G. (WETTSTEIN A., SCHMIDLIN J.): Verfahren zur Herstellung von delta$^{4}$-Halogenpregnenen, 1 065 844 (1959).

956. Jenapharm VEB (SCHUBERT A. *et al.*): Verfahren zur Einführung von Hydroxylgruppen in Steroide, 1 067 020 (1959).

957. Ciba A. G. (WETTSTEIN A. *et al.*): Verfahren zur Herstellung von neuen 14-Oxypregnanen und ihren funktionellen Derivaten, 1 067 431 (1959).

958. Merck A. G. (MANNHARDT H., METZ H.): Verfahren zur Herstellung von 11alpha-Hydroxy- und 11-Ketosteroiden, 1 069 623 (1959).

959. Am. Cyanamid Co. (BLANK R. H. *et al.*): Verfahren zur Herstellung neuer Steroidverbindungen der 1,4-Pregnadienreihe, 1 070 176 (1959).

960. Pfizer Co. (SHULL G. M.): Verfahren zur Herstellung von 3-Keto-delta$^{1,4}$-steroiden, 1 071 080 (1959).

961. Farbwerke Höchst (HÜBENER H. J., SCHMIDT-THOMÉ J.): Gewinnung einer 20-beta-Hydroxysteroiddehydrogenase, 1 075 521 (1960).

962. Schering A. G. (RASPÉ G., RICHTER H.): Vefahren zur Herstellung von 5-Androsten-3,11-diol-17-onen, 1 080 553 (1960).

963. O. Mathieson chem. Corp. (THOMA R. W. *et al.*): Vefahren zur Herstellung von 11beta-Hydroxysteroiden der Pregnanreihe, 1 081 457 (1960).

964. Upjohn Co. (ŠEBEK O. K., SPERO G. B.): Verfahren zur Herstellung von 6-Methyl-17alpha, 21-dioxy-1,4-pregnadien-3,20-dionen und deren 21-Acylaten, 1 085 873 (1960).

964*a*) Upjohn Co. (ROSSELET J. P., ŠEBEK O. K., SPERO G. B.): Verfahren zur Herstellung von therapeutisch wirksamen Steroidverbindungen, 1 087 599 (1960).

965. Merck A. G. (BORK K. H. *et al.*): Verfahren zur Herstellung von ungesättigten, fluorhaltigen Steroiden, 1 092 468 (1960).

966. Pfizer Co. (BADIA D. J. *et al.*): Verfahren zur Herstellung von 3-Keto-delta$^{1,4}$-steroiden, 1 093 792 (1960).

967. O. Mathieson chem. Corp. (KROLL H. A. *et al.*): Verfahren zur Herstellung von hydroxylierten Steroiden der 3-Keto-delta$^{1,4}$-pregnadienreihe, 1 094 743 (1960).

968. Scherico Ltd. (CARVAJAL F.): Verfahren zur Herstellung von 6beta-Hydroxysteroiden, 1 095 278 (1960).
969. Takeda pharm. Ind. (HASEGAWA T. *et al.*): Verfahren zur Herstellung von 11- oder 19-Hydroxysteroiden, 1 100 623 (1961).
970. Am. Cyanamid Co. (BERNSTEIN S. *et al.*): Verfahren zur Hydroxylierung von 7-Desoxysteroiden der Pregnanreihe, 1 101 413 (1961).
971. Vismara S. p. A. (ERCOLI A., GARDI R.): Verfahren zur Herstellung von Cortison, 1 102 149 (1961).
971*a*) Ciba A. G. (WETTSTEIN A., *et al.*): Pregnanolon Derivate, 1 102 733 (1961).
972. Farbwerke Höchst A. G. (LINDNER F. *et al.*): Verfahren zur Herstellung von 1,4,17(20)-Pregnatrienen, 1 107 225 (1961).
973. Upjohn Co. (CAMPBELL J. A. *et al.*): Verfahren zur Herstellung von Steroiden, 1 109 169 (1961).
974. Ciba A. G. (WETTSTEIN A., VISCHER E.): Verfahren zur selektiven Reduktion von Steroiden, 1 110 637 (1961).
975. Merck Co. (ARTH G. E. *et al.*): Verfahren zur Herstellung von in 11-Stellung durch sauerstoffhaltige Gruppen substituierten 16-Methyl-1,4-pregnadien-17alpha-ol-3,20-dion-Verbindungen, 1 113 690 (1961).
976. Syntex S. A. (RINGOLD H. J. *et al.*): Verfahren zur Herstellung von Hydroxypregnanverbindungen, 1 118 196 (1961).
976*a*) Lab. Franc. Chimiothérap. (MULLER G.): 9alpha-Fluor-16alpha-methyl-1,4-pregnadiene, 1 121 611 (1962).
977. Chinoin (WIX G., RADOS M.): Verfahren zur Herstellung von delta$^{1,4}$-Androstadien-3,17-dion, 1 122 945 (1962).
978. Merck Co. (MCALEER W. J., STOUDT T. H.): Verfahren zur Herstellung von 16alpha-Methyl-11alpha, 17alpha-dioxy-4-pregnen-3,20-dion, 1 122 949 (1962).
979. Upjohn Co. (HANZE A. R., MURRAY H. C., ŠEBEK O. K.): Verfahren zur Herstellung von 9alpha-Oxysteroiden der Pregnanreihe, 1 124 945 (1962).
980. Merck A. G.: Verfahren zur Herstellung von fluorhaltigen 16-Methylen-Verbindungen, 1 125 921 (1962).
981. Scherico Ltd. (CARVAJAL F. *et al.*): Verfahren zur Herstellung von 11-hydroxylierten Steroiden der Pregnanreihe, 1 128 954 (1962).
982. Ciba A. G. (WETTSTEIN A., VISCHER E.): Verfahren zur Herstellung von 11alpha-Oxy-17-alpha-methyl-testosteron, 1 130 442 (1962).
983. Merck A. G. (BORK K. H. *et al.*): Verfahren zur Herstellung von 16-Methylen-steroiden, 1 130 805 (1962).
984. Chinoin (WIX G. *et al.*): Verfahren zur Herstellung von delta$^{1,4}$-Androstadien-3,17-dion, 1 131 210 (1962).
985. Soc. Farmaceutici Italia (SPALLA C. *et al.*): Verfahren zur Herstellung von Hydrocortison, 1 132 922 (1962).
985*a*) Merck A. G. (MENNHARDT H. J. *et al.*): 16-Methylen-Steroide, 1 134 074 (1962).
986. Searle Co. (MUIR R. D., DODSON R. M.): Verfahren zur Herstellung von 9alpha-Oxy-3-keto-delta$^{4}$-steroiden der Androstan-19-Norandrostan- und Pregnanreihe, 1 134 669 (1962).
987. Chinoin (WIX G., ALBRECHT K.): Verfahren zur Herstellung von delta$^{1,4}$-Androstadien-3,17-dion, 1 135 455 (1962).
988. Schering A. G. (RASPÉ G. *et al.*): Verfahren zur Herstellung von 1,4-Dien-3-keto-steroiden, 1 135 899 (1962).
989. Ciba A. G. (WETTSTEIN A., VISCHER E.): Verfahren zur Herstellung von delta$^{4}$-6alpha-Fluor- oder -Chlor-16alpha-alkyl-11alpha, 17alpha, 21-trihydroxy-pregnen-3,20-dionen und delta$^{1,4}$-analogischen Verbindungen, 1 135 901 (1962).

990. Chinoin (WIX G., ALBRECHT K.): Verfahren zur 11alpha-Hydroxylierung von Progesteron, 1 137 436 (1962).
991. Upjohn Co. (LINCOLN J. H. j., SCHNEIDER W. P.): Verfahren zur Herstellung von 16alpha-Methyl-4-pregnen- und 16alpha-Methyl-1,4-pregnadienderivaten, 1 137 437 (1962).
992. Upjohn Co. (CAMPBELL J. A. *et al.*): Verfahren zur Herstellung von 6-Fluor-17beta-hydroxy-17alpha-alkyl-1,4-androstadien-3-onen, 1 139 838 (1962).
993. Merck A. G. (BORK K. H. *et al.*): Verfahren zur Herstellung von fluorhaltigen 16-Methylen-1,4-pregnadienen, 1 140 574 (1962).
994. O. Mathieson chem. Corp. (THOMA R. W. *et al.*): Verfahren zur Herstellung von 11alpha-Hydroxy-A-Norprogesteron, 1 143 812 (1963).
995. Upjohn Co. (PETERSON D. H. *et al.*): Verfahren zur Herstellung von Steroidverbindungen, 1 144 273 (1963).
996. Chinoin (PAUNCZ J. *et al.*): Verfahren zur Gewinnung von delta$^{1,4}$-Androstadien-3,17-dion, 1 145 608 (1963).

## German (East) Patents

997. Jenapharm VEB (SCHUBERT A. *et al.*): Verfahren zur Herstellung von delta$^{4}$-3-Keto-15alpha-Oxypregnenen, 15 599 (1958).
998. Jenapharm VEB (SCHUBERT A. *et al.*): Verfahren zur Herstellung von hydroxylierten Steroiden, 19 318 (1960).
999. Jenapharm VEB (SCHUBERT A. *et al.*): Verfahren zur Herstellung von Dihydroxy-delta$^{4}$-3-ketosteroiden, 19 651 (1960).
1000. Jenapharm VEB (SCHUBERT A., SIEBERT R.): Verfahren zur Herstellung von in 14alpha-Stellung hydroxylierten Steroiden, 19 658 (1960).
1001. Jenapharm VEB (SCHUBERT A., SIEBERT R.): Verfahren zur mikrobiellen Hydroxylierung von Steroiden in 9-Stellung, 20 643 (1961).
1002. Jenapharm VEB (SIEBERT R. *et al.*): Verfahren zur Herstellung von delta$^{4}$-11alpha, 17alpha, 21-Trihydroxypregnen-3,20-dion, 21 912 (1961).
1003. Jenapharm VEB (SCHUBERT A., SIEBERT R.): Verfahren zur Einführung von Hydroxylgruppen in Steroide, 23 995 (1962) — Suppl. to 19 318.

## U.S.S.R. Patents

1004. BARMENKOV A. S., JEROSHIN V. K.: Chromatographic determination of the progesterone fermentative transformation, 119 656 (1960).
1005. WIX G., ALBRECHT K.: Microbiological oxidation of steroids, 135 427 (1961).
1006. SUVOROV N. N. *et al.*: Preparation of epi-hydrocortisone, 145 573 (1961).
1007. SUVOROV N. N., SKRJABIN G. K. *et al.*: Preparation of prednisolone and dianabol, 154 981 (1962).

## Czechoslovakian Patents

1008. HANČ O., ČAPEK A., TADRA M.: Preparation of testosterone, 86 728 (1957).
1009. HANČ O., ČAPEK A., TADRA M.: Preparation of delta$^{4}$-androstene-3,17-dione and testosterone, 87 068 (1957).

1010. HANČ O. *et al.*: Microbial hydroxylation of steroids of the pregnane or androstane series, 89 993 (1959).
1011. HANČ O. *et al.*: Production of testosterone and its derivatives, 90 161 (1959).
1012. HANČ O., ČAPEK A., TADRA M.: Biosynthetical preparation of cortisol and its 11alpha-epimer, 95 389 (1960).
1013. HANČ O., TADRA M., ČAPEK A.: Processing mixture of epimeric delta$^4$-pregnene-11, 17alpha, 21-triol-3,20-diones to cortisol, cortisone and their esters, 105 063 (1962).
1014. ČAPEK A., HEJDÁNEK M., HANČ O., TADRA M., KAKÁČ B., CECHNER J.: Preparation of 1,4-dienes by microbial dehydrogenation of delta$^4$-3-ketosteroids by two-step method, 109 820 (1963).

## *d*) Papers dealing with isolation, identification and analysis of steroids

1015. ALDERS L.: Liquid--liquid-extraction. New York: Elsevier 1955.
1016. AMENDOLLA C., ROSENKRANZ G., SONDHEIMER F.: Synthesis of steroidal hormone analogues hydroxylated at $C_6$; J. chem. Soc. 1954, 1226.
1017. BALANT Ch. P., EHRENSTEIN M.: 6beta- and 6alpha-acetoxy and hydroxy-derivatives of progesterone and androstenedione, J. org. Chem. 17: 1587, 1952.
1018. BANES D.: The chemical assay of adrenal cortex extracts, J. am. pharm. Assoc. Sci. Ed. 42: 669, 1953.
1019. BARMENKOV A. S., EROSHIN V. K.: Method of controlling the hydroxylation of progesterone to the 11alpha-hydroxy-progesterone (in Russian), Med. Prom. 14, No. 4, p. 40, 1960.
1020. BARTON G. M., EVANS R. S., GARDNER J. A. F.: Paper chromatography of phenolic substances, Nature 170; 249, 1952.
1021. BARTON D. H. R., MCMORRIS T. C., SEGOVIA R.: The nature of the Porter-Silber reaction, J. chem. Soc. 1961, 2027.
1022. BAULD W. S.: Some errors in the colorimetric estimation of oestriol, oestrone and oestradiol by the Kober reaction, Biochem. J. 56: 426, 1954.
1023. BENNETT R. D., HEFTMANN E.: Thin-layer chromatography of corticosteroids, J. Chromatogr. 9: 348, 1962.
1024. BERGSTROM S. *et al.*: The stereochemistry of 7alpha-hydroxylation in the biosynthesis of cholic acid from cholesterol, J. am. chem. Soc. 80: 2337, 1958.
1025. BIANCHI M.: Colorimetric determination of delta$^4$-3-keto-steroids in the presence of their delta$^1$-dehydro derivatives, Chim. e Ind. 41: 33, 1959.
1025*a*) BOBBITT I. S.: Thin layer chromatography, Reinhold Co. (New York) 1963.
1026. BODÁNSZKY A., KOLLONITSCH J.: Color reagent for paper chromatography of steroids, Nature 175; 729, 1955.
1027. BOWDEN K. *et al.*: Researches on acetylenic compounds. Part I. The preparetion of acetylenic ketones by oxidation of acetylenic carbinols and glycols, J. chem. Soc. 1946, 39.
1028. BRADLEY J. G.: Pulse column for solvent extraction, U.S. patent 2 811 423 (1957).
1029. BROCKMANN H., SCHODDER H.: Aluminiumoxyd mit abgestuftem Adsorptionsvermögen zur chromatographischen Adsorption, Chem. Ber. 74: 73, 1941.
1030. BROOKS C. J. W., NORYMBERSKI J. K.: The oxidation of corticosteroids with sodium bismuthate, Biochem. J. 55; 371, 1953.
1031. BURSTEIN S., DORFMAN R. I.: Steroid metabolism in guinea pigs. I. Metabolism of cortisol in vivo, J. biol. Chem. 213: 581, 1955.

1032. BURSTEIN S.: Extraction of $C_{21}O_5$ and $C_{21}O_6$ steroids from aqueous media, Science 124: 1030, 1956.

1033. BURSTEIN S.: 2alpha-hydroxylation of cortisol in the guinea pig, J. am. chem. Soc. 78: 1769, 1956.

1034. BUSH I. E.: Methods of paper chromatography of steroids applicable to the study of steroids in mammalian blood and tissues, Biochem. J. 50: 370, 1951.

1035. BUSH I. E.: The chromatography of steroids, London: Pergamon Press 1961.

1036. BUTENANDT A., Schmidt-Thomé J., WEISS T.: Über ein 16-Oxytestosteron, Chem. Ber. 72: 417, 1939.

1037. BUTT W. R., HENLY A. A., MORRIS C. J. O. R.: The polarographic estimation of steroid hormones. IV. Determination of 3alpha- and 3beta-hydroxy-17-ketosteroids, Biochem. J. 42: 447, 1948.

1037*a*) CAINELLI G. *et al.*: Direkte Einführung einer Sauerstoffunktion in die Methylgruppe C-18 im intakten Steroidgerüst, Helv. chim. Acta 42; 1124, 1959.

1038. CARSTENSEN H.: Separation of corticosteroids by countercurrent distribution, Acta chem. Scand. 9: 1026, 1955.

1039. CARSTENSEN H.: Countrecurrent distribution for the identification of aldosterone, Arkiv Kemi 10: 235, 1956.

1040. CARTLAND G. F., KUIZENGA M. H.: The preparation of extracts containing the adrenal cortical hormone, J. biol. Chem. 116: 57, 1936.

1041. CAVINA G., CINGOLANI E.: Ricerca e determinazione di steroidi cortico-simili di interesse farmaceutico mediante cromatografia su carta (in Italian), Farmaco (Pavia), Ed. prat. 15: 246, 1960.

1042. ČERNÝ V. *et al.*: Chemistry of steroid compounds, Prague: Publishin House of Czech. Acad. of Science 1960.

1043. ČERNÝ V., JOSKA J., LÁBLER L.: Application of then layer chromatography without binder for rapid analytical and preparative separation of steroids, Coll. czech. chem. Com. 26: 1658, 1961.

1044. Chen C., Wheeler J., Tewell M. E.: Methods of estimation of adrenal cortical steroids with tetrazolium salts, J. Lab. clin. Med. 42: 749, 1953.

1045. CHINOIN: Improvements in or relating to the recovery of delta$^{1,4}$-androstadiene-3,17-dione, Brit. patent 881 490 (1961).

1046. CINGOLANI E., CAVINA G., AMORTIMO V.: Sul controllo dei corticosteroidi nei preparati fermaceutici (in Italian), Farmaco (Pavia), Ed. prat. 15: 301, 1960.

1047. CLARK J.: A colorimetric reaction for the estimation of cortisone, hydrocortisone, aldosterone and related steroids, Nature 175: 123, 1955.

1048. CLARKE R. L. *et al.*: The reaction of lead tetraacetate with progesterone and testosterone, J. am. chem. Soc. 77: 661, 1955.

1048*a*) COPIUS PEEREBOOM J. W.: Chromatographic sterol analysis as applied to the investigation of milk fat and other oils and fats, Versl. nadbouwk onderz, No. 69.8 (Wageningen) 1963.

1049. CRAIG L. C.: Identification of small amounts of organic compounds by distribution studies, J. biol. Chem. 155: 519, 1944.

1050. CRAIG L. C.: Partition chromatography and counter-current distribution, Anal. Chem. 22: 1346, 1950.

1051. CRAIG L. C., CRAIG D., SCHEIBEL E. G.: Extraction and purification. In A. Weissberger (Ed.), Technique of organic chemistry, Ed. 2., Vol. III., p. 149–393. New York: Interscience Publ. 1956.

1052. DEYL Z., PAVLÍČEK M., ROSMUS J.: Paper chromatography in centrifugal field, Chem. Listy 57; 479, 1963 (in Czech).

1053. Djerassi C. *et al.*: Correlation of digitogenin with progesterone, J. am. chem. Soc. 77: 3673, 1955.
1054. Djerassi C. *et al.*: The constitution and stereochemistry of digitogenin, J. am. chem. Soc. 78: 3166, 1956.
1055. Djerassi C.: Optical rotatory dispersion. New York: McGraw-Hill 1960.
1056. Dobriner K., Lieberman S., Rhoads C. P.: Studies in steroid metabolism. I. Methods for the isolation and quantitative estimation of neutral steroids present in human urine, J. biol. Chem. 172: 241, 1948.
1057. Dorfman L.: Ultraviolet absorption of steroids, Chem. Rews. 53: 47, 1953.
1058. Dowlatabadi H.: Eine neue Methode zur Bestimmung der Steroide mit einer delta$^4$-3-Keto-Gruppe mit Blautetrazolim, Naturwiss. 45: 517, 1958.
1059. Dreiding A. C., Pummer W. J., Tomasewski A. J.: The dienonephenol rearrangement with mineral acids, J. am. chem. Soc. 75: 3159, 1953.
1060. Drieding A. S., Voltman A.: The structure of Inhoffen's phenolic steroids. I. Dehydrogenation, J. am. chem. Soc. 76: 537, 1954.
1061. Edwards R. W. H., Kellie A. E.: The determination of 17-ketogenic steroids, Acta Endocrinol. 27: 262, 1958.
1062. Ehrenstein M.: 6-Oxoprogesterone and the stereochemical configuration of several 3,5,6-triols, J. org. Chem. 4: 506, 1939.
1062*a*) Ehrhart G. *et al.*: Neue Synthesen in der Steroidreihe, Z. angew. Chem. 52: 363, 1939.
1063. Eik-Nes K.: Determination of 17,21-dihydroxy-20-ketosteroids in blood plasma, J. clin. Endocrinol. 17: 502, 1957.
1064. Eisenbrand J., Picher H.: Über den polarographischen Nachweis von biologisch wichtigen Ketonen der Steringruppe, Z. physiol. Chem. 260: 83, 1939.
1065. Eisenlohr H.: A new centrifugal solvent extractor, Ind. Chemist 27: 271, 1951.
1066. Engel L. L., Slaunwhite W. R. *et al.*: A relation between phase composition and partition coefficients for the estrogens, J. biol. Chem. 191: 621, 1951.
1067. Engel L. L., Baggett B.: Non-ketonic steroids. Estimation of alcoholic steroids, Rec. Progr. Hormone Res. 9; 251, 1954.
1068. Engel L. L. *et al.*: Relations between phase composition and partition coefficients for some neutral steroids, Anal. Chem. 26: 639, 1954.
1068*a*) Engel L. L.: Physical properties of the steroid hormones, Pergamon Press (London) 1963.
1069. Evans R. M. *et al.*: Studies in the synthesis of cortisone. Part XVIII. The preparation of cortisone esters from 4,5alpha-dihydrocortisone, J. chem. Soc. 1956, 4356.
1070. Fieser L. F., Fieser M.: Natural products related to phenanthrene. Ed. 3. New York: Reinhold 1949.
1071. Fieser L. F., Fieser M.: Streoids. New York: Reinhold 1959.
1072. Fieser L. F., Romero M. A.: Rearrangement of 6beta-bromo-delta$^4$-cholestene-3-one to 2alpha-acetoxy-delta$^4$-cholestene-3-one, J. am. chem. Soc. 75: 4716, 1953.
1073. Fried J., Thoma R. W. (O. Mathieson chem. Co.). Synthesis of steroids, Austral. patent 220 072 (1959).
1074. Girard A., Sandulesco G.: Sur une nouvelle série de réactifs du groupe carbonyle, leur utilisation à l'extraction des substances cétoniques et à la caractérisation microchimique des aldéhydes et cétones, Helv. chim. Acta 19: 1095, 1936.
1075. Glaxo Lab.: Procédé perfectionné de séparation de dérivés de 21-acyloxy-prégnane ou alloprégnane à partir de mélange en contenant, French patent 1 179 221 (1959).
1076. Gornall A. G., MacDonald P. M.: Ouantitative determination of the steroid hormones with 2,4-dinitrophenylhydrazine, J. biol. Chem. 202: 279, 1953.

1077. GRIMMER G.: Reaktionskinetische Studien an den Hydroxyl-Gruppen der Steroid-Molekel (Beiträge zur Konstitutionsermittlung unbekannter Steroide im Mikromasstab mit 0,3 bis 0,4 mg Substanz), Angew. Chemie 69: 400 1957.

1077*a*) GUERCIO P. A.: Perfectionnements apportés aux procédés pour obtenir des produits de fermentation du type corticostéroide, Fr. patent 1 209 063 (1960).

1077*b*) GUERCIO P. A.: Recovery of steroids fro m fermentation broth, U.S. patent 2 935 520(1960).

1078. HAMILTON J. G. jr., HOLMAN R. T.: Displacement analysis of lipides. VIII. Preliminary studies with steroids, Arch. Biochem. Biophys. 36: 456, 1952.

1079. HASKINS A. L., SHERMAN A. I., ALLEN W. M.: Paper chromatographic separation and ultraviolet analysis of commercially prepared progesterone, J. biol. Chem. 182: 429, 1950.

1080. HECKER E.: Verteilungsverfahren im Laboratium. Monographien zur Angewandten Chemie und Chemie-Ingenieur-Technik. Weinheim/Bergstr.: Verlag Chemie 1955.

1081. HEFTMANN E.: Chromatography. New York: Reinhold 1961.

1082. HELLER K., WENTZKE M.: Ein Spektropolarimeter — aufgebaut aus Geräten des VEB Carl Zeiss Jena, Jenaer Rundschau 4: 143, 1961.

1083. HEŘMÁNEK S., SCHWARZ V., ČEKAN Z.: Chromatographie neutraler Steroide an einer dünnen Aluminiumoxydschicht, Coll. czech. chem. Com. 26: 1669, 1961.

1084. HEŘMÁNEK S., SCHWARZ V., ČEKAN Z.: Measurement of activity of alumina. by means of thin layer chromatography, Coll. czech. chem. Com. 26: 3170, 1961.

1085. HEŘMÁNEK S., SCHWARZ V., ČEKAN Z.: Über die Chromatographie einiger organischen Verbindungen auf dünnen Aluminiumoxydschicht, Pharmazie 16: 566, 1961.

1086. HERSHBERG E. B., WOLFE J. K., FIESER L. F.: Polarographic determination of dehydroisoandrosterone and other 3-hydroxy-delta$^5$-steroids, J. biol. Chem. 140: 215, 1941.

1087. HERZOG H. L. *et al.*: Partial synthesis of 11-ketotestosterone and of adrenosterone, J. am. chem. Soc. 75: 266, 1953.

1088. HERZOG H. L.: 1-Hydroxycortisone, 1-dyhroxyhydrocortisone, and esters thereof, U.S. patent 2 737 518 (1956).

1089. HESSE G., DANIEL J., WOHLLEBEN G.: Aluminiumoxyde für die chromatographische Analyse und Versuche zu ihrer Standardisierung, Angew. Chem. 64: 103, 1952.

1090. HONEGGER C. G.: Preparative thin-layer chromatography, Helv. chim. Acta 45: 1409, 1962.

1091. HOSOI M.: A new color reaction and colorimetric estimation of testosterone and its esters, J. Pharm. Soc. Japan 75: 1506, 1955.

1092. HOUGH L.: Application of paper partition chromatography to the separation of the polyhydric alcohols, Nature 165; 400, 1950.

1093. HUANG-MINLON *et al.*: Steroid 17alpha-acetates, J. am. chem. Soc. 74: 5394, 1952.

1094. HÜTTENRAUCH R.: Farbreaktionen der kunjugiert ungesättigten Keto-Steroide, Z. physiol. Chem. 326: 166, 1961.

1095. HÜTTENRAUCH R.: Über die Bildung von Azomethin-Farbsalzen aus Ketosteroiden, Arch. Pharm. 295: 721, 1962.

1096. INHOFFEN H. H., HUANG-MINLON: Transitions of steroids in aromatic compounds. III. Aromatization of 1,4-cholestadien-3-one, Naturwiss. 26: 756, 1938.

1097. JACQUES J., KAGAN H. B.: Séparation chromatographique des stéréoisomeres. In: E. Lederer, Chromatographie en chimie organique et biologique, 584—634. Paris: Masson et C$^{ie}$ Éditeurs, Vol. I. 1959.

1098. JEANLOZ R. W. *et al.*: Chemical transformations of steroids by adrenal perfusion. III. delta$^4$-androstene-3,17-dione, J. biol. Chem. 203: 453, 1953.

1099. JULIAN P. L. *et al.*: 17alpha-hydroxy-11-desoxycorticosterone (Reichstein's substance S), J. am. chem. Soc. 72: 5145, 1950.

1100. KABASAKALIAN P., DE LORENZO S., MCGLOTTEN J.: Polarographic analyses of mixtures of prednisone and cortisone, Anal. Chem. 28: 1669, 1956.
1101. KABASAKALIAN P., MCGLOTTEN J.: The polarographic reduction of prednisone, prednisolone and their precursors, J. am. chem. Soc. 78: 5032, 1956.
1102. KABASAKALIAN P., MCGLOTTEN J.: Polarographic reduction of nonconjugated steroidal ketones, Anal. Chem. 31: 1091, 1959.
1103. KAHNT F. W., NEHER R., WETTSTEIN A.: Bildung von 18- und 19-Oxy-Cortexon durch Einwirkung von Nebennieren-Homogenaten auf Cortexon, Helv. chim. Acta 38: 1237, 1955.
1104. KAKÁČ B.: Photometrical determination of prednisone in mixture with cortisone, Česk. Farm. 8; 437, 1959 (in Czech).
1105. KAKÁČ B.: Determination of prednisone and his 20beta-hydroxy derivative in fermentation medium in the presence of cortisone (in Czech), Česk. Farm. 12: 177, 1963.
1106. KAKÁČ B., VEJDĚLEK Z.: Handbuch der Kolorimetrie, Bd. II. Jena: Fischer-Verlag 1963.
1107. KEIL B.: Extraction and countre-current distribution. In: B. KEIL *et al.*: Laboratory technique of organic chemistry (in Czech), 481—547. Ed. 2. Prague: Publishing house of Czech. Acad. of Science 1963.
1108. KELLIE A. E., WADE A. P.: Analysis of urinary 17-oxosteroids by gradient elution, Biochem. J. 66: 196, 1957.
1109. KELLIE A. E.: Determination of 17-ketosteroids and 17-ketogenic steroids, Analyst 82: 722, 1957.
1110. KOBER S.: Eine kolorimetrische Bestimmung des Brusthormons (Menformon), Biochem. Z. 239: 209, 1931.
1111. KUBOTA T., MIYAZAKI H.: Fluorescence spectra of phosphoric acid solutions of steroids, Ann. Report Shionogi Res. Labor. 11: 155, 1961.
1112. KUZNĚCOVA L. A., ROMANČUK M. A.: The use of activated charcoal for isolation of epicortisol from fermentation medium (in Russian), Med. Prom. 17, No. 8, p. 32, 1963.
1113. LAKSHMANAN T. K., LIEBERMAN S.: An improved method of gradient elution chromatography and its application to the separation of urinary ketosteroids, Arch. Biochem. Biophys. 53: 258, 1954.
1114. LANGBEIN G., MEYER M.: Verfahren zur Herstellung von Cortisonestern, German (East) patent 22 448 (1961).
1115. LEDERER E., LEDERER M.: Chromatography. A review of principles and applications. New York: Elsevier 1957.
1116. LETTRÉ - INHOFFEN - TSCHESCHE: Über sterine, Gallensäuren und verwandte Naturstoffe. Bd. 2. Stuttgart: Enke 1959.
1117. LEVESLEY P., WATERS W. A., WRIGHT A. N.: The mechanisms of oxidative fission of stereoisomeric 1,2-glycols, J. chem. Soc. 1956, 840.
1118. LEVY H., KUSHINSKY S.: Methods of isolating corticosteroids from blood, Rec. Progr. Hormone Res. 9: 357, 1954.
1119. LINFORD J. H.: Experiments on the use of silic acid for steroid chromatography, Canad. J. Biochem. and Physiol. 34: 1153, 1956.
1120. LOMBARDO M. E. *et al.*: A method for the extraction of steroids from blood, J. biol. Chem. 212: 345, 1955.
1121. LONG A. G., WYMAN L. J.: Separation of steroid mixtures. Brit. patent 815 832 (1959).
1122. MADER W. J., BUCK R. R.: Colorimetric determination of cortisone and related ketol steroids, Anal. Chem. 24: 666, 1952.
1123. MANCERA O., ROSENKRANZ G., SONDHEIMER F.: Synthesis of 11beta-hydroxytestosterone and 11-ketotestosterone, J. chem. Soc. 1953, 2189.

1124. MATTHEWS J. S., DE LOS ANGELES CERVANTES M.: Centrifugally accelerated paper partition chromatography of steroids, J. Chromatogr. 9: 195, 1962.

1125. MATTHEWS J. S., PEREDA A. L., AGUILERA A.: The quantitative analysis of steroids by thin-layer chromatography, J. Chromatogr. 9: 331, 1962.

1126. MATTOX V. R., KENDALL E. C.: Adrenal cortical hormones; introduction of a double bond, J. biol. Chem. 188: 287, 1951.

1127. MATTOX V. R.: The formation of a glyoxal side chain from steroids with dihydroxyacetone and delta$^{16}$-ketol side chains, J. am. chem. Soc. 74: 4340, 1952.

1128. MATTOX V. R., MASON H. L.: Chromatographic fractionation of beef adrenal extract. I. Isolation of aldosterone, J. biol. Chem. 223: 215, 1956.

1129. MEYER A. S.: Chemical transformations of steroids by adrenal perfusion. V. Blood incubations and perfusions with cortisone, J. biol. Chem. 203: 469, 1953.

1130. MEYER A. S., JEANLOZ R. W., PINCUS G.: Chemical transformations of steroids by adrenal perfusion. IV. Dehydroepiandrosterone, J. biol. Chem. 203: 463, 1953.

1131. MEYER A. S., LINDBERG M. C.: Optical-analytical studies on steroids. Reducing characteristics of hydroxylated and ketone steroids toward blue tetrazolium, Anal. Chem. 27: 813, 1955.

1132. MEYER A. S.: Light absorption between 230-500 mμ of alpha, beta-unsaturated and 3,6-diketo steroids in alkaline ethanol, J. org. Chem. 20: 1240, 1955.

1133. MIESCHER K., FISCHER W. H.: Zur partiellen Reduktion von Androstendion zu Testosteron, Helv. chim. Acta 22: 158, 1939.

1134. MILLS J. S. *et al.*: 6alpha-Fluoro-16alpha-hydroxy cortical hormones, J. am. chem. Soc. 81; 1264, 1959.

1135. MIZSEI A., SZABÓ A.: Analytical control of the biological oxidation in the production of delta$^{1,4}$-androstadiene-3,17-dione, J. Biochem. Microbiol. Technol. Engng. 2: 411, 1960.

1136. MIZSEI A., SZABÓ A.: Rapid determination of 17alpha-oxy-11-desoxy-corticosterone in the course of biological fermentative oxidation, J. Biochem. Microbiol. Technol. Engng. 3: 21, 1961.

1137. MIZSEI A., SZABÓ A.: A rapid estimation of prednisolone produced by fermentative oxidation, J. Biochem. Microbiol. Technol. Engng. 3: 119, 1961.

1138. MOORE J. A., HEFTMANN E.: Chemistry of the adrenocortical steroids. In: Eichler O., Farah A.: Handbuch der experimentellen Pharmakologie, Ergänzungswerk, Bd. 14/1, p. 186—410, Berlin: Springer 1962.

1139. MORRIS C. J. O. R., WILLIAMS D. C.: The polarographic estimation of steroid hormones. VI. Determination of individual adrenocortical hormones in human peripheral blood, Biochem. J. 54: 470, 1953.

1140. MORRIS C. J. O. R.: The polarographic estimation of ketosteroids, Rec. Trav. chim. 74: 476, 1955.

1140*a*) MOTTIER M., POTTERAT M.: De l'extraction des colorants pour denrées alimentaires avec la quinoléine et de leur identification par chromatographie sur plaque d'alumine, Anal. chim. Acta 13: 46, 1955.

1141. NEHER R.: Chromatography of sterols, steroids, and related compounds (389 ref.), J. Chromatogr. 1: 122—165, 205—258, 1958 (in German).

1142. NEHER R., WETTSTEIN A.: Farbreaktionen mit Steroiden, insbesondere Corticosteroiden im Papierchromatogramm, Helv. chim. Acta 34: 2278, 1951.

1143. NEHER R., WETTSTEIN A.: Papierchromatographie von schwach polaren Steroiden, Helv. chim. Acta 35: 276, 1952.

1144. NEHER R., WETTSTEIN A.: Isolierung und Konstitutionsermittlung weiterer Pregnanverbindungen aus Nebennieren, Helv. chim. Acta 39: 2062, 1956.

1145. Nielsen A. T.: Colorimetric assay of testosterone, Acta Endocrinol. 1: 362, 1948.
1146. Norymberski J. K., Woods G. F.: Selective reduction of oxosteroids with sodium borohydride, Chemistry Industry 1954, 518.
1147. Norymberski J. K.: The partial reduction of steroid hormones and related substances, J. chem. Soc. 1956, 517.
1148. Norymberski J. K., Stubbs R. D.: Indirect analysis of corticosteroids: the determination of steroidal dihydroxyacetones, Biochem. J. 64: 168, 1956.
1149. Oertel G. W., Eik-Nes B. K.: Determination of delta$^5$-3beta-hydroxy steroids, Anal. Chem. 31: 98, 1959.
1150. Oertel G. W.: Quantitative determination of estrogens, Naturwiss. 47: 183, 1960.
1151. Olivetto E. P., Hershberg E. B.: The selective reduction of steroidal 11,20-diketones with sodium borohydride, J. am. chem. Soc. 75: 488, 1953.
1152. Oughton J. F., Stephenson L. (Glaxo Lab.): Verfahren zur Isolierung von 1,4-3-keto steroiden, Swiss patent 346 541 (1960).
1153. Parsons J., Beher W. T., Baker G. D.: Steroid X-ray diffraction powder data, Anal. Chem. 28: 1514, 1956.
1154. Pfeffer K. H. *et al.*: Separate determination of 11-hydroxy and 11-desoxycorticoids, Arch. exptl. Path. Pharmacol. 214: 165, 1952.
1155. Picher H., Hajdu P.: Über die Bestimmung von kleinen Mengen Cortison neben predni-son, Medizin und Chemie 6: 259, 1958.
1156. Pincus G., Pearlman W. H.: Fractionation of neutral urinary steroids, Endocrinology 29: 413, 1941.
1157. Podbielniak W. J.: Apparatus for the centrifugal fractionation of materials such as in oil refining, U.S. patent 2 003 308 (1935).
1158. Poos G. J. *et al.*: 4b-Methyl-7-ethylenedioxy-1,2,3,4,4*a*alpha, 4beta, 5,6,7,8,10,10*a*beta-dodecahydrophenarthrene-4beta-ol-1-one and related tricyclic derivatives, J. am. chem. Soc. 75: 422, 1953.
1159. Porter C. C., Silber R. H.: A quantitative color reaction for cortisone and related 17,21-dihydroxy-20-keto steroids, J. biol. Chem. 185: 201, 1950.
1160. Prelog V., Häfliger O.: Über das polarographische Verhalten der Cyclanon-betainyl-hydrazone und die polarographische Bestimmung von Cyclanonen, Helv. chim. Acta 32: 2088, 1949.
1161. Procházka Ž.: Paper partition chromatography of steroid acids (in Russian), Coll. czech. chem. Com. 19: 98, 1954.
1162. Procházka Ž.: Thin-layer chromatography without a binding agent, Chem. Listy 55: 974, 1961.
1163. Randerath K.: Dünnschicht-Chromatographie. Weinheim/Bergstr. Verlag Chemie 1962.
1164. Rao N. P., Goldberg H. R., Axelrod L. R.: Synthesis of 2beta-hydroxy steroids, J. org. Chem. 28: 270, 1963.
1165. Reich H., Nelson D. H., Zaffaroni A.: Isolation of 17-hydroxycorticosterone from blood obtained from adrenal veins of dogs, J. biol. Chem. 187: 411, 1950.
1166. Reichstein T.: Über Cortin, das Hormon der Nebennierenrinde, Helv. chim. Acta 19: 29, 1936.
1167. Reichstein T.: Über Bestandteile der Nebennieren-Rinde. VI. Trennungsmethoden, sowie Isolierung der Substanzen F. a H. und J, Helv. chim. Acta 19: 1107, 1936.
1168. Reichstein T., v. Euw J.: Isolierung der Substanzen Q (desoxycorticosteron) und R sowie weiterer Stoffe, Helv. chim. Acta 21: 1197, 1938.
1169. Reichstein T., v. Euw J.: Über Bestandteile der Nebennierenrinde. Die Verseifung von Estern reduzierender Ketale mit schwachen Alkalien, Helv. chim. Acta 21: 1181, 1938.

1170. REICHSTEIN T., SHOPPEE C. W.: Chromatography of steroids and other colorless substances by the method of fractional elution, Discussions Faraday Soc. 7: 305, 1949.
1171. REINEKE L. M.: Paper chromatography in steroid determination, Anal. Chem. 28: 1853, 1956.
1172. ROMETSCH R.: Über die Fraktionierung von Gemischen durch Gegenstromextraktion I., Helv. chim. Acta 33: 184, 1950.
1173. ROSENFELD R. S., GALLAGHER T. H.: Deacetoxylation of steroid ring C ketol acetates, J. am. chem. Soc. 77: 4367, 1955.
1174. ROSENKRANZ G., MANCERA O., SONDHEIMER F.: Synthesis of C-2 oxygenated derivatives of Reichstein's substance S and of cortisone, J. am. chem. Soc. 77: 145, 1955.
1174*a*) RÖSSLER H., BRÜCKNER K.: Eine einfache Methode zur Bestimmung von 1,4-Dien-3-keto-steroiden neben 4-En-3-keto-steroiden. Naturwiss. 48: 695, 1961.
1175. RUDOLPH H.: A photoelectric polarimeter attachment, J. opt. Soc. Amer. 45: 50, 1955.
1176. SAVARD K.: Paper partition chromatography of $C_{19}$- and $C_{21}$-ketosteroids, J. biol. Chem. 202: 457, 1953.
1177. SCHEIBEL E. G.: Fractional liquid extraction, Chem. Eng. Progress 44, No. 9: 681, 1948.
1178. SCHRIEFFERS H., KORUS W.: Bestimmung von Cortison, Hydrocortison, Prednison, Prednisolon und 9alpha-Fluor-hydrocortison als 17-Ketosteroide nach Wismutatoxydation, Z. physiol. Chem. 313: 66, 1958.
1179. SEKI T.: Chromatographic separation of oestriol, oestrone and oestradiol-17beta, Nature 181: 768, 1959.
1180. SEKI T.: Chromatographic separation of some delta$^4$-3-ketosteroids, J. Chromatogr. 2: 667, 1959.
1181. SEKI T., MATSUMOTO K.: Chromatographic separation of 17-keto-steroids, J. Chromatogr. 10: 400, 1963.
1182. SHOOLERY I. N., ROGERS M. T.: Nuclear magnetic resonance spectra of steroids, J. am. chem. Soc. 80: 5121, 1958.
1183. SIBLÍKOVÁ O., HAIS I. M.: A rewiew on some colour and fluorescence detections of steroids, Česk. Farm. 7, 1, 1958 (in Czech).
1184. SIBLÍKOVÁ O., ČERNÁ M., HAIS I. M.: A rewiew on some colour and fluorescence detections of steroids. II. Česk. Farm. 11: 187, 1962 (in Czech).
1185. SILBER R. H., PORTER C. C.: The determination of 17,21-dihydroxy-20-ketosteroids in urine and plasma, J. biol. Chem. 210: 923, 1954.
1186. SJÖSTRÖM E., NYKÄNEN L.: Behavior of estradiol and nonphenolic sex hormones in ion-exchange columns containig strongly basic resins, Suomen Kemistilehti 29B: 23, 1956.
1187. SJÖSTRÖM E., NYKÄNEN L.: Reaction mechanism between estradiol benzoate and a strongly basic anion exchange resin in methanol and ethanol, J. org. Chem. 21: 1321, 1956.
1188. SJÖSTRÖM E., NYKÄNEN L.: Ion exchange separation of phenolic and nonphenolic sex hormones in pharmaceutical preparations, J. am. pharm. Assoc. Sci. Ed. 46: 321, 1957.
1189. SMITH J.: Chromatographic and electrophoretic techniques, Vol. I. London: Heinemann 1960.
1190. SMITH L., HALMER M.: 16alpha-Hydroxy steroids I. Characterisation of triamcinolone, J. am. pharm. Assoc. Sci. Ed. 48: 348, 1959.
1191. SMITH L. L., FOELL T.: Differentiation of delta$^4$-3-keto steroids and delta$^{1,4}$-3-keto steroids with isonicotinic acid hydrazide, Anal. Chem. 31: 102, 1959.
1192. SMITH L. L., FOELL T.: Thin-layer chromatography of steroids on starch-bound silica gel chromatoplates, J. Chromatogr. 9: 339, 1962.
1193. SMITH L. L. *et al.*: 16alpha-Hydroxy steroids. XII. 21-Amino derivatives of 9alpha-fluorhydrocortisone, J. am. chem. Soc. 84: 1265, 1962.

1194. SONDHEIMER F. *et al.*: Synthesis of steroidal hormone analogs hydroxylated at C-2, J. am. chem. Soc. 75: 4712, 1953.
1195. SONDHEIMER F., BURSTEIN S., MECHOULAM R.: Synthesis in the cardiac aglycone field. III. The conversion of 14alpha- to 14beta-hydroxy group in the androstane series. The ultraviolet spectra of delta$^{15}$-androsten-17-ones, J. am. chem. Soc. 82: 3209, 1960.
1196. STAHL E.: Dünnschicht-Chromatographie. Berlin: Springer 1962.
1197. SLAUNWHITE W. R. *et al.*: The fluorescence and extinction and partition coefficients of estrogens, J. biol. Chem. 191: 627, 1951.
1198. SWEAT M. L.: Sulphuric acid-induced fluorescence of corticosteroids, Anal. Chem. 26: 773, 1954.
1199. SWEAT M. L.: Silica gel microcolumn for chromatographic resolution of cortical steroids, Anal. Chem. 26: 1964, 1954.
1200. TALBOT N. B., EITINGTON I. V.: Use of the periodate acid reaction in the measurement of non-ketonic steroids obtained after various types of hydrolysis, J. biol. Chem. 154: 605, 1944.
1201. TALBOT N. B. *et al.*: The colorimetric assay of urinary corticosteroid-like substances, J. biol. Chem. 160: 535, 1945.
1202. TANSEY R. P., CROSS J. M.: A quantitative colorimetric method for certain ketosteroids with a modification of the Zimmermann reaction, J. am. pharm. Assoc. Sci. Ed. 39: 660, 1950.
1203. TAYLOR W.: Isolation of oxo-steroids under neutral conditions, Nature 182: 1735, 1958.
1204. TOUCHSTONE J. C., MURAVEC T.: Enhancement of the fluorescence of progesterone (and other steroids) in sulphuric acid, Anal. Chem. 32: 822, 1960.
1205. TRAPPE W.: Die Trennung von biologischen Fettstoffen aus ihren natürlichen Gemischen durch Anwendung von Adsorptionssäulen. II. Mitt. Abtrennung der phosphor- und stickstofffreien Lipoidfraktionen, Biochem. Z. 306: 316, 1940.
1206. TREYBAL R. E.: Liquid extraction. Chemical engineering series. New York: McGraw-Hill 1951.
1207. TSUDA K., OHKI E., SUZUKI J.: Reduktion von Reichstein's Substance S mittels Zink in Essigsäure, Chem. Pharm. Bull. Japan 7: 522, 1959.
1208. TSUDA K. *et al.*: An aromatization reaction of A cross-conjugated dienone system with zinc, J. org. Chem. 26: 2614, 1961.
1209. UPJOHN CO.: Cortisone. Brit. patent 739 023 (1955).
1210. VAEDTKE J., GAJEVSKA A.: Thin-layer partition chromatography a quick method of chromatography for steroids, J. Chromatogr. 9; 345, 1962.
1211. VELLUZ L. *et al.*: Saponification de deoxycorticostérone, Bull. Soc. chim. France 1947, 123.
1212. WALSER A., SCHLUNKE H. P.: Porter-Silber-reaktion mit $C_{16}$ substituierten corticosteroiden, Experientia 15: 71, 1959.
1213. WEISENBORN F. L. (O. Mathieson chem. Corp.): 9alpha-halo-A-norprogesterones, U. S. patent 2 950 289 (1960).
1214. WEISENBORN F. L., SIH Ch. J. (O. Mathieson chem. Corp.): 9alpha-hydroxy-A-norprogesterone, U. S. patent 3 056 838 (1962).
1215. WEISS-BERG E., TAMM Ch.: Über die Steroidoxydasen von Fusarium lini (Bolley). 9. Mitteilung über Reaktionen mit Mikroorganismen. Helv. chim. Acta 46:, 1166, 1963.
1216. WENDLER N. L., GRABER R. P.: Alkaline degradation of the cortical side chain, Chemistry Industry 1956, 549.
1217. WILSON H.: Chromogenic values of various ketosteroids in a micro modification of the Zimmermann reaction: Comparison with the macro procedure, Arch. Biochem. Biophys. 52: 217, 1954.

1218. WOLFE J. K., HERSHBERG E. B., FIESER L. F.: The polarographic determination of ketonic steroids, J. biol. Chem. 136: 653, 1941.
1219. ZAFFARONI A., BURTON R. B., KEUTMANN E. H.: Adrenal cortical hormones: Analysis by paper partition chromatography and occurence in the urine of normal persons, Science 111: 6, 1950.
1220. ZAFFARONI A., BURTON R. B.: Identification of corticosteroids of beef adrenal extract by paper chromatography, J. biol. Chem. 193: 749, 1951.
1221. ZIMMERMANN W.: Zur Methodik der Bestimmung der 17-Ketosteroide, Z. Vitamin-, Hormon- und Fermentforsch. 4: 456, 1951.
1222. ZUMAN P., TENYGL J., BŘEZINA M.: Polarographie der Steroide. I. Direkt reduzierbare Ketosteroide: delta$^4$-cholestenon, Methyltestosteron, Testosteron, Progesteron und Desoxycorticosteron, Coll. czech. chem. Com. 19: 46, 1954.

## *e)* Papers dealing with general problems in microbiology and fermentation:

1223. ANNEAR D. J.: A method for the routine preservation of bacteria, Austr. J. exp. Med. Sci. 40: 1, 1962.
1224. BARMENKOV A. S.: About preservation of microbe cultures (in Russian), Mikrobiologya 28: 444, 1959.
1225. FABIÁN J. *et al.*: Determination of residual water in freeze-dried bacterial cultures, Folia microbiol. 7: 247, 1962.
1226. HEROLD M. *et al.*: Antibiotics (in Czech), Publ. House Czechoslov. Acad. Sci. (Prague) 1957.
1227. KAROW E. *et al.*: Effect of physical variables on microbiological transformation of steroids, Ind. Eng. Chem. 48: 2213, 1956.
1228. MAZUR P. *et al.*: The effect of spray-drying on the viability of fungous spores, J. Bacteriol. 71: 257, 1956.
1229. MILLER R. E., SIMONS L. A.: Survival bacteria after twenty-one years in the dried state, J. Bacteriol. 84: 1111, 1962.
1230. PRESCOTT S. C., DUNN C. G.: Industrial Microbiology, McGraw Hill (New York) 1949.
1231. RAPER K. B. *et al.*: Preservation of molds by the lyophil process, Mycologia 37: 499, 1945.
1232. STÁRKA J.: Physiology and biochemistry of microbes (in Czech), Stát. pedagog. nakl. (Prague) 1959.

# SUBJECT INDEX

For a detailed *list of microorganisms* see page 129, *steroid metabolites* are listed on page 147.

*Absidia* 129
Acetonides, preparation 107
Acetylation of steroids 40, 98, 116
*Actinomyces* 130
Adsorption of steroids 80
Aldehyde group, reduction 36
Aldosterone 16, 61, 128
Androgens 13
  preparation 118
$\Delta^{1,4}$-androstadiene-3,17-dione 29, 73, 91, 92, 108, 120
Androstane-3β,17β-diol 37, 38
Androstan-17β-ol-3-one 38
$\Delta^5$-androstene-3β,17β-diol 35, 38
$\Delta^4$-androstene-16β,17β-diol-3-one 107
$\Delta^1$-androstene-3,17-dione 37
$\Delta^4$-androstene-3,17-dione 18, 20, 31, 35, 37, 45, 49, 54, 56, 58, 61, 62, 63, 73, 80, 87, 91, 109, 119, 120
$\Delta^4$-androsten-1α-ol-3,17-dion 20
$\Delta^4$-androsten-2β-ol-3,17-dion 20
$\Delta^4$-androsten-15α-ol-3,17-dione 24
$\Delta^4$-androsten-19-ol-3,17-dione 30
$\Delta^5$-androsten-3α-ol-17-one 18, 69, 80, 87
$\Delta^5$-androsten-3β-ol-17-one 18, 35
$\Delta^4$-androsten-17β-ol-3-one 76, 80, 98
Aromatization of steroids 30, 105, 109, 115
*Aspergillus* 130

Carboxyl group, formation 33
Cholestanol 36
Cholestenone 19, 33, 36
Cholesterol 12, 13, 19, 21, 34, 56, 58
  degradation 51, 117
Cholic acid 13, 38
Chromatography, column 89
  paper 70, 88, 95
  thin-layer 72, 89, 95
Colorimetric determination of steroids 73
  with $H_2SO_3$ 73, 74
  with tetrazolium salts 74
  Zimmermann method 74, 94
  Porter-Silber method 74, 93, 101
  with 2,4-dinitrophenylhydrazine 75, 94
  with the hydrazide of isonicotinic acid 75
  with 2,6-tert-butyl-p-cresol 75
Colour reaction 93
  with Lugol solution 69, 70
  with 2,4-dinitrophenylhydrazine 70
  with p-fenylenediamine 70, 71
  with tetrazolium salts 71
  with $SbCl_3$ 71
  with $K_3Fe(CN)_6 + FeCl_3$ 71
  Liebermann reaction 71
  with $H_2SO_4$ 73, 94
Column chromatography 89
  adsorbents 89, 90
  artefacts formation 89
  elution, solvents for 90
  distribution 90
  separation of isomers 90
Compounds S — Reichstein 16, 23, 27, 31, 36, 50, 61, 74, 77, 84, 87, 95, 126
Configuration, determination
  epimer hydroxy groups 103, 107
  side-chain 103, 104
  20-hydroxy groups 106
Counter-current distribution 90, 91
Cortexolone 16, 23, 27, 31, 36, 43, 50, 61, 74, 77, 84, 87, 95, 126
Cortexone 16, 28, 43, 58, 61, 74, 76, 83, 84, 94, 105
Corticoids 14
  preparation 124
Corticosterone 16, 28, 74, 94

Cortisol 15, 16, 23, 39, 43, 56, 61, 67, 74, 80, 84, 87, 88, 91, 94, 99, 126
Cortisol 11$\beta$,21-diacetate 39
Cortisone 16, 29, 61, 69, 74, 77, 79, 84, 91, 94, 95, 125, 126, 127
*Corynebacterium* 134
Crystal structure, study of 96
Crystallization of steroid metabolites 86, 87, 88
Cultivation of microorganisms 65
*Cunninghamella* 134
*Curvularia* 134
Cyclopentanoperhydrophenanthrene 11

Deacetylation 99
Dehydratation
  by strong acids 99, 100, 106
  by alcalies 106
  by $Al_2O_3$ 98, 107
  by zinc in acetic acid 107
Dehydroandrosterone 18
7-dehydrocholesterol 34
Dehydrocorticosterone 84
Dehydroepiandrosterone 18, 56, 62, 69, 80, 87, 119
Dehydrogenases of steroids 45, 47, 48, 53
1-2-dehydrogenation of steroids 29, 33, 42, 70, 105, 109, 114, 118
  mechanism 47, 49
  of androstenes 120, 121
  of pregnenes 127, 128
1-dehydrotestololactone 29, 33, 120
1-dehydrotestosterone 29
Deoxycholic acid 38
Deoxycorticosterone, see cortexone
*Dermatophyta,* influence of steroids 59
"Detoxication" reaction 64
Digitoxigenin 20, 21, 24, 26
3$\alpha$,7$\alpha$-dihydroxy-12-ketocholanic acid 35
6$\beta$,11$\alpha$-dihydroxyprogesterone 109
7$\alpha$,14$\alpha$-dihydroxyprogesterone 22
7$\alpha$,15$\beta$-dihydroxyprogesterone 98
12$\beta$,15$\beta$-dihydroxyprogesterone 24
Dische's reaction 94
Double bonds, formation
  by microorganisms 70, 109
  by $SeO_2$ 105
  by chloranil 105
DPN 45

Emulsions 83
  breaking of 83
  formation of 78, 83
"Enzymatic transcrystallization" 42
Enzyme action, specificity 42
11$\alpha$-epicorticosterone 28, 94, 125
11$\alpha$-epicortisol 23, 84, 88, 91, 94, 103, 126
17$\alpha$-epitestosterone 45, 61
Epoxidation 114
9$\alpha$,11$\alpha$-epoxy-$\Delta^4$-pregnene-17$\alpha$,21-diol-3,20-dione 34
14$\alpha$,15$\alpha$-epoxy-$\Delta^4$-pregnene-17$\alpha$,21-diol-3,20-dione 34
Epoxides, formation 34
Ergosterol 12, 55, 58, 61
Esterification of steroids 40, 116
Estradiol 13, 15, 19, 58, 61, 94
  preparation 121
Estriol 122
Estrogens 13
  preparation 121
  separation 72
  isolation 91
Estrone 19, 30, 121, 122
Extraction of steroids 71, 78—88
  solvents 81
  extraction coefficient 82
  extractors, types 82, 83
  selective 83, 84
Extracts,
  purification of 84—86
    from lipids 86
  decolourizing 86
    with $Al_2O_3$ 84
    with charcoal 84
  drying 84, 85, 71
  evaporation 85

Fermentation course, analyses 69, 70, 71, 72, 73
Fermentation medium, 65
  isolation of steroids 79
Fermentation process,
  single phase- 65
  double phase- 65
*Flavobacterium* 135
Fluorescence reactions 93, 94, 95
9$\alpha$-fluoro-$\Delta^{1,4}$-pregnadiene-11$\beta$,16$\alpha$,17$\alpha$,21-tetraol-3,20-dione 107
*Fusarium* 135

Gestagens 13
preparation 123
Girard hydrazones 76, 77
Glycoside bond, cleavage 116
Grimmer's route 103

D-Homo steroids 38, 98
Hydrogenation of steroids 37, 116
Hydrolysis 39, 116
7β-hydroxycholesterol 21
14α-hydroxycortisol 27
1β-hydroxydigitoxigenin 20, 21
5β-hydroxyditoxigenin 26
12β-hydroxydigitoxigenin 24
Hydroxy groups, removal of
by acids 107
by alcalies 106
by $Al_2O_3$ 107
Hydroxylation of steroids 20, 50, 69, 111, 118, 123
10-hydroxy-19-nortestosterone 27
20β-hydroxyprednisone 29
6β-hydroxyprogesterone 109
7α-hydroxyprogesterone 22
7β-hydroxyprogesterone 22
9α-hydroxyprogesterone 26, 52
11α-hydroxyprogesterone 23, 62, 69, 80, 109, 118, 123
14α-hydroxyprogesterone 22, 27
15β-hydroxyprogesterone 25, 98
16α-hydroxyprogesterone 25
17α-hydroxyprogesterone 27, 39
9α-hydroxysolasodin 26
6β-hydroxytestosterone 21
14α-hydroxytestosterone 27
16β-hydroxytestosterone 107

Identification of steroids 92—109
by chemical methods 97—108
by esterification 98
by microbiological methods 108
by colour reactions 93
by fluorescence reactions 94, 95
by physical methods 95—97
Isolation of steroids 78
from mycelium 87, 80
Isomerase of steroids 45, 46
Isomerization of steroids
$\Delta^5 \rightarrow \Delta^4$, mechanism 46
of side chain 99, 100, 101
Isomerization by acids
by alcalies 100, 101
into D-Homosteroids 105

3-Ketobisnor-$\Delta^4$-cholen-22-al 37
3-ketobisnor-$\Delta^4$-cholen-22-ol 37
$\Delta^4$-3-ketoetienic acid 33
Keto group, formation 30, 31
reduction 35
20-keto group of pregnenes, reduction 48
Kober reaction 73

Melting point, determination 95
17aα-methyl-D-Homo-11α-hydroxytestosterone 39
17α-methyl-testosterone 119
Microbial binding of steroids 56
Microorganisms, maintenance 67
list of 129
Molecular rotation, determination 96
*Mucor* 138
*Mycobacterium* 139

*Neurospora crassa* 57
influence of steroids 58
*Nocardia* 140
19-norsteroids 46, 109, 121
19-nortestosterone 27
Nuclear magnetic resonance 97

Optical rotation, determination 95
Optical rotatory dispersion 97
Oxidation
$CrO_3$ 102—103
α-diol groups 102, 104
α-ketol groups 102, 104
selective 103, 105
$HJO_4$ 103—104
$NaBiO_3$ 104, 103
lead tetraacetate 104, 103
of the methylene group 30
Oxidation
manganese dioxide 104
Oppenauer 105, 76
N-bromacetamide 105
N-bromsuccinimide 105
chloranil 105
$SeO_2$ 105
Oxidative processes 18, 110
Oxygen, consumption and steroids 61

Paper chromatography 70, 95
detection of spots 70, 72
preparative 88
$R_F$ values 70
solvent systems 70
Patent literature 205
*Penicillium* 141
Polarography 76
Prednisolone 74, 80, 127
Prednisone 29, 30, 68, 74, 77, 79, 88, 91, 127, 128
$\Delta^{4,9(11)}$-pregnadiene-17α,21-diol-3,20-dione 34
$\Delta^{4,14}$-pregnadiene-17α,21-diol-3,20-dione 34
$\Delta^{1,4}$-pregnadiene-17α,21-diol-3,11,20-trione 74, 77, 79, 88, 91
$\Delta^{4,14}$-pregnadiene-3,20-dione 107
$\Delta^{1,4}$-pregnadiene-11β,17α,21-triol-3,20-dione 74, 80
Pregnane-17α,21-diol-3,11,20-trione 30
$\Delta^{5}$-pregnene-3β,11α-diol-7,20-dione 31
$\Delta^{4}$-pregnene-6β,11α-diol-3,20-dione 109
$\Delta^{4}$-pregnene-7α,15β-diol-3,20-dione 98
$\Delta^{4}$-pregnene-11α,21-diol-3,20-dione 94
$\Delta^{4}$-pregnene-11β,21-diol-3,20-dione 74, 94
$\Delta^{4}$-pregnene-17α,21-diol-3,20-dione 74, 77, 84, 87, 95
$\Delta^{4}$-pregnene-17α,21-diol-3,11,20-trione 69, 74, 77, 79, 84, 91, 94, 95
$\Delta^{4}$-pregnene-3,20-dione see progesterone
Pregnenes, cleavage of side chain 49, 120
$\Delta^{4}$-pregnene-6β,17α,21-triol-3,20-dione 23
$\Delta^{4}$-pregnene-11α,17α,21-triol-3,20-dione 23, 84, 88, 91, 94, 103
$\Delta^{4}$-pregnene-11β,17α,21-triol-3,20-dione 23, 74, 80, 84, 87, 88, 91, 94, 99
$\Delta^{4}$-pregnene-15β,17α,21-triol-3,20-dione 98
$\Delta^{4}$-pregnene-17α,20β,21-triol-3-one 36
$\Delta^{4}$-pregnen-6β,-ol-3,20-dione 109
$\Delta^{4}$-pregnen-11α-ol-3,20-dione 69, 80, 109
$\Delta^{4}$-pregnen-15β-ol-3,20-dione 98
$\Delta^{4}$-pregnen-21-ol-3,20-dione 74, 76, 83, 84, 94, 105
$\Delta^{5}$-pregnen-3β-ol-20-one 13, 19, 31, 56, 58, 123
$\Delta^{4}$-pregnen-21-ol-3,11,20-trione 84
*Prim.* alcohol group, formation 28, 36
Progesterone 13, 15, 19, 22, 23, 24, 25, 26, 28, 29, 31, 32, 33, 49, 52, 53, 58, 61, 62, 73, 76, 80, 120, 123, 124, 125
"Pseudocrystallo-fermentation" 42
*Pseudomonas* 142
Purification of extracts 84, 85
lipids, removing of 86

Racemates, microbial cleavage 128
Rearrangement
by acids 74, 101
of the side chain 74, 101
dienone-phenol 108
Reduction
metalhydride ($NaBH_4$) 104, 105
of 17-keto group 45, 105, 119, 121
of 20-keto group 48
double bonds 106
microbiological 106
selective 105, 106
Meerwein-Ponndorf 106
Zn-AcOH 106
catalytical 106
Reduction processes 35, 115
Reviews 181
*Rhizopus* 143

Saponification of steroid esters 39, 116
Savard's rule 95
*Sec.* alcohol group, oxidation 18
formation 19, 32, 35
elimination 38
9,10-*seco*-3-hydroxy-1,3,5(10)-androstatrien-9,17-dione 109
*Seco* steroids 51, 52, 53, 109
Side chain of steroids, cleavage 31, 32, 33, 70, 74, 115
configuration of 103
β-sitosterol 56
Solasodin 26
Spectrophotometry 75
Sterane 11
Steroid acetates, formation 40, 116
Steroid esters, saponification 39, 116
Steroid hormones,
biogenesis 13
metabolism 14
degradation 51
influence on microbes 55
preparation 117
Steroid hydrocarbons 12
Steroid hydroxylation 50, 123
Steroid metabolites 147

Steroid metabolites list of 147
  acetylation 98
  adsorption on charcoal 80
  aromatization 105, 109
    by alcalies 105
    by microorganisms 109
  paper chromatography of 70—72
  thin-layer chromatography of 72
  colorimetry 73—75
  spectrophotometry 75—76
  polarography 76—77
  separation (mixtures) 88—92
  isolation 79—92
  identification 92—109
  extraction 81—86
  crystallization 86—88
Steroids, acetonides, preparation 107
  acetylation 98
  adsorption 80
  analysis 69
  aromatisation 105
  chemistry 11
  chromatography 89
  colorimetry 73
  colour reactions 71, 93
  degradation 51
  identification 78
  influence on microbes 55
  isolation of metabolites 78
  isolation, analyses, literature 224
  mechanism of transformations 41
  and microbial metabolism 61
  and microbial respiration 63
  microbial binding 56
  solubility 41
Steroids mixtures,
  separation of, by chemical methods 91
  separation by esterification 91
  as Girard hydrazones 91, 92
  as derivatives 92
  by fractional crystallization 88
  as digitonides 92
  as bisulfite compounds 92
  as cyclo-borate complex 92
Steroid side chain, cleavage 49
Sterols 12
  degradation 51, 117
  influence on microbes 55
  microbial binding 56
Stigmasterol 56
*Streptomyces* 144

*Tert.* alcohol group, formation 25
Testolic acid 54
Testololactone 32, 48, 49, 53, 54, 120
Testosterone 13, 15, 21, 32, 35, 38, 40, 45, 49, 56, 58, 61, 63, 76, 80, 98
  preparation 118
Testosterone acetate 40
Thin-layer chromatography 72, 95
  adsorbents 72, 89
  preparative 89
  detection of spots 72
Transformations of steroid 65
  conditions 66
  types of reactions 110
Triamcinolone 107
*Trichoderma* 146
*Trichothecium* 146
3,7,12-triketocholanic acid 35
Types of reactions 110

Zimmermann reaction 94

# SUPPLEMENT TO LITERATURE

## (*including the year 1965*)

## *a*) Reviews


1233. ACHREM A. A., TITOV J. A.: Microbiological transformations of steroids (in Russian), p. 504, Moskva: Nauka 1965.

1234. DORFMAN R. I., UNGAR F.: Steroid transformations by microorganisms. In: Dorfman R. I., Ungar F:. Metabolism of steroid hormones, p. 224—288, New York: Academic Press 1965.

1235. CHARNEY W., HERZOG H.: Microbiological transformation of steroids – A handbook, p. 670, New York: Academic Press 1965.


## *b*) Individual papers


1236. AARONSON S.: A role for a sterol and a sterol precursor in the bacterium *Rhodopseudomonas palustris*, J. gen. Microbiol. 37: 225, 1964.

1237. BALLET P., TRUONG VAN THUONG: Transformation de la digitoxine par *Streptomyces platensis* McGuire (NRRL 2364), Ann. pharm. franç. 23: 57, 1965.

1238. BOLT C. C. *et al*: Microbiological conversion of 19-nortestosterone (VI), 1beta-hydroxylation, Rec. Trav. chim. 84: 626, 1965.

1239. BRANNON D. R. *et al*: Transformation of progesterone and related steroids by *Aspergillus tamarii*, J. org. Chem. 30: 760, 1965.

1240. ČAPEK A., KAKÁČ B., TADRA M.: The preparation of prednisone from cortisone acetate by a microbial procedure (in Czech), Čs. Farmacie 12: 408, 1963.

1241. ČAPEK A., TADRA M., TŮMA J.: Separation of androstane 17-hydroxyepimers, Folia Microbiol. 9: 380, 1964.

1242. Čapek A., HANČ O., TADRA M., TŮMA J.: The effect of substituents on the direction of the transformation of steroids by *Beauveria bassiana*, Folia Microbiol. 11: 159, 1966.

1243. ČAPEK A., HANČ O., TADRA M., TŮMA J.: Microbial transformations of steroids XXVI. Improvement of the preparation of cortisone from cortexolone (in Czech), Čs. Farmacie 15: 198, 1966.

1244. CASAS-CAMPILLO C., BAUTISTA M.: Microbiological aspects in the hydroxylation of estrogens by *Fusarium moniliforme*, Appl. Microbiol. 13: 977, 1965.

1245. CHEN J. W. *et al*: 1-Dehydrogenation of steroids at levels above their solubilities, Ind. Eng. Chem. (Proc. design & develop.) 4: 421, 1965.

1246. CORONELLI C. *et al*: An epoxidation by *Corynebacterium simplex*, Experientia 20: 208, 1964.

1247. CRABBÉ P., CASAS-CAMPILLO C.: Microbiological hydroxylation of estrane derivatives with *Fusarium moniliforme*, J. org. Chem. 29: 2731, 1964.

1248. CURTIS P. J.: The transformation and resolution of 8-azasteroids by fungi, Biochem. J. 97: 148, 1965.

1249. DODSON R. M. *et al*: Identifizierung der als 6beta,15alpha-Dihydroxyprogesteron, 11alpha,15beta-Dihydroxyprogesteron und 12beta,15beta-Dihydroxyprogesteron beschreibenen Steroide mit 12beta,15alpha-Dihydroxyprogesteron, Helv. chim. Acta 48: 1933, 1965.

1250. EL-TAYEB O. *et al*: Steroid epoxide cleavage by *Cylindrocarpon radicicola*, Biochim. biophys. Acta 93: 402, 1964.

1251. EL-TAYEB O. *et al*: Substrate specificity in steroid side-chain degradation by microorganisms Biochim. biophys. Acta 93: 411, 1964.

1252. GIBSON D. T. *et al*: Microbiological degradation of steroids, In: Abstracts of papers, 150th Meeting Am. Chem. Soc. p. 100C, Washington 1965.

1252a. GIBSON D. T. *et al*: On the mechanism of ring A cleavage in the degradation of 9,10-secosteroids by microorganisms, J. biol. Chem. 241, 551, 1966.

1253. GREENSPAN G. *et al*: Microbiological hydroxylation of alkaloids from *Funtumia latifolia*, J. org. Chem. 30: 4215, 1965.

1254. GVOZDIAK J. P., KOLESNYKOV H. D.: Microbial transformations of cardinoglucosides (in Ukrainian), Farm. Ž. (Kyjev) 19: 2, 37, 1964.

1255. HAYAKAWA S., SAMUELSSON B.: Transformation of cholic acid in vitro by *Corynebacterium simplex*. Bile acids and steroids 132, J. biol. Chem. 239: 94, 1964.

1256. HOLMLUND CH. E. *et al*: Microbiological transformations of 21-acetoxy-17alpha-hydroxy-16alpha-methylpregn-4-ene-3,20-dione, Steroids 5: 459, 1965.

1257. IRMSCHER K. *et al*: Darstellung von 7alpha-Hydroxy- und 7alpha-Methoxytestosteron-Derivaten, Chem. Ber. 97: 3363, 1964.

1258. KOGAN L. M. *et al*: Microbiol hydroxylations of steroids by *Actinomycetes* (in Russian), Doklady Akad. Nauk USSR, 160: 346, 1964.

1259. KOGAN L. M. *et al*: Reduction of 17alpha,21-deoxysteroids by *Actinomyces albus* 3006 (in Russian), Izv. Akad. NaukUSSR (chim.) 1964; 2008.

1260. KOGAN L. M. *et al*: Oxidation of pregnenolone by *Rhizopus nigricans* VNICHFI — 7 (in Russian), Izv. Akad. Nauk USSR (chim.) 1964: 2016.

1261. KOGAN L. M. *et al*: Transformation by *Actinomycetes* of estadiol to estrone (in Russian), Izv. Akad. Nauk USSR (biol.), 1965: 285.

1262. KOGAN L. M. *et al*: Substrate specificity of microbial transformation of steroids by *Actinomyces albus* ATCC 3006 (in Russian), Chem. prirod Sojed. 1965: 100.

1263. KONDO E., TORI K.: Microbiological 9alpha,12alpha- and 9alpha,18-dihydroxylations of androst-4-ene-3,17-dione, J. Am. Chem. Soc. 86: 736, 1964.

1264. KOŠČENKO K. A., SKRJABIN G. K.: Deacetylation by molds of cortisone acetate (in Russian), Mikrobiologyia 34: 252, 1965.

1265. LÁBLER L., ŠORM F.: Microbiological hydroxylation of 18-substituted steroids, In: Pasqualini J. R. — Jayle M. F.: Structure and metabolism of corticosteroids, p. 59—64, New York: Academic Press, 1964.

1266. LASKIN A. I. *et al.*: Microbial hydroxylations of estrone and estradiol in the 6beta,7alpha- and 15alpha-positions, J. org. Chem. 29: 1333, 1964.

1267. LASKIN A. I. *et al*: 2beta-Hydroxylation of steroids by *Gnomonia fragariae*, Steroids 5; 57, 1965.

1268. LEE S. S., SIH CH. J.: Mechanism of steroid oxidation by microorganism. VI. Metabolism of 3beta-hydroxy-5alpha,6alpha-oxiandrostan-17-one, Biochemistry 3: 1267, 1964.

1269. MANSON A. J. *et al*: A new microbiological steroid reaction, J. org. Chem. 30: 307, 1965.

1270. MARTÍNKOVÁ J.: Three-steps microbial transformation of 9-fluorocortisol-21-acetate to 9alpha-fluoro-16alpha-hydroxyprednisolone (triamcinolone) (in Czech), Dissertation Tech. University Prague 1965.

1271. Naito A. *et al*: Transformation of steroids by *Pseudomonas chloraphis*, Steroids 3: 327, 1965.
1272. Noguchi S., Fukushima D. K.: Microbiological hydroxylation of saturated 17-keto-steroids, J. org. Chem. 30: 3552, 1965.
1273. Nozaki Y. *et al*: Microbiological 12beta-hydroxylation of digitoxin, a steroidal glycoside, Agricult. biol. Chem. (Japan) 29: 783, 1965.
1274. Okada M., Saito Y.: Microbiological formation of 3beta,7alpha,15alpha-trihydroxy-androst-5-en-17-one, Steroids 6: 651, 1965.
1275. Okada M. *et al*: Microbiological transformation of dehydroepiandrosterone, progesterone, desoxycorticosterone, and testosterone by *Gibberella saubinetti*, J. pharm. Soc. Japan 85: 9, 816, 1965.
1276. Peterson G. E., Davis J. R.: Cholesterol utilization by *Streptomyces spp.*, Steroids 4: 677, 1964.
1277. Procházka Ž., Šorm F.: On steroids. XCII. Microbial hydroxylation of 14beta,18-cyclo-progesterone, Collection 30: 1874, 1965.
1278. Protiva J. *et al.*: Transformation of steroids by microorganisms of the genus *Flavobacterium*, Folia microbiol. 9: 218, 1964.
1279. Raman P. B., Peron F. G.: Microbiological transformation of 18-hydroxy-11-desoxy-corticosterone into 18-hydroxycorticosterone, Steroids 5: 249, 1965.
1280. Rosenfeld R. S., Gallagher T. F.: Further studies of the biotransformation of cholesterol to coprostanol, Steroids 4: 515, 1964.
1281. Ryžkova V. M. *et al.*: Deacetylation by *Bacillus megaterium* of steroid acetates (in Russian) Mikrobiologyia 34: 407, 1965.
1282. Sato Y., Hayakawa S.: The microbial hydroxylation of tomatidine, J. org. Chem. 29: 198, 1964.
1283. Sato Y. *et al*: Microbial hydroxylation of 5,6-dihydrosolasodine, J. org. Chem. 20: 3732, 1964.
1284. Sax K. J. *et al.*: Microbiological 16-oxidation of estr-4-en-3-one, J. org. Chem. 29: 2351, 1964.
1285. Sax K. J. *et al.*: Formation of 1alpha, 2alpha-dihydroxy-steroids, Steroids 5: 345, 1965.
1286. Schubert K. *et al.*: Stereospezifische Hydrierung von delta$^{1,4}$-Androstendion(3,17) zu delta$^{1}$-5beta-Androstendion-(3,17) und 5beta-Androstanol-(3alpha)-on-(17) mit *Clostridium paraputrificum* unter anaeroben Bedingungen, Z. physiol. Chem. 332: 310, 1964.
1287. Schubert K. *et al.*: Stereospezifische Hydrierung konjugierter Doppelbindungen in Steroiden durch Darmbakterien, Angew. Chem. 76: 796, 1964.
1288. Schubert K. *et al.*: Formation of low molecular degradation products from progesterone by microorganisms, Steroids 4: 581, 1964.
1289. Schubert K. *et al.*: Stereospecific hydrogenation of delta$^{4}$-, delta$^{1,4}$-, delta$^{4,6}$-, and delta$^{1,4,6}$-ketosteroids by *Clostridium paraputrificum*, Steroids, Suppl. I, 175, 1965.
1290. Schubert K. *et al.*: Mikrobieller Abbau von Cholesterin, Naturwiss. 52: 20, 1965.
1291. Schubert K., Kaufmann G.: Bildung von Sterinestern in der Bakterienzelle, Biochim. biophys. Acta 106: 592, 1965.
1292. Schwarz V. *et al.*: The isolation and the characterization of 1beta-hydroxy-11-deoxy-cortisol as a by-product of a microbial 11-hydroxylation, Steroids 4: 645, 1964.
1293. Sih C. J., Wang K. C.: A new route to estrone from sterols, J. Am. Chem. Soc. 87: 1387, 1965.
1294. Sih C. J. *et al.*: 3,4-Dihydroxy-9,10-secoandrosta-1,3,5(10)-triene-9,17-dione. An intermediate in the microbiological degradation of ring A of androst-4-ene-3,17-dione, J. Am. Chem. Soc. 87: 1385, 1965.
1295. Sih C. J. *et al.*: On the mechanism of ring A cleavage in the degradation of 9,10-seco steroids by microorganisms, J. Am. Chem. Soc. 87: 1386, 1965.

1296. SIH C. J. *et al.*: 3,4-dihydroxy-9,10-secoandrosta-1,3,5(10)-triene-9,17-dione, an intermediate in the microbiological degradation of ring A of androst-4-ene-3,17-dione, J. biol. Chem. 241, 540, 1966.
1297. SKRJABIN G. K. *et al.*: Transformation by *Mycobacterium sp.* 193 of cortisol, cortisone and derivatives (in Russian), Izv. Akad. Nauk USSR (biol.) 1964 715.
1297a. SMIRNOV P. N.: Polarographic control of microbial transformation of methyltestosterone to dianabol (in Russian), Med. Prom. 19: 11, 38, 1965.
1298. TAKAHASHI T.: Transformation of $C_{19}$-steroids by *Pellicularia filamentosa f. sp. microsclerotia* IFO 6298, Agr. biol. Chem. 28: 1, 38, 1964.
1299. TSONG Y. Y. *et al*: Mechanism of steroid oxidation by microorganisms. V. Reduction of the delta$^{6,7}$ double bond, Biochim. biophys. Acta 93: 398, 1964.
1300. VONDROVÁ O., TADRA M., ČAPEK A.: Hydroxylation of the steroid molecule by strains of *Streptomyces fradiae*, Folia microbiol. 8: 176, 1963.
1301. WHITMARSH J. M.: Intermediates of microbiological metabolism of cholesterol, Biochem. J. 90: 23P, 1964.
1302. ZAKRZEWSKI Z. *et al*: Preparation of 17alpha-methyl-1,4-androstadien-17beta-ol-3-one by microbial transformation (in Polish), Przemysl chem. 43: 564, 1964.
1303. ZVJAGINCEVA I. S., SKRJABIN G. K.: Dehydrogenation of steroids by *Mycobacterium* (in Russian), Izv. Akad. Nauk USSR (biol.) 1964: 525.

## *c*) Patent literature

U.S. patents

1304. Am. Cyanamid Co. (HOLMLUND C. E., EVANS R. H.): 16alpha-Hydroxylation of steroids by *Staurophoma species*, 3 071 516 (1963).
1305. O. Mathieson chem. Corp. (LASKIN A. I., FRIED J.): Process for microbial 6beta-hydroxylation of steroids containing aromatic A-rings, 3 115 443 (1963).
1306. O. Mathieson chem. Corp. (LASKIN A. I., FRIED J.): Process for microbial hydroxylation of steroids containing aromatic A-ring, 3 115 444 (1963).
1307. Am. Cyanamid Co.: Microbiological 16alpha-hydroxylation of steroids, 3 116 219 (1963).
1308. Upjohn Co.: Process for the 9alpha-hydroxylation of steroids with *Ascochyta*, 3 116 220 (1963).
1309. Eli Lilly & Co. (MALLETT G. E.): delta$^{1}$-Dehydrogenation of steroids by fermentation with *Actinoplanaceae*, 3 128 238 (1964).
1310. Ciba Corp. (URECH J. *et al.*): Process for the microbiological oxidation of steroids using *Mucor griseocyanus*, 3 147 195 (1964).
1311. Syntex Corp. (CASAS-CAMPILLO C.:): Process for the production of 11alpha-hydroxylation of steroids with *Panaeolus*, 3 149 050 (1964).
1312. Syntex Corp. (CASAS-CAMPILLO C.:): Process for the preparation of 11alpha-hydroxy delta$^{4}$-3-keto steroids from 11-unsubstituted delta$^{5}$-3-hydroxy and 3-acyloxy steroids using *Psilocybe caerulescens* var. *mazatecorum*, 3 190 809 (1965).
1313. Sterling Drug Co.: Microbial reduction process and steroid products thereby, 3 198 809, (1965).
1314. Upjohn Co. (HANSON F. R., MAXON W. D.): Process for the microbiological oxygenation of progesterone 3 201 324 (1965).

British Patents

1315. Merck Co.: Steroids compounds, 931 221 (1963).
1316. Roussel - Uclaf: New steroid compounds and processes for their production and conversion into substituted corticosteroids, 934 710 (1963).
1317. Roussel - Uclaf: New steroid compounds and processes for their preparation and conversion into substituted corticosteroids, 934 711 (1963).
1318. Merck Co.: Steroid compounds, 934 802 (1963).
1319. Merck Co.: New anabolic steroid and the preparation thereof, 935 334 (1963).
1320. Am. Cyanamid Co.: 17-Desoxysteroids and methods of preparing the same, 935 701 (1963).
1321. Boots Pure Drug Co.: Improvements in steroid processes, 936 590 (1963).
1322. Syntex S. A.: Process for the manufacture of fluorine compounds of the pregnane series, 937 988 (1963).
1323. Boots Pure Drug Co.: Improvements in steroid processes, 938 440 (1963).
1324. Am. Cyanamid Co.: delta$^{1}$-Unsaturated steroids and process for preparation, 939 820 (1963).
1325. Roussel - Uclaf: New steroid compounds and processes for their preparation, 943 951 (1963).
1326. Ciba A. G.: Process for the manufacture of 11alpha-hydroxy-6alpha-halogen-pregnenes, 944 321 (1963).
1327. Syntex S. A.: New 6-halo-steroids and process for the preparation thereof, 944 906 (1963).
1328. O. Mathieson chem. Corp.: Steroids, 945 105 (1963).
1329. Merck Co.: Steroid compounds, 946 860 (1964).
1330. Schering Corp.: Process for the manufacture of 1,4-diene-3-keto-steroids and new 1,4-keto-steroids, 948 188 (1964).
1331. Ciba Ltd.: Process for the manufacture of 11alpha-18-dihydroxyprogesterone, 950 744 (1964).
1332. O. Mathieson chem. Corp.: A-Nor steroids, 951 108 (1964).
1333. Roussel - Uclaf: New steroid compounds and processes for the preparation thereof, 952 491 (1964).
1334. Ciba Ltd.: Process for the microbiological oxidation of steroids, 953 486 (1964).
1335. Societa Farmaceutici Italia: 16alpha-hydroxy-steroids, 957 392 (1964).
1336. Schering Corp.: 16alpha-Halomethyl-derivatives of steroids of the 17-alpha-hydroxy-pregnane series and a process for their manufacture, 957 436 (1964).
1337. Eli Lilly Co.: 6alpha-Chloro-16alpha-methylprogesterone, 963 421 (1964).
1338. General Mills Inc.: Improvements in or relating to steroids, 967 606 (1964).
1339. Merck A. G.: New unsaturated 16-methylene-3-keto-steroids and process for their manufacture, 968 255 (1964).
1340. Schering Corp.: 11beta- and 14alpha-Hydroxylated 16alpha-methyl steroids and process for their manufacture, 969 904 (1964).
1341. Brinkmann A. G.: 20-Esters of 6beta-bromo-delta$^{4}$-pregnen-20beta-ol-3-one and the manufacture of delta$^{4}$-pregnene-6,20beta-diol-3-one and allopregnan-20beta-ol-3,6-dione, 973 916 (1964).
1342. Lepetit S. p. A.: Steroids, 974 064 (1964).
1343. Merck A. G.: 6alpha-Methyl-16-methylene-steroids and process for their manufacture, 987 064 (1965).
1344. Syntex S. A.: Preparation of 11alpha-hydroxy-delta$^{4}$-3-keto steroids, 989 918 (1965).
1345. Schering A. G.: New delta$^{1,4}$-16alpha-methylsteroids and a process for their manufacture, 990 384 (1965).

1346. Upjohn Co.: Improvements in or relating to steroids and the manufacture thereof, 997 167 (1965).
1347. O. Mathieson chem. Corp.: Synthesis of steroids, 997 380 (1965).
1348. Hoffmann - La Roche: The manufacture of 17-oxygenated retrosteroids, 1 000 889 (1965).
1349. Richter Corp.: Procedure for the preparation of delta$^{1}$-dehydrosteroids, 1 003 852 (1965).
1350. Upjohn Co.: Improvements in or relating to steroids and the manufacture thereof, 1 004 667 (1965).
1351. Upjohn Co.: Improvements in or relating to the preparation of steroids, 1 006 818 (1965).
1352. O. Mathieson chem. Corp.: Phenanthrene derivatives, 1 012 491 (1965).

French Patents

1353. Schering A. G.: Nouveaux delta$^{1,4}$-16alpha-méthylstéroides et leur préparation, 1 315 629, Supplement No. 84 738 (1963).
1354. Upjohn Co.: Procédé de fabrication de 6alpha- et 6beta-fluoro steroïdes, 1 326 323 (1963).
1355. Schering A. G.: Nouveaux composés dérivant du prégnane et leur préparation, 1 328 103 (1963).
1356. Societá Farmaceutici Italia (Amici A. M. et al.): Procédé de préparation de 16alpha-hydroxystéroïdes, 1 328 731 (1963).
1357. O. Mathieson chem. Corp.: Procédé de préparation de la 11alpha-hydroxy-A-norprogestérone, 1,328 910 (1963).
1358. Ciba A. G.: Procédé pour la préparation de nouveaux prégnènes, entre autres du delta$^{4}$--3,20-dioxo-6alpha-fluoro-16alpha-méthyl-11beta,14alpha,17alpha,21-tétrahydroxy-prégnène et de son 21-triméthylacétate, 1 331 345 (1963).
1359. Ciba A. G.: Procédé de préparation de nouveaux prégnadiénes entre autres du delta$^{1,4}$-3,20-dioxo-6alpha-fluoro-16alpha-méthyl-11beta,14alpha,17alpha,21-tetrahydroxy-prégnadiéne, 1 331 346 (1963).
1360. Merck A. G.: Procédé de fabrication de 16-méthylène-stéroïdes, 1 333 715 (1963).
1361. Farbwerke Hoechst A. G.: Dérivés de progéstérone substitués en position-16 et leur procédé de préparation, 1 334 933 (1963).
1362. Ciba A. G.: Procédé de préparation de nouveaux stéroïdes, entre autres de delta$^{1,4}$-3-oxo-18-nor-androstadiènes oxygénés en position 11, 1 344 973 (1963).
1363. Upjohn Co.: Procédé de préparation de composés 6-fluoro de la série de l'androstane et de la série de l'oestrane, 1 347 653 (1963).
1364. Merck Co.: Procédé de préparation de stéroïdes par fermentation, 1 362 314 (1964).
1365. Lepetit S. p. A. (Kluepfel D., Coronelli C.): Nouveau procédé de préparation de stéroïdes, 1 364 069 (1964).
1366. Philips Co.: 11,17alpha,21-Trihydroxy-9beta,10alpha-pregnen-3,20-dione et sa préparation, 1 364 261 (1964).
1367. Roussel - Uclaf: Nouveau composé cortisonique et procédé de préparation, 1 365 265 (1964).
1368. Merck A. G.: Procédé de préparation de 16-méthylène stéroïdes, 1 367 410 (1964).
1369. Am. Cyanamid Co.: Méthode de fermentation pour la conversion des 20-dihydroxystéroides de la série du prégnane en 20-cétostéroides correspondants, 1 369 309 (1964).
1370. Roussel - Uclaf: Dérivés de la cortisone et procédé de préparation, 1 369 316 (1964).
1371. Philips Co.: Procédé de préparation de 10alpha-méthyl-stéroïdes oxygénés en 16, 1 372 850 (1964).
1372. Philips Co.: Procédé d'oxygénation de stéroïdes a l'aide de la famille *Sphaerioidaceae*, 1 372 851 (1964).

1373. Philips Co.: Procédé de préparation de 17beta-hydroxy-rétrostéroides, 1 372 852 (1964).
1374. Schering A. G.: Procédé de préparation de la 16alpha-chlorométhyl-delta$^{1,4}$-prégnadiène-11beta,21-diol-3,20-dione et de ses 21-esters, 1 374 605 (1964).
1375. Am. Cyanamid Co. (HOLMLUND CH. E. *et al.*): Procédé pour la préparation de nouveaux composés stéroides plus spécialement des prégnénes et des prégnadiénes, 1 375 659 (1964).
1376. VNICHFI: Procédé de préparation de la delta$^{1}$-déhydrocortisone (prednisone), 1 381 682 (1964).
1377. Richter Corp.: Procédé de préparation par voie microorganique de delta$^{4}$-3-céto-stéroïdes, 1 383 288 (1964).
1378. Richter Corp.: Procédé de préparation de délta$^{1}$-déhydro-stéroïdes, 1 397 299 (1965).
1379. Boots Pure Drug Co.: Procédé perfectionné pour la préparation de stéroïdes, 1 397 547 (1965).
1380. Hoffmann - La Roche: Procédé pour la préparation de dérivés de rétroandrostane, 1 404 545 (1965).
1381. Upjohn Co.: Procédé d'oxygénation de progestérone par voie microbiologique, 1 406 102 (1965).

## Swiss Patents

1382. Ciba A. G. (WETTSTEIN A., NEHER R.:) Verfahren zur Herstellung von Allopregnan-3beta,16alpha-diol-20-cn, 372 049 (1963).
1383. Hoffmann - La Roche: Verfahren zur Erzeugung vcn Vitamin $B_{12}$ durch Propionsäurebakterien, 373 865 (1963).
1384. Merck Co. (MCALEER W. J., STOUDT T. H.:) Verfahren zur Herstellung von 11-Oxysteroiden, 374 659 (1964).
1385. Syntex S. A. (BATRES E. *et al.*): Verfahren zur Herstellung neuer Pregnadiene, 374 660 (1964).
1386. Merck Co. (ARTH G. E. *et al.*): Verfahren zur Herstellung von 11beta,21-Dioxy-16alpha-alkyl-4,17(20)-pregnadien-3-cn-21-acylaten, 375 002 (1964).
1387. Chinoin (WIX G., RADOS M.): Verfahren zur Herstellung von delta$^{1,4}$-Androstadien-3,17-dion, 376 499 (1964).
1388. Ciba A. G. (WETTSTEIN A., VISCHER E.): Verfahren zur mikrobiologischen Reduktion von Steroiden, 376 910 (1964).
1389. Ciba A. G. (WETTSTEIN A. *et al.*): Verfahren zur Herstellung von 11alpha,18-Dihydroxyprogesteron, 376 912 (1964).
1390. Ciba A. G. (WETTSTEIN A. *et al.*): Verfahren zur Herstellung neuer delta$^{1,4}$-Pregnadiene, 377 812 (1964).
1391. Merck A. G.: Verfahren zur Herstellung von Reichsteins-Substanz-S, 378 880 (1964).
1392. Takeda Pharm. Ind. (TAKEDA R. *et al,*): Verfahren zur Herstellung von Steroidverbindungen, 381 226 (1964).
1393. Upjohn Co. (ROSSELET J. P. *et al.*): Verfahren zur Herstellung von Steroidverbindungen, 381 679 (1964).
1394. Upjohn Co. (MAGERLEIN B. J. *et al.*): Verfahren zur Herstellung von Steroidverbindungen, 382 156 (1964).
1395. Upjohn Co. (MAGERLEIN B. J. *et al.*): Verfahren zur Herstellung von Steroidverbingungen, 382 157 (1964).
1396. Chinoin (WIX G. *et al.*): Verfahren zur Herstellung von delta$^{1,4}$-Androstadien-3,17-dion aus delta$^{4}$-3-Ketosteroiden, 382 733 (1964).

1397. Upjohn Co. (MAGERLEIN B. J. *et al.*): Verfahren zur Herstellung von Steroidverbindungen, 382 741 (1964).
1398. Upjohn Co. (LINCOLN F. H., JR., *et al.*): Verfahren zur Herstellung neuer Steroidverbindungen, 382 742 (1964).
1399. Chinoin (WIX G., ALBRECHT K.): Verfahren zur mikrobiologischen Hydroxylierung von Steroiden 383 358 (1964).
1400. Chinoin (PAUNCZ J. *et al.*): Verfahren zur Gewinnung von Androstadien-(1:2,4:5)-dion--(3,17) aus seinen Lösungen, 383 363 (1964).
1401. Ciba A. G. (WETTSTEIN A., VISCHER E.): Verfahren zur Darstellung von 11alpha-Oxy-6alpha-halogensteroiden der Pregnanreihe, 383 364 (1964).
1402. Upjohn Co. (HANZE A. R. *et al.*): Verfahren zur Herstellung von 9alpha-Oxysteroiden 384 574 (1965).
1403. Takeda Pharm. Ind. (TAKEDA R. *et al.*): Verfahren zur Herstellung von Steroidverbindungen, 385 204 (1965).
1404. Upjohn Co.: Verfahren zur Herstellung neuer 6-Fluorsteroide, 387 026 (1965).
1405. Merck Co.: Verfahren zur Herstellung von 16-Alkyl-9alpha-halogen-11beta,17alpha,21-trioxy-4-pregnen-3,20-dion-21-acylat-Verbindungen, 390 912 (1965).
1406. Ciba A. G. (WETTSTEIN A. *et al.*): Verfahren zur Herstellung von Dehydroverbindungen der Steroidreihe, 394 168 (1965).
1407. Merck A. G. (JAHN U. *et al.*): Verfahren zur Herstellung von 1-Dehydro-progesteronderivaten, 395 082 (1965).
1408. Chinoin (WIX G., ALBRECHT K.): Verfahren zur Herstellung von Androstadiendion, 395 981 (1966).
1409. Ciba A. G. (URECH J. *et al.*): Verfahren zur Herstellung von delta$^4$-3,20-Dioxo-6alfa-fluor-16alpha-methyl-11beta,14alpha,17alpha,21-tetrahydropregnen und dessen 21-Ester, 396 897 (1966).

## German (West) Patents

1410. Schering A. G. (KIESLICH K., RASPÉ G.): Verfahren zur Gewinnung antiphlogistisch wirksamer Steroide, 1 147 226 (1963).
1411. O. Mathieson chem. Corp. (FRIED J.): Verfahren zur Herstellung von Steroiden, 1 149 003 (1963).
1412. Schering A. G. (KIESLICH K., RASPÉ G.): Verfahren zur Gewinnung von antiphlogistisch wirksamen Steroiden, 1 150 074 (1963).
1413. Ciba A. G. (WETTSTEIN A. *et al.*): Verfahren zur Herstellung von 6alpha-Halogen- bzw. 6alpha,9alpha-Dihalogen-pregnenen, 1 150 385 (1963).
1414. Upjohn Co. (FONKEN G. S., MURRAY H. C.: :Verfahren zur 1(2)-Dehydrierung von 3-Keto-, delta$^4$-3-Keto- bzw. delta$^5$-3-Hydroxy-20-ketosteroiden der Pregnanreihe, 1 152 412 (1963).
1415. Merck Co. (STOUDT T. H.): Verfahren zur Herstellung von 1,4-Bis-dehydro-3-ketosteroiden, 1 153 017 (1963).
1416. Ciba A. G. (WETTSTEIN A. *et al.*): Verfahren zur Herstellung von 11alpha,18-Dihydroxy-progesteron, 1 154 468 (1963).
1417. Takeda Pharm. Ind. (TAKEDA R. *et al.*): Verfahren zur Umwandlung von Steroiden der Pregnanreihe mit Hilfe von Mikroorganismen, 1 154 802 (1963).
1418. Roussel - Uclaf (AMIARD G., HEYMES R.): Verfahren zur Herstellung von 18-nor-Cortison bzw. 21-Estern desselben, 1 155 445 (1963).

1419. O. Mathieson chem. Corp. (WEISENBORN F. L., LASKIN A. I.): Verfahren zur Herstellung von Steroiden, 1 156 408 (1963).
1420. Merck A. G. (MANNHARDT H. - J. *et al.*): Verfahren zur Herstellung von 16-Methylensteroiden, 1 156 804 (1963).
1421. Upjohn Co. (MAGERLEIN B. J. *et al.*): Verfahren zur Herstellung von therapeutisch wirksamen Steroidverbindungen, 1 158 063 (1963).
1422. BRINKMANN A. G. (KASPAR E., KIESLICH K.): Verfahren zur Herstellung von delta$^4$-Pregnen-6,20beta-diol-3-on bzw. des isomeren Allopregnan-20beta-ol-3,6-dions sowie ihrer Ester, 1 159 440 (1963).
1423. Merck A. G. (BORK K. - H. *et al.*): Verfahren zur Herstellung von ungesättigten 16-Methylen-3-keto-steroiden, 1 159 945 (1963).
1424. Schering A. G.: Verfahren zur Herstellung von 21-Methyl-delta$^4$-pregnen-11beta,17alpha-diol-3,20-dion, 1 167 829 (1964).
1425. O. Mathieson chem. Corp.: Verfahren zur Herstellung von 11alpha-Hydroxy-6alpha-halogensteroiden, 1 168 901 (1964).
1426. Schering A. G. (KIESLICH K. *et al.*): Verfahren zur Herstellung von delta$^{1,4}$-16alpha-Methylsteroiden, 1 169 444 (1964).
1427. Merck Co.: Verfahren zur Herstellung von 6alpha,16alpha-Dimethyl-11beta,17alpha,21-trihydroxy-4-pregnen-3,20-dion bzw. dessen 21-Acetat, 1 170 948 (1964).
1428. Schering A. G.: Verfahren zur Herstellung von 1alpha-Methylcorticoiden, 1 172 259 (1964).
1429. Societá Farmaceutici Italia: Verfahren zur Herstellung von 16alpha-Hydroxysteroiden, 1 173 090 (1964).
1430. Upjohn Co. (MURRAY H. C., REINEKE L. M.): Verfahren zur Herstelllung von 11alpha, 17alpha-Dihydroxyprogesteron, 1 175 670 (1964).
1431. Merck A. G. (BORK K. H. *et al.*): Verfahren zur Herstellung von ungesättigten 16-Methylen-17alpha-acetoxy-progesteron-derivaten, 1 176 132 (1964).
1432. Merck A. G., (BORK K.-H. *et al.*): Verfahren zur Herstellung von 1-Dehydro-16-methylen-17alpha-acetoxy-progesteron-derivaten, 1 176 134 (1964).
1433. Upjohn Co. (FONKEN G. S., MURRAY H. CH.): Verfahren zur Herstellung von 16alpha-Hydroxy-5beta-pregnan-3,11,20-trion, 1 176 135 (1964).
1434. Ciba A. G. (URECH J. *et al.*): Verfahren zur mikrobiologischen Herstellung von 11beta-Hydroxysteroiden der Pregnanreihe, 1 179 547 (1964).
1435. Richter Corp. (WIX G. *et al.*): Verfahren zur Herstellung von Prednisolon, 1 180 747 (1964).
1436. Wisconsin Alumni Research Foundation (KNIGHT S. G.): Verfahren zur mikrobiologischen Umwandlung von Steroiden und Fettsäuren, 1 181 703 (1964).
1437. Upjohn Co.: Verfahren zur Herstellung von therapeutisch wirksamen Steroidverbindungen, 1 185 608 (1965).
1438. Upjohn Co. (HANSON F. R.): Verfahren zur Herstellung von 1-Dehydro-6alpha-methylhydrocortison, 1 187 239 (1965).
1439. Philips Co.: Verfahren zur Herstellung von 11alpha,17alpha,21-Trihydroxy-9beta,10alpha-pregn-4-en-3,20-dion, 1 200 815 (1965).
1440. Philips Co.: Verfahren zur Herstellung von 17beta-Hydroxyretro-(10alpha-methyl-9beta)-steroiden, 1 203 261 (1965).
1441. Merck A. G.: Verfahren zur Herstellung von im Ring A gesättigten 3-Keto-steroiden auf mikrobiologischem Wege, 1 205 092 (1965).

German (East) Patents

1442. STACHOWIAK A. *et al.*: Verfahren zur Herstellung von delta$^{1,4}$-3-Keto-steroiden der Pregnanreihe mit einer Sauerstofffunktion in 11-Stellung, 33 325 (1964).
1443. Jenapharm VEB (SCHUBERT A. *et al.*): Verfahren zur stereospezifischen Hydrierung von Steroiden, 33 406 (1964).
1444. MÜLLER P. *et al.*: Verfahren zur Umwandlung von Steroiden, 33 875 (1964).
1445. Jenapharm VEB (SCHUBERT A. *et al.*): Verfahren zur partiellen selektiven Hydrierung von Steroiden, 41 964 (1965).

U. S. S. R. Patents

1446. SUVOROV N. N. *et al.*: Preparation of prednisolone from cortisol acetate, 152 656 (1964).
1447. SUVOROV N. N. *et al.*: Preparation of delta$^{1,4}$-androstadiene-3,17-dione, 157 682 (1964).
1448. SUVOROV N. N. *et al.*: Preparation of cortisol, 166 926 (1964).
1449. KOGAN L. M. *et al.*: Preparation of 17alpha-hydroxyprogesterone, 170 981 (1965).

Czechoslovakian Patents

1450. PROTIVA J. *et al.*: Preparation of delta$^{4}$-3-ketosteroids by microbial oxidation, 110 080 (1964).
1451. PROTIVA J. *et al.*: Preparation of cortisol and its 11alpha-epimer possibly of cortisone, 110 082 (1964).
1452. ŠMARIN - ALPĚJEV V., HILBERT O.: Isolation of steroid compounds, 110 097 (1964).
1453. ČAPEK A. *et al.*: Preparation of prednisona from cortisone acetate, 119 072 (1966).

GPSR Compliance

*The European Union's (EU) General Product Safety Regulation (GPSR) is a set of rules that requires consumer products to be safe and our obligations to ensure this.*

*If you have any concerns about our products, you can contact us on ProductSafety@springernature.com*

In case Publisher is established outside the EU, the EU authorized representative is:

Springer Nature Customer Service Center GmbH
Europaplatz 3
69115 Heidelberg, Germany

**Batch number: 10584556**

Printed by Printforce, the Netherlands